THE LOEB CLASSICAL LIBRARY

FOUNDED BY JAMES LOEB

THEOPHRASTUS

DE CAUSIS PLANTARUM

III

LCL 475

THEOPHRASTUS

DE CAUSIS PLANTARUM

BOOKS V–VI

EDITED AND TRANSLATED BY

BENEDICT EINARSON

AND

GEORGE K. K. LINK

HARVARD UNIVERSITY PRESS
CAMBRIDGE, MASSACHUSETTS
LONDON, ENGLAND
1990

Library of Congress Cataloging-in-Publication Data

Theophrastus.
De causis plantarum.
(The Loeb classical library: Greek authors; 471, 474, 475)
1. Botany—Pre-Linnean works I. Title. II. Series.
QK41.T23 1976 581 76–370781
ISBN 0–674–99519–8 (v. 1)
ISBN 0–674–99523–6 (v. 2)
ISBN 0–674–99524–4 (v. 3)

Typeset by Chiron, Inc.
Printed in Great Britain by St. Edmundsbury Press Ltd,
Bury St. Edmunds, Suffolk, on acid-free paper.
Bound by Hunter & Foulis Ltd, Edinburgh, Scotland.

CONTENTS

SIGLA

(see volume I, pages lix–lxi)

U	Vatican City, Urbinas graecus 61 (11th century)
U^d	the diorthotes of U
u	correctors (more probably an Italian corrector) of the 15th century
N	Florence, Laurentian Library, desk 85, 22 (15th century)
v	Venice, Library of St Mark 274 (dated 3 Jan. 1443)
Gaza	Theodorus Gaza's Latin translation (completed 1451)
M	Florence, Laurentian Library, desk 85, 3 (15th century)
C	Oxford, Corpus Christi College 113 (15th century)
H	Harvard College Library 17 (15th century)
a	Aldine Aristotle, fourth volume
P	Paris, National Library 2069 (15th century)
B	Vatican City, Vaticanus graecus 1305 (15th century)

U^c	a correction by the first hand
U^{cc}	such a correcton made in the course of writing
U^{ac}	the reading before correction by the first hand
U^r	a reading due to erasure
U^{ar}	the reading before erasure
U^m	a reading or note in the margin by the first hand
U^t	a reading in the text
U^{ss}	a superscription
U^1	a reading by the first hand

ΠΕΡΙ ΦΥΤΩΝ ΑΙΤΙΩΝ

Ε–Ζ

E[1]

1.1 τῶν ἐν τοῖς δένδροις καὶ φυτοῖς τὰ μὲν αὐτόματα γίγνεται, τὰ δ' ἐκ παρασκευῆς καὶ θεραπείας.

ἑκάτερα δὲ αὐτῶν ἔχει διαίρεσιν· τῶν γὰρ αὐτομάτων τὰ μέν ἐστι φύσει, τὰ δὲ παρὰ φύσιν (καὶ γὰρ ἐν τοῖς φυτοῖς ὑπάρχει τι τοιοῦτον, ὥσπερ καὶ ἐν τοῖς ζῴοις· οἷον ἐάν τι μὴ τὸν οἰκεῖον ἐνέγκῃ βλαστὸν ἢ καρπόν, ἢ μὴ κατὰ τὴν τεταγμένην ὥραν, ἢ μὴ ἐκ τῶν εἰωθότων μερῶν, ἤ τι τοιοῦτον ἕτερον, ἅπαντα γὰρ ταῦτα παρὰ φύσιν)· τῶν δ' ἐκ τέχνης καὶ θεραπείας τὸ μέν ἐστι συνεργοῦν τῇ φύσει πρὸς εὐκαρπίαν καὶ πλῆθος, τὸ δὲ εἰς ἰδιότητα καὶ τὸ περιττὸν τῶν καρπῶν (οἷον τὸ ποιῆσαι βότρυν ἀγίγαρτον, ἢ ἐκ τοῦ αὐτοῦ κλήματος μέλανα καὶ λευκὸν φέρειν, ἢ καὶ ἐν αὐτῷ τῷ βότρυϊ, καὶ ὅσα δὴ τοιαῦθ' ἕτερα ποιοῦσι, πλείω γάρ ἐστιν).

[1] αρχ(η) τοῦ ε̄ U.

BOOK V

A Fourfold Division

Some occurrences in trees and plants are spontaneous [A], others come from human provision and care [B]. 1.1

Each class is subdivided: spontaneous occurrences [A] are either natural [A1] or unnatural [A2] (for in plants, as in animals, there is such a thing as an unnatural occurrence; for example a tree or plant may not bear its proper shoot or fruit, or may bear it not at the regular season or from the usual parts, or may do some other such thing, all of these occurrences being unnatural). Among the procedures of art and care [B], on the other hand, there is one set [B1] that collaborates with the plant's nature to bring about good and abundant fruit, and another [B2] directed to producing fruit of a special and extraordinary character (for instance to grow a grape cluster that has no stones, or a black and a white cluster on the same twig, or black and white grapes in the same cluster, and all such other results as are obtained, for there are a number of them).

1.2 περὶ μὲν οὖν τῶν κατὰ φύσιν, καὶ τῶν συνεργούντων τῇ φύσει, πρότερον εἴρηται·

περὶ <δὲ>[1] τῶν παρὰ φύσιν, καὶ ὅσα πρὸς τὸ περιττὸν ἥκει, νῦν λεκτέον· καὶ πρῶτον μὲν[2] ὑπὲρ τῶν παρὰ φύσιν ἢ ὄντων ἢ δοκούντων, καὶ γὰρ τούτων ἐστί τις αἰτία, καθάπερ ἐν τοῖς πρώτοις (οἷον ὅσαι[3] πάρωροι καρποτοκίαι, καὶ μὴ ἐκ τῶν αὐτῶν μερῶν[4] γίγνονται· λέγω δ' οἷον εἴ ποτε συκῆ κατὰ χειμῶνα, τοῦ ἦρος ὑποφαίνοντος, ἤνεγκε καρπόν, καὶ ῥόα καὶ ἄμπελος ἐκ τῶν ἀκρεμόνων[5] καὶ ἐκ τοῦ στελέχους, καὶ εἰ δή τι παραπλήσιον τουτοισί[6]).

1.3 πρὸς[7] δὴ ταῦτα λαβεῖν δεῖ πρῶτον μέν, ὅτι συμβαίνει τὰ τοιαῦτα τοῖς πολυκάρποις καὶ ὑγροῖς (οἷον ἀμέλει καὶ τοῖς εἰρημένοις· ἅπαντα γὰρ [τὰ][8] ὑγρὰ ταῦτα καὶ πολύκαρπα), ξηρῷ δ' οὐδενὶ καὶ ὀλιγοκάρπῳ.

δεύτερον δέ, πρός τε τὰς βλαστήσεις καὶ καρποτοκίας τὰς παρώρους, ὅτι πάντων ἔχουσίν τινας ἀρχὰς ἐν αὐτοῖς βλαστητικὰς ἢ κλῶνες, ἢ πτόρθοι, ἢ ἀκρεμόνες, ἢ ὅ τι χρὴ καλεῖν τὰ ἐσχα-

[1] a. [2] Wimmer : ἢ U^{c} (ἡ U^{ac}) N : aP omit.
[3] u a : ὅσα U N P. [4] Wimmer : ωρων U.
[5] u : μονων U^{t} : κρεμονων U^{c} (κρε ss.).
[6] U : τούτοισι Heinsius. [7] u : πόρος U. [8] Itali.

We have already discussed natural occurrence[1] [A1] and procedures that collaborate with the nature of the plant[2] [B1]. 1.2

The Unnatural [A2] : *Three Premises*

We must now discuss occurrences contrary to nature[3] [A2] and procedures that achieve the extraordinary[4] [B2], and first [A2] the occurrences that are either unnatural or regarded as such, since these too, like the matters first discussed,[5] have some causation (as for instance the cases of fruit borne out of season or not borne from the same parts: I mean for example if a fig tree bears fruit in winter, when there is a hint of approaching spring, and if a pomegranate or vine bears from its boughs and trunk, and any other cases resembling these).

1.3 To explain these matters we must set up first the point that such things occur in trees that are abundant bearers and fluid (as in fact the ones mentioned,[6] all of them fluid and all abundant bearers), but in none that is dry and a scanty bearer.

A second point, to explain unseasonable sprouting and fruiting, is that in all of them the "twigs" (*klônes*) or "shoots" (*ptórthoi*) or "boughs" (*akre-*

[1] *CP* book 2. [2] *CP* book 3.
[3] *CP* 5 1–4.
[4] *CP* 5 5–7.
[5] [A1] in *CP* 2 and [B1] in *CP* 3.
[6] *CP* 5 1. 1 (fig, pomegranate and vine).

τεύοντα τῶν δένδρων (οὐδὲν γὰρ ἐπὶ πᾶσι κοινόν, ἀλλ' ἔνια προσηγορίας <ἰδίας>[1] τινὰς ἔχει, καθάπερ θαλλία, κράδη, κλῆμα· καλοῦσι δέ τινες καὶ

1.4 κλῶνας). ὅτι δὲ ἔχουσιν ἀρχάς τινας τοῦ βλαστάνειν φανερὸν ἐκ τῶν καταπηγνυμένων, ἃ πολλάκις, ἀρριζώτων ὄντων τῶν κάτω, μεθίησι βλαστὸν (οἷον κλήματα καὶ κράδαι[2] καὶ ἐλάας χάρακες[3]), ἐνίοτε δὲ καὶ μὴ καταπαγέντα, ἀλλ' ἀφῃρημένα (καθάπερ τά τε κλήματα καὶ τὰ τῆς ἐλαίας ξύλα).

τρίτον δ' ὅτι[4] συμβαίνει τὴν πάρωρον βλάστησιν γίνεσθαι χειμώνων μαλακῶν ἢ νοτίων γενομένων.

ἐκ τούτων γὰρ οὐκ ἂν δόξειεν ἄλογον εἶναι τὸ συμβαῖνον, ἠθροισμένης τῆς γονίμου δυνάμεως καὶ ὑγρότητος, ὁτὲ μὲν καθ' αὑτήν, ὁτὲ δὲ λειπομένης τῆς ἔμπροσθεν ἀφ' ἧς ὁ καρπὸς ἦλθεν, ὥσπερ

§1. 4–5: Athenaeus iii. 2 (77 C).

[1] added by Heinsius after ἔχει, placed here by me.
[2] Heinsius (κράδας Scaliger) : κλάδη U.
[3] Schneider : χάρακας U.
[4] Wimmer : δέ τι U.

mónes), or whatever one is to call the extremities of trees, contain in themselves certain starting-points capable of sprouting. (There is no name for these extremities that applies to all trees, but some have certain special names, as *thallía* [live-twig], *krádē* [fig-twig] and *klêma* [vine-twig]; some persons use for twig the word *klônes*).[1] That the twigs have certain starting-points of sprouting is evident from those that are set in the ground, which often send out a shoot, although their lower parts have not rooted, as twigs of vine and fig and sticks of olive wood[2]; sometimes they do so even when not set in the ground, but merely removed from the tree, as with vine twigs[3] and olive wood.[4] 1.4

A third point is that unseasonable sprouting occurs when the winter has been mild or had winds from the south.

The Unnatural [A2] : *Unseasonable "Fruit"*

In the light of these points the occurrence would not appear unreasonable, once the generative power and fluid had been collected, sometimes independently, sometimes when the power and fluid that was there before and produced the fruit of the previous year has left a remainder, as occurs with the

[1] In *HP* 1 1. 9 the twig is called *kládos*.

[2] *Cf. CP* 1 12. 9 (for vine, fig and olive); *CP* 1 3. 3 (for olive). [3] *Cf. CP* 1 13. 9. [4] *Cf. CP* 1 7. 4.

συμβαίνει καὶ ἐπὶ τῆς συκῆς· ὑπολειπομένης γὰρ πλείονος τῆς τοιαύτης ὑγρότητος, ὅταν ἀὴρ ἐπιγένηται μαλακὸς καὶ ὑγρὸς καὶ θερμός, ἐξε-

1.5 καλέσατο τὴν βλάστησιν. ὅτι[1] δὲ τοῦτο συμβαίνει φανερόν, ἐκ[2] τούτου[3] <γὰρ>[4] τοῦ μέρους ὁ καρπὸς ἀνίετο[5] ὅθεν καὶ οἱ πρόδρομοι. τάχα δὲ καὶ οἱ πρόδρομοι μὲν διὰ τὴν μαλακότητα τοῦ ἀέρος οὐ προτεροῦσί τε καὶ πέττονται μᾶλλον· εἰ δ' ἄρα μὴ τοῦτο, ἀλλὰ τἆλλά γε πάντα διὰ τὰς αὐτὰς καὶ παραπλησίους αἰτίας ἐστίν, ἤτοι γενομένων τινῶν ὑπολειμμάτων, ἢ καθ' αὑτὰ[6] συνισταμένων ἀμφοῖν.

[1] u^{c} P : ὅτε U : ὁτὲ u^{ac} N a.
[2] U : ὅτι ἐκ u : εἰ ἐκ Wimmer. [3] u : του U.
[4] ego. [5] ego : άνιεται U.
[6] Scaliger (*ex integro* Gaza) : καὶ ταῦτα U.

---

[1] Here the true fig or the fruiting shoot from which it is produced.

[2] The fruit (*sŷkon* or true fig) is produced in front of the leaf (that is, in the axil of the new leaf), and not, like the premature arrivals (*pródromoi*) and the *ólynthoi*, behind it (that is, in the axils of the leaves of previous years): *cf. CP* 51. 7; 51. 8.

[3] *Pródromoi* in the Greek, literally "fore-runners" or "headlong runners." *Sŷkon* (the true fig) is neuter, *pródromos* masculine; presumably *ólynthos* (masculine) was once understood. Four words are used of the true or

fig-tree. For if a good amount of such fluid is left over, ensuing mild, moist and warm weather calls forth sprouting. That this occurs is evident, since the fruit[1] was put forth from the part[2] from which the "premature arrivals"[3] come. Perhaps in the case of the premature arrivals at least the earlier production and concoction[4] is not so much due to the mildness of the air[5]; but even if this is not so, the rest of the imperfect productions[6] are all due to the same and similar causes: either a remainder has been left over, or both collections[7] are made independently. 1.5

false fig: (1) *sŷkon*: the true fig; see also (4) below; (2) *ólynthos*: these come out in season, remain on the tree, and are sometimes edible (*CP* 5 1. 6–8); the term covers the varieties intermediate between *sŷkon* and *pródromos*; (3) *pródromos* ("first arrival"): these come out far in advance of the season (although some come out later), are irregular in their production, soon fall off, and are never edible; (4) *erinón*: properly the fruit of the caprifig, but the term is also used of the "premature arrivals" and the unripe *sŷkon*.

[4] Only in colour: *CP* 5 1. 6.

[5] It is due more to the state of the tree: *cf.* *CP* 5 1. 8 *ad fin.* The premature arrivals usually come so early that it is not easy to speak of them as called forth by the "mildness of the air."

[6] The *ólynthoi* of various types: *CP* 5 1. 6 *ad fin.* and 5 1. 7.

[7] That which produced last year's figs (*sŷka*), and that which produces this year's "premature arrivals" (*pródromoi*) and *ólynthoi*.

1.6 οὐκ ἄλογον δ᾽ οὐδετέρως·[1] προϋπάρχει γὰρ ἡ τοῦ δένδρου πρὸς ἄμφω φύσις ὑγρά τις οὖσα καὶ προβλαστής,[2] ὥστ᾽ ἂν εὐδίαι πλείω χρόνον ἐπιγένωνται, καὶ τοῦ χρώματος ποιεῖν τὴν μεταβολήν (ἐπεὶ τόν γε[3] χυλὸν οὐ δύναται πέττειν οὐδαμῶς, ἀλλ᾽ ἡ φαντασία σχεδὸν γίνεται κατὰ τὴν χρόαν). οἱ δ᾽ ὄλυνθοι, συνεργούσης ἤδη τῆς ὥρας, πεπαίνονται μέχρι τινός· ἔτι δ᾽ οὗτοί γε κατὰ φύσιν πώς εἰσιν, ἀεὶ γὰρ γίνονται τὴν αὐτὴν καὶ τεταγμένην ὥραν, ὥσπερ καὶ ὁ τῶν διφόρων

1.7 συκῶν λεγομένων[4] καρπός. ἀλλ᾽ ἐπὶ τοσοῦτον τὸ ὅμοιον ληπτέον, ὅτι παραπλησία τις ἡ αἰτία πάντων, ἐκ τοῦ συνίστασθαι τὴν τοιαύτην ὑγρότητα καὶ δύναμιν. διὸ καὶ ὄπισθεν τοῦ φύλλου πάντα ταῦτα γίγνεται, καὶ οὐχ, ὥσπερ τὸ σῦκον, ἐμ-

[1] οὐδ᾽ ἑτέρως aP : οὐδ᾽ ἕτερος U[c] (ἔ- U[ac]) N : οὐδ᾽ ἕτερον u.
[2] ego (*cf.* προαυξής) (πρωΐβλαστος Schneider : πρωϊβλαστής LSJ) : προβλάς U N P : πρόβλαστος a. [3] Schneider : τε U.
[4] Gaza, Scaliger : λεγόμενος U.

---

[1] The white *ólynthoi* of *CP* 5 1. 8; the black are meant by the "fruit of the so-called double-bearing 'fig-trees'" in the following sentence. Only the true fig (*sŷkon*) is ever fully ripe. [2] *Cf. HP* 3 7. 3, cited in note 2 on *CP* 5 1. 8.

[3] A reference to the trees that bear both figs and black *ólynthoi* (*CP* 5 1. 8), the same as the "wild" fig-trees of *CP* 2 9. 13. There was some doubt whether they deserved

Production of premature arrivals is unreasonable from neither source: the nature of the fig-tree is to begin with well fitted for either production, being fluid and given to early sprouting. So that if fair weather follows the production for any considerable time the tree goes on to change the colour of the premature arrival; as for the juice, the tree is quite unable to concoct it, the appearance of concoction being confined, one may say, to the coloration. The *ólynthoi*, on the other hand, do get ripened up to a point,[1] with the season already lending its aid. But these moreover are in a way natural,[2] since they are always produced at the same regular season, as is also the fruit of the so-called double-bearing "fig-trees."[3] But we can press the resemblance[4] only so far as to say that the causation is similar in all, coming from the accumulation of this sort[5] of fluid and power. This moreover is why all these grow from behind the leaf, and not, like the fig, from 1.6 1.7

the name of "fig-trees": *cf. HP* 3 7. 3: ". . . and if after all some fig-trees bear *ólynthoi*." The same tree is called an *ólynthos* at *HP* 1 14. 1: "Bearing fruit from both old and new wood are any apple-trees there may be that bear two crops . . . ; and further the *ólynthos* that concocts its fruit [that is, its *ólynthoi*] and bears figs on the new wood."

[4] Between the black *ólynthoi*, the white *ólynthoi* and the premature arrivals.

[5] Presumably both collections were made before the new wood had developed far enough to have a collection of its own, producing the true fig.

προσθεν. ἀλλ' αἱ μὲν φύουσαι τοὺς προδρόμους ἀσθενέστεραι τυγχάνουσιν, ὥστε οὐδ' ἐπιμένειν δύνανται, τῆς βλαστήσεως γινομένης ἐκεῖσε.

1.8 ὅλως δὲ πολυειδές τι τὸ τῶν συκῶν ἐστιν· αἱ μὲν γὰρ ὀλυνθοφόροι μόνον, σῦκα δ' οὐ φέρουσιν (οἷον αἱ τοὺς λευκοὺς ὀλύνθους φέρουσαι τοὺς ἐδωδίμους)· ἕτεραι δὲ φέρουσι καὶ σῦκα καὶ ὀλύνθους μέλανας, ἀβρώτους δὲ τούτους ἢ καὶ ἐδωδίμους (φέρουσι[1] δὲ τὸν ὄλυνθον ὄπισθεν τοῦ θρίου· γίνονται δ' ἐν αὐτοῖς καὶ ψῆνες)· αἱ <δ'>[2] ὅλως οὐ φέρουσιν ὀλύνθους. πάλιν δὲ τοὺς προδρόμους αἱ μὲν φέρουσιν (οἷον ἥ τε Λακωνικὴ καὶ <ἡ>[3] λευκομφάλιος καὶ ἕτεραι πλείους), αἱ δ' οὐ φέρου-

§1.8: *Cf.* Athenaeus, iii. 12 (77 F); *ibid.* 77 C; Pliny *N. H.* 16. 113.

[1] Gaza, Scaliger : -σαι U.
[2] u^{ss}? (δὲ Itali after Gaza).
[3] Athenaeus.

---

[1] To the position in front of the leaf, from which position the true fig is produced. The occurrence mentioned in *HP* 2 3. 3 is therefore monstrous: "a fig-tree has been known to grow its figs from behind its *eriná* [that is, its premature arrivals]."

[2] *Cf. HP* 3 7. 3: "Whereas other trees bear both the

before it. But there is the difference that the fig-trees producing the premature arrivals are weaker, so that the premature arrivals are unable even to remain on the tree when the sprouting is taking the other direction.[1] And in general fig-trees are a much varied group: some bear *ólynthoi* only and no figs[2] (like the trees that bear the white and edible *ólynthoi*); others bear both figs and black *ólynthoi*, the latter inedible or edible too, and borne behind the leaf[3] (fig-insects moreover are found in them); still others bear no *ólynthoi* at all. Again there are the premature arrivals: some trees bear them, as the Laconian and white-naveled variety and several others, some do not; and in those that bear them the

1.8

same fruit and the familiar annual increments, leaf, flower and shoot (some also bearing a catkin or tendril), some bear an even greater variety of products, as . . . the fig-tree both its *eriná* [that is, *pródromoi*] that drop off before the other crop and there are some, if they are fig-trees, that bear *ólynthoi*, but these perhaps are in a way the fruit."

[3] *Cf.* Athenaeus, iii. 12 (77 C): "In the second book of his History of Plants the philosopher says: 'There is another variety in Greece and Cilicia and Cyprus that bears *ólynthoi*, which bears its fig in front of the leaf, its *ólynthos* behind; and some bear the *ólynthoi* all from last year's shoot and not from this year's. They bear the *ólynthos* before they do the figs, and it is ripe and sweet and not like the *ólynthos* of our country. And it also gets much larger than the figs; and its season is not long after the tree sprouts.'"

σιν· τῶν δὲ φερουσῶν οὐ συνεχής, ἀλλὰ παραλλάττουσα ἡ φορὰ[1] *κατὰ τὴν διάθεσιν· ὁτὲ μὲν γὰρ ἤνεγκαν, ὁτὲ δ' οὐκ ἤνεγκαν, ἐὰν δὲ κακῶς, οὐκ*[2] *ἤνεγκαν.*

1.9 *φέρουσι δὲ οὐ παρ' αὐτὸ τὸ θρῖον μόνον, ἀλλὰ καὶ κατωτέρω πολύ, καὶ ἔνιαί γε ἐκ τῶν παχέων σφόδρα. συμβαίνει δ' ἐνίας*[3] *καὶ ἐκπέττειν τινὰς τῶν ὕστερον· οὐ γὰρ ἅμα πάντας, ἀλλὰ τοὺς μὲν πρότερον φέρουσιν, τοὺς δὲ μικρὸν πρὸ τῶν ἐρινῶν,*[4] *ἐγγύτεροι δὲ οὗτοι τοῦ βλαστοῦ καὶ τῶν σύκων,*[5] *ἀποπίπτουσι δ' ἐκεῖνοι πρότερον καὶ ἧττον ἐκπέττονται.*

1.10 *τὰ μὲν οὖν συμβαίνοντα ταῦτ' ἐστίν, οὐκ ἀλόγως δ' ἂν δόξειεν οὐδέν.*

τὸ γὰρ μὴ πάσας φέρειν, ἀλλ' ὅσαι ὑγρότεραι καὶ ἰσχὺν ἔχουσαι εἰς τὴν ἐπίσπασιν, εὔλογον· ἐκ γὰρ τῆς ὑπολοίπου[6] *τροφῆς ἐβλάστησεν ἡ σύστασις, ἐφελκυσαμένῃ*[7] *δὲ πλέον, οὐκ ἄλογον τὸ μὴ πᾶσαν καταπέψαι*[8] *τότε.*

τὸ <*δὲ*>[9] *μὴ ἐνδελεχὲς αὐτῶν τούτων διὰ τὸ*

[1] Scaliger : *διαφορὰ* U. [2] U : *δὲ, κακῶς* [*οὐκ*] Heinsius.
[3] u : *ἐνίους* U N aP. [4] U N aP : *ἐρινεῶν* u.
[5] u : *συκῶν* U N aP. [6] u : *-λή-* U. [7] *-μένῃ* (that is, *-ῃ*) U : *-μένην* Schneider. [8] u aP : *καταπεμψαι* U (*-έμψαι* N). [9] Gaza (*autem*), Scaliger : *τε* Wimmer.

bearing is not constant, but varies with the condition of the tree: sometimes it bears them, sometimes not, and if its condition is poor it does not bear them.

1.9 The trees bear the premature arrivals not only right behind the leaf but also a good distance below it, some bearing them even from the branch where it is very thick. It happens that some trees even bring a few of the later premature arrivals to concoction, for they do not bear all of them at the same time, but bear some earlier, others a little in advance of the unripe figs.[1] The late premature arrivals are closer to the new shoot and to the true figs, but the earlier ones fall off first and their concoction is less complete.

1.10 Such then are the facts, and none would appear to occur unreasonably.

For it is reasonable that not all fig-trees should bear premature arrivals,[2] but only such as are more fluid and have strength to attract food, since the shoot comes from an accumulation of left-over food; and when the tree has attracted a greater than ordinary amount of food it is not unreasonable that it should not concoct[3] all of it at the time.

Interrupted bearing in the trees that do bear

[1] "Unripe figs" renders *eriná*: see note 3 on *CP* 5 1. 5.

[2] *Cf. CP* 5 1. 8.

[3] That is, turn it into fruit.

μὴ ὁμοίας εἶναι τὰς διαθέσεις· ἀσθενεστέρως γὰρ ἔχουσα καὶ χεῖρον θεραπευθεῖσα δῆλον ὡς ἐλάττω λήψεται τὴν ὑγρότητα, μὴ λαβοῦσα δὲ πλῆθος, οὐδ' ὑπόλειμμα[1] ποιήσεται·[2] ἐπεὶ γεωργηθεῖσα ὁμοίως, καὶ τῶν ἐκ τοῦ ἀέρος συνεργησάντων, 1.11 ἀποδώσει καὶ τῷ[3] ὕστερον. (ἐνίοτε δὲ κἀκεῖνο συμβαίνει, μὴ τυχούσης θεραπείας, ὥστε τοὺς μὲν προδρόμους ἐνεγκεῖν, τὸν δὲ καρπὸν μὴ δύνασθαι. τοῦτο δ' ὅτι τὸ μὲν ὑπῆρχε προκατειργασμένον, θάτερον δ' ὥσπερ ἐκ καινῆς ἔδει ποιεῖν.)

ἀλλὰ μὴν οὐδὲ τὸ μὴ ἐκπέττειν τοὺς ἐκ τῶν παχέων,[4] ἀλλὰ τοὺς ἄνω μᾶλλον· ἡ τροφὴ[5] γὰρ ἅμα καὶ ἡ θερμότης ἄνω φέρεται, καὶ ὅλως ἡ τοῦ δένδρου ὁρμὴ πρὸς τὴν βλάστησιν, ὥσθ' ἅμα συναναφερομένου καὶ τοῦ θερμοῦ, δι' ἀμφοτέρων ἡ ὑπόλειψις[6] (διὸ καὶ αἱ βλαστήσεις, καὶ αἱ κατὰ μέρος ἀνθήσεις, καὶ αἱ καρποτοκίαι τῶν ἀνθῶν[7]

[1] u : -λημ- U.
[2] ego : ποιήσαι U (ποιῆσαι u N) : ποιήσει aP.
[3] U : τὸ N aP.
[4] παχέων ἄλογον Schneider.
[5] N : στροφῆ U (-ῇ u aP).
[6] Gaza, Heinsius : ὑπόληψις U.
[7] U : αὐτῶν Wimmer.

premature arrivals[1] is due to the dissimilar condition of the tree: when the tree is in a weaker state and has been worse tended it will evidently get less fluid, and getting no great amount, will leave nothing over; since if the tendance has been as good as it was before, and the weather has done its share, the tree will be forthcoming with premature arrivals in the following year. (Occasionally, when the tree has not been tended, another situation occurs: the tree bears its premature arrivals but is unable to bear its fruit. This is because the supply for the first had already been prepared and was available, whereas the supply for the true figs had to be produced as it were from scratch.) 1.11

Nor yet is it unreasonable that the fig-tree should fail to concoct fully the premature arrivals growing from the thick part of the branches, but should succeed better with those growing further up.[2] For the food moves upward and with it the heat, and indeed the whole drive of the tree is toward new growth, so that, since the heat accompanies the food in this upward movement, the lagging behind of these premature arrivals is due to their being bypassed by both. (This moreover is why a tree's process of sprouting, and progressive flowering, and later production of fruit from the flowers advance

[1] *Cf. CP* 5 1. 8.
[2] *Cf. CP* 5 1. 9.

ὕστερον, ἄνω[1]).

καὶ περὶ μὲν τούτων ἱκανῶς εἰρήσθω.

1.12 παραπλήσιον δὲ καὶ τὸ συμβαῖνον καὶ ἐπὶ τῶν ἀμπέλων ἐστίν· καὶ γὰρ αὗται προβλαστάνουσιν ἐνίοτε, συνηθροισμένης τῆς γονίμου καὶ βλαστητικῆς ἀρχῆς. ὁ δὲ τόπος ὡρισμένος αὐταῖς, ἐκ τῶν ὀφθαλμῶν γὰρ ἡ βλάστησις· ὁποῖα δ' ἂν καὶ τὰ τοῦ ἀέρος ἐπιγένηται, καὶ ἐπὶ πόσον χρόνον, οὕτω καὶ τὰ τῆς αὐξήσεως ἀκολουθεῖ.

σχεδὸν δὲ καὶ ὅσα πρωϊβλαστῆ καὶ πρωϊανθῆ φύσει, ταὐτὸ τοῦτο πάσχει· κυϊσκόμενα γὰρ ἐν ὥρᾳ, προκαλουμένων τῶν εὐδιῶν, πολλάκις ἀποτίκτει πρὸ τῶν καιρῶν, καὶ προάγει μέχρι οὗ ἂν ὁ ἀὴρ ὁμαλίζῃ.

τὰς μὲν οὖν παρώρους βλαστήσεις ἐν ταύταις ὑποληπτέον ταῖς αἰτίαις εἶναι.

2.1 τὰς δὲ μὴ ἐκ τῶν οἰκείων μερῶν, ἀλλ' οἷον ἀκρεμόνων καὶ στελέχους, οὐ πόρρω ταύτης, ὅταν ἀθροισμὸς εἰς ταῦτα γένηται τῆς γονίμου δυνά-

[1] ego : ἀλλὰ U N P : a omits.

upward.)

Let this discussion suffice for these matters.

The Unnatural [A2] : *Premature Sprouting in the Vine*

1.12 Similar too is what occurs in the vine: the vine too sometimes sprouts prematurely when the starting-point of generation and sprouting has been collected. In the vine however the place is fixed, the sprouting coming from the eyes[1]; and the extent of growth depends on the character and duration of the ensuing weather.

Indeed one might say that all trees that naturally sprout and flower early are affected like this, for being pregnant early they often bring forth prematurely, invited by the fine weather, and continue to develop the shoot so long as the weather lasts.

We must suppose, then, that the cases of sprouting out of season are covered by these causes.

The Unnatural [A2] : *Sprouting from a Strange Part*

2.1 We must suppose that sprouting that is not from the proper parts, but for instance from the boughs or trunk, is not far removed from sprouting out of season. It occurs when the generative power collects in

[1] The "eye" of the grape vine is the bud in its axil.

μεως· ἀρχῆς γὰρ ἐπιστάσης,[1] *ἡ ἐπιρροὴ τῆς τροφῆς ἐξομοιοῦται τῷ ὑποκειμένῳ, καθάπερ ἐν τοῖς ἄλλοις. ἡ δὲ σύστασις οὐκ [ἂν] ἄλογος,*[2] *εἰ τοιαύτην τινὰ κατὰ μόριον ἔλαβε κρᾶσιν, ἀλλ' εὐλογωτέρα πολὺ*[3] *τῆς*[4] *ἐν ζῴοις γινομένης, οἷον εἰ*[5] *κέρας ἐκ τοῦ στήθους ἤ τι τοιοῦτον ἕτερον· αὕτη γὰρ μείζων ἡ παραλλαγὴ τῶν ἐν τοῖς φυτοῖς ὅσῳ μᾶλλον ὁμοιομερῆ τὰ φυτὰ τῶν ζῴων.*

2.2 *ἐὰν δὲ δὴ καὶ προβλάστημά τι γένηται τῷ βλαστῷ,*[6] *ἔλαττον ἔτι τὸ ἄτοπον (ἐνίοτε γὰρ καὶ φύεται κλῆμα ἐκ τοῦ στελέχους)· ἰσχύσαντα δὲ καρπολογῆσαι*[7] *ταῦτ', οὐκ ἄλογον. ἀλλ' ὅταν αὐτὸς ὁ καρπὸς ἐκ τοῦ ἀκρεμόνος ἢ τοῦ στελέχους, ἀτοπώτερον, ὥσπερ ἐπὶ τῶν ῥοῶν· αἰτίαν δὲ τὴν εἰρημένην ὑποληπτέον.*

[1] u aP : *-στασεις* U : *-στάσις* N : *-συστάσης* Schneider (taking this to be the reading of U).
[2] Scaliger (*temere* Gaza) : *ἂν ἄλογος* U N : *ἀνάλογος* aP.
[3] u : *πολλοὶ* U : *πολλῷ* N aP.
[4] u aP : *τοῖς* U N.
[5] N (*οἱονεὶ* [*οἱ-* a]P) : *οἷν εἰ* U : *οἶν εἰ* u.
[6] U : *fructui* Gaza : *καρπῷ* Wimmer.
[7] U : *fructum prestet* Gaza : *καρπογονῆσαι* or *καρποτοκῆσαι* Scaliger : *καρποφορῆσαι* Heinsius.

---

[1] The nature and power are teleological, and have a goal: *cf. CP* 6 4. 4; 6 6. 7; 6 17. 2. Otherwise the starting-

these parts: when a starting-point has become fixed, the food that flows to it becomes assimilated, just as in the other parts, to the thing that is to be produced.[1] The formation of such a starting-point is not unreasonable, if the tree has acquired in some part a tempering that is favourable, but is a good deal more reasonable than the formation that occurs in animals, as when a horn[2] grows from the chest or something else of the kind occurs, since here the departure is greater than that in plants to the extent that plants are more undifferentiated than animals.[3]

2.2 Now if the misplaced sprouting has in addition a preliminary shoot from which it sprouts, the oddity is still less (sometimes in fact a vine-twig grows from the trunk); and it is not unreasonable that these sproutings should gain strength and acquire fruit; the case is however odder when the fruit comes directly from the branch or trunk, as in pomegranates.[4] We must take the causation to be the one mentioned.[5]

point would produce not fruit, but just more trunk.

[2] *Cf.* Aristotle, *On the Generation of Animals*, iv. 4 (770 b 36–37): "An instance has occurred of a goat born with a horn next to its leg (*πρὸς τῷ σκέλει*)."

[3] And so a fruit is closer to a piece of trunk than a horn is to an animal's chest.

[4] *Cf. HP* 2 3. 3: ". . . the pomegranate and the vine have grown fruit from the trunk . . ."

[5] *CP* 5 2. 1.

αἱ δὲ τοιαῦται παραλλαγαὶ τῶν τόπων ἐλάττους εἰσίν, οἷον εἴ ποτε συκῆ τις ἤνεγκεν ἐκ τοῦ ὄπισθεν τοῦ θρίου τὸν καρπόν, μικρὰ γὰρ ἡ μεταβολή, καὶ ὁ τόπος ὅλως οὐκ ἀλλότριος τοῦ συμβαίνοντος.

2.3 *αἴτιον δέ, ὅτι τὴν μὲν ἐπιβλάστησιν ἀσθενῆ, τὴν δ' ἐνταυθα συρροὴν ἰσχυροτέραν ἔσχεν· ὥστ', ὀψιαιτέρας γενομένης τῶν προδρόμων, ἐπένειμέ*[1] *τε καὶ ἐξέπεψεν.*[2]

φαίνεται δὲ καὶ τῇδε ἧττον ἄτοπον, ὅτι γένος τί φασιν εἶναι τοιοῦτον <ὃ> *δὴ*[3] *καλοῦσιν ὀπισθοκάρπιον·*[4] *εἰ γὰρ ὅλως τι πέφυκεν οὕτω καρποτοκεῖν τῶν ὁμογενῶν, οὐκ ἄτοπον συμβῆναί ποτε καὶ τοῖς μὴ πεφυκόσι, τῆς ὁμοίας διαθέσεως γινομένης.*[5]

[1] U : *ἐπέμεινέ* Schneider.
[2] U r : *ἐξέπεμψεν* U ar : *ἐξεπέφθη* Schneider.
[3] Schneider : δὴ ὃ u.
[4] U c from ὁ- : *ὀπισθόκαρπον* Heinsius.
[5] U : *γενομένης* Schneider.

---

[1] *Cf. HP* 2 3. 3: ". . . thus it has been known to happen that a fig-tree grew its figs from behind the place where it bore its *eriná* [that is, its premature arrivals]." If the figs grew behind the *eriná*, and the *eriná* grew behind the leaf, the figs presumably also grew behind the leaf.

Such deviations in location as the following are smaller: so if an occasional fig-tree has borne its fruit from the region behind the leaf,[1] since the change is slight and the place is in any case not inappropriate to the occurrence.[2]

2.3 The cause is this: the new growth acquired by the fig-tree was weak, whereas the collection of power and food that it acquired here[3] was unusually strong; as a result, since the new growth occurred later than the production of the premature arrivals, the tree assigned their food to its figs and concocted them completely.

The following circumstance moreover makes the occurrence appear less odd: experts say that there is a variety of fig-tree of this description (and actually call it *opisthokárpion*).[4] For if a variety of the same tree is in general so constituted as to bear fruit in this position naturally, it is not odd that such bearing should occasionally occur in the trees too that do not naturally bear in this way, when they are in the same condition.

[2] It is still a twig, not a branch or the trunk, and the purpose of the leaf is to shelter the fruit (Aristotle, *Physics*, ii. 7 [199 a 25–26]; *On the Soul*, ii. 1 [412 b 2]).

[3] That is, on the old wood behind the leaf.

[4] "With fruit behind (the leaf)"; the name is not found elsewhere. Perhaps the fig-trees of *CP* 5 1. 3 are meant, which bear figs in front of the leaves and black *ólynthoi* behind them.

2.4 ἐπεὶ καὶ <τὸ>[1] ἐκ τῶν ἀκρεμόνων ἐνεγκεῖν τινα ὁμοιότητα ἔχει τῇ ἐν Αἰγύπτῳ συκαμίνῳ· φέρει γὰρ δὴ κἀκείνη τὸν καρπὸν ἐκ τοῦ στελέχους, οὐκ ἐκ τῶν ἀκρεμόνων, εὐβλαστής τις οὖσα, καὶ εὔτονος, ὡς ἔοικεν, πρὸς καρποτοκίαν (σημαίνει δέ, τὸ πλεονάκις ἀπὸ τῶν αὐτῶν φέρειν, ἀφαιρουμένων), οὐ πεπαίνει[2] δὲ καλῶς μὴ ἐπικνισθέντων[3] καὶ περιαλειφθέντων ἐλαίῳ (καθάπερ ἐλέχθη), <διὰ>[4] τὴν εὐτροφίαν καὶ τὸ πλῆθος τῆς ἐπιρροῆς· ἀφαίρεσίς τε γὰρ γίνεται τῇ ἐπικνίσει,[5] καὶ τὸ ἔλαιον ἅμα διαθερμαῖνον, καὶ ὁ ἥλιος, ἀνεῳγμένων, ταχεῖαν ποιεῖ[6] τὴν πέψιν.

2.5 ἔοικε <δὲ>[7] παραπλησίῳ[8] τρόπον τινὰ τὸ συμβαῖνον τοῖς κατὰ μέρος ἀνθοῦσιν ἀπὸ τῶν κάτωθεν ἀρχομένοις.[9] ἐκείνων τε γὰρ τὰ μὲν ἰσχύοντα καὶ οἷον[10] τετελεωμένα καὶ ἀνθεῖ καὶ καρπογονεῖ, τὰ δὲ ἀσθενέστερα καὶ ἐπάνω προαύξεται[11] καὶ βλαστάνει καὶ ἀεὶ τὸ τελεούμενον ἀνθεῖ, τὴν δὲ τῆς τροφῆς ἐπιρροὴν ἕκαστον εἰς

[1] Schneider. [2] aP : -ειν U N.
[3] U N aP : -κνη- u. [4] aP.
[5] N aP : -κνή- U. [6] ego : ποιεῖται U. [7] aP.
[8] ego (παραπλήσιον Schneider) : παραπλησίως U.
[9] u aP : -ων U N.
[10] N aP : οἷο U. [11] U : προσ- Schneider.

Indeed bearing from the branch has a certain similarity to the case of the fig-mulberry in Egypt, since this tree too bears its fruit from the trunk (but not from the branches),[1] since it sprouts readily and has a sustained vigour, it appears, for fruit production, as is shown by its bearing repeatedly from the same parts when the fruit is removed. But it does not ripen the fruits properly unless they are scratched and smeared with oil (as we said),[2] owing to its good feeding and the great influx of food. For the amount is reduced by the scratches and at the same time the heating effect of the oil and the sun on the fruit thus laid open makes its concoction rapid. 2.4

What happens in plants that flower progressively from the lower parts upward[3] resembles in a way what happens here.[4] Thus in the progressive flowerers the parts that are strong and (as it were) completed both flower and bear fruit, whereas the parts that are weaker and above these both grow in length and sprout, and it is only as it is approaching completion that each of these flowers; and each part, 2.5

[1] *Cf. HP* 1 1. 7 (the fig-mulberry bears even on the trunk); *HP* 1 14. 2 (the fig-mulberry bears on the trunk; some say on the branches as well).

[2] *CP* 1 17. 9; 2 8. 4.

[3] *Cf. CP* 4 10. 2–3 with the notes.

[4] In the fig-trees of *CP* 5 2. 2, last paragraph.

<τὴν>[1] ἑαυτοῦ φύσιν καταμερίζεται καὶ δύναμιν· καὶ τούτων ὁμοίως <ἣ>[2] μὲν εἰς τὴν βλάστησιν φέρεται τῆς τροφῆς, ἡ δ' εἰς τὸ ἰσχυρὸν ἤδη καὶ καρπογονοῦν, ἐκεῖνο δὲ οὔπω δύναται διὰ τὴν ὑγρότητα καὶ τὴν ἀτέλειαν.

καὶ περὶ μὲν τούτων ἀρκείτω τὰ εἰρημένα.

3.1 τὰς δὲ μεταβολὰς τῶν καρπῶν εἴ[3] τινες [εἰ][4] ἐκ λευκῶν μέλανες ἢ ἐκ γλυκέων[5] ὀξεῖς, ἢ ἀνάπαλιν (καθάπερ φασὶν ἐπί τε συκῆς καὶ ἀμπέλου καὶ ῥόας τοῦτο συμβαίνειν), ἐνιαχοῦ μὲν οὐδὲ θαυ-

[1] aP. [2] ego.
[3] Gaza (*si*), Itali : ἀεί U.
[4] Scaliger (ἢ Itali).
[5] u : γλυκαίων U[c] (γ ss.).

---

[1] That is, what is sprouting turns the food to sprout, what is flowering turns it to flower, and what is fruiting turns it to fruit.

[2] The fig-trees of *CP* 5 2. 2, last paragrapah.

[3] Here the premature arrivals.

[4] With this paragraph *cf. HP* 2 3. 1–2: "At all events it is reported that there is in such matters a change that is spontaneous, sometimes of the fruit, sometimes including the tree, and soothsayers take these changes for signs, for instance that an acid pomegranate tree has borne a sweet pomegranate and a sweet tree an acid, and again that the trees themselves change with no such limitation and turn sweet from acid and acid from sweet . . . ; again that a fig-

above and below, devotes its share of the influx to developing its own nature[1] and power. So too in our trees[2]: some of the food is carried to the new growth, some to the part that is already strong and producing fruit,[3] but the new growth is not yet able to produce fruit because of its fluidity and incomplete development.

Let the foregoing discussion suffice for these matters.

The Unnatural [A2] : *Changes in the Character of the Fruit:* (1) *the Vine*

3.1 As for the changes of the fruit,[4] when it turns from white to black or from sweet to acid or the reverse, as is reported of the fig, the vine, and the pomegranate—that they occur in some regions is

tree turns from white to black and from black to white, this last occurring similarly with the black vine. Now these cases are taken to be portents and unnatural, but all such cases as are familiar are not even viewed with surprise, for instance that the so-called 'smoky' vine should bear a white cluster where it bore a black and a black where it bore a white. For not even the soothsayers interpret such occurrences. Indeed the people living where the country is of a nature to produce the changes (as we said the pomegranate changed in Egypt [*HP* 2 2. 7]) are not surprised at the changes there; it is the change in Greece that is surprising."

μαστόν, οὐδ᾽ ὅλως ἄτοπον φαίνεται, τῷ τὴν φύσιν ὁτὲ μὲν οὕτως, ὁτὲ δ᾽ ἐκείνως,[1] καὶ τοῖς μέρεσιν ἀνομοίως, φέρειν, οἷον τὴν κάπνεον[2] ἄμπελον καλουμένην· αὕτη γὰρ δοκεῖ τοὺς μὲν λευκοὺς ἐνίοτε, τοὺς δὲ μέλανας, καὶ ὁτὲ μὲν πάντας τοιούτους, ὁτὲ δέ, τοιούτους φέρειν· διὸ καὶ οὐδ᾽ οἱ μάντεις εἰώθασιν τοῦτο κρίνειν ὡς τέρας· τὸ γὰρ εἰωθὸς οὐ τέρας.

3.2 αἴτιον[3] τῆς παραλλαγῆς, ὅτι τὸ χρῶμα τοιοῦτον[4] <ἔχει>[5] τῶν βοτρύων, οὔτε μέλαν οὔτε λευκόν [ἔχει],[5] ἀλλὰ καπνῶδες[6] (ὅθεν καὶ τὴν προσηγορίαν ἔσχεν). ἐὰν οὖν μικρά τις ἐναλλαγὴ γένηται, δοκεῖ μεταβεβληκέναι τὴν χρόαν. οὐδέτερον <δ᾽>[7] εἰλικρινές, οὔτε τὸ μέλαν οὔτε τὸ λευκόν, ἔχει· διὸ καὶ ἐνίοτε τῶν βοτρύων ἑτερόχροοί τινες, οἱ μὲν εἰς τὸ λευκόν, οἱ δ᾽ εἰς τὸ μέλαν μᾶλλον ἀποκλίναντες. αἰτία δ᾽ ἡ τροφή, καὶ ἡ

[1] u aP : ἐκεινος U.
[2] Wimmer (*cf.* κάπνειον *HP* 2 3. 2 : κάπνεων Schneider) : καπνέων U. [3] αἴτιον <δὲ> aP.
[4] ego : τοιούτων U : τούτων u. [5] ego.
[6] aP : καπνωνσε U : καπνώσε u : καπνῶσε N. [7] aP.

---

[1] *Cf.* Aristotle, *On the Generation of Animals*, iv. 4 (770 b 17–24): "This is why people speak of portents neither here nor in other matters where something customarily

not regarded as even wonderful nor yet as in general odd, because it is the nature of the tree that bears the fruit now in this way, now in that, and differently on different parts of the tree, as with the so-called "smoky" vine; for this vine is held to bear at some times some of its clusters of a white colour, others of a black, and sometimes to bear all of them of the one colour, sometimes of the other. This is why even the soothsayers are not accustomed to interpret the matter as a portent, the customary being no portent.

3.2 The cause of the variation is that the vine has a colour in the clusters of this sort: neither black nor white, but smoky (from which circumstance it took its name). So if a small variation in this smokiness occurs, the vine is held to have changed its coloration. But the vine has neither shade, neither the black nor the white, unmixed with the other; and for this reason at some times some of the clusters differ in shade from the rest, some inclining more to white, some to black.[1] The cause is the food and

occurs, as in *pericarpia*. For there is a certain vine which some call 'smoky'; and if it bears black clusters the event is not adjudged a portent, because the vine has the custom of doing this very often indeed. The cause is that the vine in its nature is intermediate between a white vine and a black, so that the transition covers no great distance nor yet is it as it were unnatural, for it is not a transition to another nature."

διάθεσις αὐτῆς τῆς ἀμπέλου.

3.3 τοῦτο μὲν οὖν (ὥσπερ ἐλέχθη) συγχωρεῖται τῷ μὴ θαυμαστῷ.

τὸ δ' ἐπὶ τῶν ῥοῶν, εἰ μὲν ὅλων τῶν δένδρων γίνονται μεταβολαί, παρόμοιον ἄν τι τὸ συμβαῖνον ἦν [1] τοῖς περὶ Κιλικίαν καὶ Αἴγυπτον (εἴπερ αὐτόματος ἡ μεταβολή), τῆς γὰρ χώρας τὸ πάθος καὶ ἡ δύναμις.

εἰ δὲ μὴ ὅλων, ἀλλά ποτε τοῦτο συνέβη, τῇ τροφῇ καὶ τῇ διαθέσει τὴν αἰτίαν ἀπολογιστέον, [2] ὡς ἐκ τοιαύτης γεγενημένης, [3] καὶ κρατηθείσης [4]

[1] ἦν U : εἴη Schneider.
[2] ego (*imputanda est* Gaza : ἀποδοτέον Heinsius) : ἀπολογητέον U.
[3] U[ar] : -νη U[r] : γενομένη N aP.
[4] U[cc] (from -θήσης) : κρατηθείση U[r] N aP.

---

[1] *CP* 5 3. 1.

[2] *Cf.* the language of *HP* 2 3. 1 (cited in note 4 on *CP* 5 3. 1): the entire tree is changed if it becomes a sweet pomegranate instead of an acid one or vice versa, whereas only the fruit is changed if a sweet tree produces acid fruit or an acid tree sweet fruit. Perhaps the language as well as the distinction came from the soothsayers.

[3] The apodosis is contrafactual in form: Theophrastus inclines to discredit such a change: *cf.* *CP* 5 3. 6.

[4] *Cf.* *HP* 2 3. 2 (cited in note 4 on *CP* 5 3. 1); *cf.* also *HP* 2 2. 7 (of the inferiority of trees produced from seed):

the condition of the vine itself.

This change in the vine, then, is (as we said)[1] conceded to belong to what is no occasion for surprise. 3.3

The Unnatural [A2] : *Changes in the Character of the Fruit:* (2) *The Pomegranate*

As for the case of the pomegranates,[1] supposing that the change is a mutation of the entire tree,[2] what happens would be similar[3] to what happens to the trees in Cilicia and Egypt[4] (that is, if the mutation is spontaneous), since the result and the power that effects it belong to the country.

If the change does not affect the entire tree, but occurred only on one occasion, one must assign the causation to the food and the condition of the tree, on the ground that the pomegranate-fruit has been produced by a tree in this condition or that,[5] and

"Regions and climates differ as well in this matter. For in some places the country is held to bring forth trees of equal excellence with the parent, as at Philippi. Few trees in few places undergo the opposite change and produce from a wild seed a cultivated tree, or from an inferior tree a tree that is simply an improvement, for we have heard of this occurring only with the pomegranate in Egypt and Cilicia: that in Egypt the acid pomegranate . . . becomes sweet after a fashion . . . , and that at Soli . . . all pomegranates are produced without kernels." *Cf.* also *CP* 1 9. 2; 2 13. 5; 2 14. 2; 6 18. 6–7.

[5] One condition (a non-permanent state) producing acid fruit, the other sweet fruit.

ὅ τι ἂν ἐθίσῃς.[1]

εἰ[2] δ' ἀνωμάλους ἤνεγκεν, <ἐν>[3] ταῖς ῥίζαις ἡ αἰτία καὶ εἴ τι μέρος ἄλλο κύριον, τῷ τὰ[4] μὲν τοιαύτην, τὰ δὲ τοιαύτην, λαβεῖν, ἢ καὶ ποιῆσαι καὶ ἀναδοῦναι, τὴν τροφήν· ἅμα δὲ καὶ τὸν ἀέρα συνεργεῖν τι, καὶ γὰρ ὑπὸ τούτων γίγνονται διαφοραί.

3.4 δόξειε δ' ἂν ἀτοπώτατον ἐνταῦθα εἶναι τὸ μερίζεσθαι τὰς ῥίζας· ἐνίοτε γὰρ ἐκ θατέρου μέρους οὐκ ἔχει τὸ δένδρον, ἢ ἐλάττους, ὁ δὲ καρπὸς ὁμοίως πανταχόθεν, ὡς ἀναμιγνυμένης δῆλον ὅτι πανταχοῦ τῆς τροφῆς.

οὐ μὴν ἀλλ' ἴσως, ἐὰν μὲν ἐκλίπωσίν[5] τινες, ἀπὸ τῶν δένδρων[6] ἡ τροφὴ πᾶσιν· ἐὰν δὲ πανταχόθεν ἔχῃ, τὸ καθ' αὑτὴν ἑκάσταις[7] οὐ κατὰ λόγον (ὡς ἄν τις φήσειεν), εὐθυπορούντων πως τῶν πόρων; ἔνια γὰρ συμβαίνει καὶ ζῆν[8] τῶν φυτῶν

[1] ego (ἐν τῇ ἀνθήσει Wimmer) : ὅτι ἂν ἐθησης U : ὅτι ἂν ἐθίσῃ u.
[2] u : ἡ U.
[3] Schneider.
[4] U : τὰς Schneider.
[5] U : ἐκλείπωσί u a (-σι N P).
[6] U : ἑτέρων Schneider.
[7] u N (ἐ- U) (ἑκάστην Wimmer) : ἑκάστῃ aP.
[8] U^{c} : ζη U^{ac}.

been mastered by whatever you accustom it to.

If the tree has borne fruit of two different characters, the causation is in the roots and any other part that may determine such matters, because some of the roots or other part have received (or prepared or distributed) food of one sort, others food of another sort (the air too contributing to the result), since differences also arise from these matters.[1]

An Objection Answered

3.4 In this last explanation the point might appear very odd that the roots divide their office and each serve a separate part of the tree, since the tree sometimes has no roots on the one side (or fewer), but the fruit is nevertheless borne equally on all sides, the implication being that the food is thrown together throughout the tree.

Nevertheless, whereas it is perhaps true that when a portion of the roots is missing the food for all parts comes from the whole tree, on the other hand, when the tree has roots on all sides, is it not reasonable (someone might urge) that each set of roots should feed its own side of the tree, the passages being more or less straight? Since some young trees

[1] The three sources of change are (a) internal to the tree (discussed in this paragraph, with the air thrown in); (b) from the country (discussed in the second paragraph of § 5 3. 3); (c) from the art of agriculture (discussed in the third).

ἐκ θατέρου[1] μέρους· αὐξανόμενα γὰρ συνεκπληροῦν πως τὸ ὅλον, τὰ μὲν ἐμπεριλαμβάνοντα,[2] τῶν δὲ καὶ ἀποπιπτόντων διὰ τὸ αὖον.

εἰ δέ τις καὶ αὕτη[3] πίστις ἐστίν, ἐκ τῶν συντιθεμένων κλημάτων μὴ ὁμογενῶν (ὅτι μία μὲν ἡ ἄμπελος, φέρει δὲ ἑκάτερον[4] τῶν μερῶν τὸν οἰκεῖον καρπόν), ὡς οὐ μιγνυμένης, ἀλλ' εὐθυπορούσης τῆς τροφῆς, εἴη δ' ἂν καὶ ταύτῃ χρῆσθαι.

τοῦτο μὲν ἥκιστα γινόμενον καὶ ἥκιστα πίστιν ἔχον ἐστίν.

3.5 αἱ δ' ἄλλαι μεταβολαὶ μάλιστ' ἂν[5] τὰς εἰρημένας ἔχοιεν αἰτίας, ὁμοίως ἐπὶ πάντων, εἴτε χρώμασιν,[6] εἴτε χυλοῖς, εἴτε ἄλλῳ τινὶ μεταβάλλουσιν· αἱ γὰρ τροφαὶ <τὰς>[7] διαθέσεις ἀλλοιοῦσι καὶ μεθιστᾶσιν,[8] ὁτὲ μὲν κρατούμεναι,

[1] u : -ους U. [2] Scaliger : εκπ- U.
[3] Scaliger : αὐτῇ U N : αὐτὴ u aP.
[4] u (ἑκατέρῳ Gaza) : εκάτερων U : ἑκατέρων N aP.
[5] ego (ἂν inserted before ἔχοιεν aP): μάλιστα U N.
[6] Scaliger (*colore* Gaza) : κράμασιν U. [7] Schneider.
[8] Heinsius (*permovere . . . possunt* Gaza) : μεθίασιν U.

---

[1] They wither because they get no food from the roots on their own side, which have been killed, and there is no common store in the tree from which they could be fed. The enclosed parts, if they survive, are fed from the parts enclosing them.

even come to live from the side that remains, for as they grow they in a way fill out the whole circumference, enclosing some of the other parts, whereas some of these actually drop off because they are withered.[1]

Further, if the following argument has any force as evidence that the food is not thrown together in the tree, but moves in a straight line—the argument from vine twigs spliced together, since a single vine results and each of the two parts bears its own type of fruit[2]—one could appeal to this evidence too.

This splicing however is of the rarest occurrence and has the least force as evidence.

The Unnatural [A2] : *Changes in the Fruit:* (3) *The Other Changes*

3.5 But the other changes[3] would mainly have the causations mentioned[4] in all matters alike, whether the change is in colour or flavour or something else[5]: differences of food alter and shift the condition of a tree, the food sometimes being mastered

[2] *CP* 5 5. 1.

[3] Mentioned in *CP* 5 3. 1 and *HP* 2 3. 1–2.

[4] *CP* 5 3. 2–4.

[5] *HP* 2 3. 1 also mentions changes from wild-fig to fig and vice versa and from olive to wild-olive and vice versa.

ὁτὲ δὲ κρατοῦσαι, καθάπερ καὶ ἐν τοῖς ἀπὸ τῶν σπερμάτων φυομένοις, πλὴν ἐνταῦθ' ἡ ἔκστασις ἀεὶ πρὸς τὸ χεῖρον, ἀλλ' ἐπὶ τῶν ἐξειργασμένων καὶ ἡμερωμένων ἔχει τὴν ὁμοίωσιν.

3.6 ἁπλῶς δ' (ὥσπερ ἐλέχθη) μικράν τινα χρὴ νομίζειν ἐν τούτοις εἶναι τὴν μεταβολήν, ὅπου μὴ ἡ χώρα μεταβάλλει, τὰ δὲ τοιαῦτα χρόνῳ γίνεται, καὶ οὐκ ἐπέτεια, καθάπερ οὐδὲ τὰ ἡμερούμενα καὶ ἀγριούμενα. μέγα δ' οὖν πρὸς πίστιν τῆς μεταβολῆς τὰ παρὰ[1] τὰς χώρας, καὶ ὅτι θεραπεῖαί τινες τῶν ῥιζῶν καὶ τῆς τροφῆς ποιοῦσι μεταβολάς· ἀλλὰ τὸ οὕτως ἐν βραχεῖ χρόνῳ καὶ ταχέως ἄπιστον, εἰ μή τις ἄμφω ταῦτα συνθήσει· καὶ τὸν χρόνον λανθάνειν, καὶ τὴν μεταβολὴν εἶναι βρα-

[1] ego (τῆς παρὰ Schneider) : τὰ περὶ U.

---

[1] *Cf. HP* 2 2. 4 (of trees): "Those grown from cuttings are all held to produce fruit like that of the parent, whereas those grown from the fruit [*i.e.* the seed] (among those able to grow in this way too) are all more or less inferior"; *CP* 2 15. 2.

[2] Such cases as that of the trees at Philippi (*cf. HP* 2 2. 7, cited in note 4 on *CP* 5 3. 3) are meant.

[3] *CP* 5 3. 2 ("a small variation"); but Theophrastus is also referring to the preceding discussion (*CP* 5 3. 1–5 3. 4), where the only major change is that of the whole tree (5 3. 3, second paragraph), which is there compared to,

and sometimes mastering (as it does in trees grown from seed, except that here the departure is always for the worse[1]; still, when the ground is carefully worked and brought under cultivation the tree may be as good as the parent).[2]

3.6 Broadly speaking we must take the change in these cases to be a small one (as we said),[3] except where a new country brings it about[4]; and changes due to the country take time and do not occur in a year (no more than does the change of a tree from wild to cultivated or from cultivated to wild).[5] The credibility of the occurrence of change, at any rate, is greatly supported by the difference made by a difference of country and by the fact that change is brought about by certain agricultural procedures affecting the roots[6] and the food.[7] But that the change should occur in so short a time and so suddenly[8] is unconvincing, unless one is to combine these two circumstances: that the lapse of time is

and here equated with, a change of country.

[4] As with the pomegranate in Cilicia and Egypt (*cf.* note 4 on *CP* 5 3. 3).

[5] *Cf.* note 5 on *CP* 5 3. 5.

[6] *CP* 5 3. 3, last paragraph; *cf.* the manures of *CP* 2 14. 2; 3 9. 3.

[7] *CP* 5 3. 3, last two paragraphs; *cf.* *CP* 2 14. 2 (watering with plenty of cold water).

[8] As is implied by its being taken for a portent: *cf.* *HP* 2 3. 1–3 and *CP* 5 3. 1.

3.7 χεῖαν. ἐπεὶ ἔν γε τοῖς ἐπετείοις οὐκ ἄλογον, οὐδ' ὁμοίως ἐφ' ὧν συμβαίνει θαυμαστόν (εἴπερ ἄρα συμβαίνει), καθάπερ ὅταν ἐκ πυρῶν αἶρα γένηται, διά τε τὸ πολλάκις γίνεσθαι, καὶ διὰ τὸ τὴν αἰτίαν οἴεσθαί πως (αἵ γὰρ[1] ὑπερομβρίαι ποιοῦσιν)· εἰ δὲ καὶ ἐκ τῶν αἰρῶν εἰς πυροὺς πάλιν ἀποκαθίσταται, τοῦτο θαυμασιώτερον· εἴη δ' ἂν ὥσπερ ἀσθένειά τις ἢ νόσος γεγενημένη τοῦ σπέρματος, ἥτις ἀπολύεται, μεθισταμένων ἄρα τῶν περὶ τὸν ἀέρα καὶ τὴν τροφήν.

ἀλλὰ γὰρ ταῦτα μὲν ἐπὶ πλέον εἴρηται, συμπαραλαμβάνοντι[2] πρὸς τὴν τῆς αἰτίας πίστιν τὰ ὁμολογούμενα.

4.1 τὸ δ' ἐνεγκεῖν ἄμπελόν ποτε καρπὸν ἄνευ φύλλων, ὡς μὲν ὅλως εἰπεῖν οὐ πιθανόν, ὡς δὲ μικρᾶς τινος γινομένης βλαστήσεως, καὶ ταύτης διὰ τὴν ἀσθένειαν ἀπορρυείσης, μᾶλλον πιθανόν, ὡς ἐν-

[1] Schneider (*namque* Gaza) : αἵτε U.
[2] Schneider : -λάμβανοντα U.

---

[1] *Cf.* *CP* 2 16. 3.
[2] *Cf.* *HP* 4 4. 5. It is rarer and an improvement, and nobody has any notion of the cause.

3.7 not noticed, and the change small. Indeed in annuals the change is not unreasonable or taken for such a wonder in the cases where it occurs (if it does occur), as when darnel comes from wheat, both because it is frequent and because people have some notion of the causation, which is heavy rains.[1] What is more of a wonder is if darnel changes back to wheat.[2] The cause would be some indisposition (as it were) or disease that had arisen in the wheat seed, and that is cured (one would suppose) by a shift of conditions in the air and food.

But we have discussed[3] the present topic at some length, including in the discussion matters that are conceded,[4] in order to establish the plausibility of the causation.

The Unnatural [A1] : *Fruit Without Leaves*

4.1 That a vine once bore fruit but no leaves[5] is not credible when you take it to mean "no leaves whatever." But if you take it to mean that sprouting occurred and that there was little of it, and the little dropped off later, owing to its weakness, the report

[3] *CP* 5 3. 1–7.

[4] That is, the case of the smoky vine, which is conceded by the soothsayers to be no portent or unnatural occurrence (*CP* 5 3. 3, first sentence).

[5] *Cf. HP* 2 3. 3: ". . . a vine has borne fruit without having any leaves."

ταῦθα πλείονος καὶ σφοδροτέρας τῆς ὁρμῆς γινομένης· ὅπερ ἐπί γέ τινων δένδρων καὶ συμβαίνει σχεδόν, ὅταν εὐκαρπήσωσιν, ὥσπερ ἐπὶ τῆς ἀμυγδαλῆς· διὰ γὰρ τὸ πλῆθος οὐδὲ φαίνεται τὰ φύλλα, καὶ ἄλλως [1] μικρὰ καὶ ἀσθενῆ γίνεται, τῆς φύσεως ἐνταῦθα ὡρμηκυίας, ὥσπερ καὶ ἐν τῷ ἀνθεῖν ἡ ἄμπελος ὅταν ὀψίσῃ καὶ διατηρῇ τὸν καρπὸν ἄνευ τῶν οἰνάρων, ὥσπερ καὶ ἡ συκῆ τῶν θρίων. 4.2 ἰσχυρότεροι γάρ, ὅταν ἅπαξ συστῶσι, γίνονται [2] τῶν φύλλων, οἱ μὲν διὰ τὸν μίσχον [3] καὶ τὴν προσάρτησιν, οἱ δὲ διὰ τὸ πιλοῦσθαι τῷ ψύχει· τῶν μὲν γὰρ φύλλων ἐξαιρεῖται τὴν ὑγρότητα καὶ τὸν ὀπόν, τῶν δὲ καρπῶν, διὰ τὸ πλῆθος καὶ τὸ μᾶλλον ἔχειν θερμότητα, τοῦτο μὲν οὐ [4] δύναται, συμπυκνῶσαι <δὲ> [5] καὶ πιλῶσαι.

παραπλήσιον δὲ τούτῳ καὶ εἴ ποτέ τις ἐλαία τὰ μὲν φύλλα ἀπέβαλεν, τὸν δὲ καρπὸν ἐξήνεγκεν· ἀσθενέστερον γὰρ ὄν, μᾶλλον ὑπήκουσεν τοῦ ψύχους ἢ εἴ τι πάθος ἦν [6] ἕτερον, πεπανθεὶς μὲν γὰρ

[1] Gaza (*alias*) : ἄλλα U N aP : ἀλλὰ u.
[2] u N P (-ωνται a) : γινον U.
[3] Gaza (*petioli*), Scaliger : σμιχον U.
[4] U^{r} N aP : μὲν οὖν U^{ar} : μόνον Wimmer.
[5] Dalecampius (*sed* Gaza).
[6] ego (Hindenlang deletes) : εἰή (εἰῆ?) U.

is more credible, since this means that the growing impulse was greater and more vigorous in the fruit. The same phenomenon in fact occurs (one might say) in certain trees when they have borne a heavy crop, as with the almond. Here the fruit is so abundant that the leaves are not even visible, and in any case come out small and weak, since the nature of the tree has directed itself to the production of fruit. So too when the vine is late in its flowering, and keeps its fruit without keeping its leaves,[1] as the fig-tree does too. For the fruit, once it has set, comes to be stronger than the leaves, some of it because of the pedicle and the attachment to the tree, some from compression by the cold (for although cold deprives the leaves of their fluid and juice,[2] it cannot do this to the fruit, since the fluidity is here more abundant and possesses more heat, but can only condense and compress it). 4.2

Similar to this is the case where an olive shed its leaves but brought forth its fruit[3]: the leaf, being weaker than the fruit, was more responsive to the cold (or whatever it was that happened); for

[1] *Cf. HP* 2 3. 3: "This [that is, fruit on the tree but no leaves, as with a vine and an olive] occurs because of wintry weather . . ."

[2] *Opós* in the Greek. It refers to the fig-juice; "fluid" refers to the vine.

[3] *Cf. HP* 2 3. 3: "And an olive tree lost its leaves but brought forth its fruit; and this is said to have happened to Thettalus son of Pisistratus."

ὁ καρπὸς αὐτόματος ἀπορρεῖ, πρὸ δὲ τοῦ πεπανθῆναι, μᾶλλον ἰσχύει καὶ προσήρτηται (διὸ καὶ ῥαβδίζουσιν τὰς ἐλαίας).

4.3 ταῦτα μὲν οὖν καὶ ὅσα ἄλλα τοιαῦτα φυσικάς τινας ἀρχάς, τὰ μὲν ἐξ αὐτῶν,[1] τὰ δ' ἐκ τοῦ περιέχοντος ἔχει.

ἐπεὶ καὶ τὰ αὐτόματα διαβλαστάνοντα ξύλα (καθάπερ τὰ ἐλάϊνα καὶ εἴ τι ἄλλο τοιοῦτον), ἅπερ εἰς τέρα[2] καὶ σημαῖα ἀνάγουσιν, οὐκ ἔστιν ἄλογον· φύσει τε γὰρ εὔζωα καὶ βλαστητικὰ διὰ πυκνότητα καὶ τὸ ἔγχυμον,[3] καὶ ὅταν ἔξωθεν ἰκμάδα τινὰ λάβῃ, ταχὺ δίδωσιν·[4] ὡς <δ'>[5] ἐπὶ τὸ πολὺ κατορυττόμενα καὶ ἐν ὑγρῷ τόπῳ βλαστάνει (πλὴν εἴ τι κοπὲν ὕστερον μικρῷ διεβλάστησεν, ἔχον ἐν ἑαυτῷ συνηθροισμένην τὴν γόνιμον ὑγρότητα, ἅμα δὲ καὶ τῆς ὥρας ὑπογύου[6] τῆς βλα-

[1] U : αὐ- Scaliger.
[2] U : τέρατα u (as at *CP* 5 4. 4).
[3] aP : ἔγιχυμον U : ἐγίχυμον u N.
[4] U : *germina edunt* Gaza : τὰς βλάστας ἀποδίδωσιν Schneider : ἀναδίδωσιν Wimmer.
[5] aP.
[6] u (-γείου N aP) : ὑποζυγίου U.

---

[1] *Cf.* *CP* 1 3. 3 (olive, myrtle, and wild olive wood); *CP* 1 7. 4, 1 12. 9, 5 1. 4 (olive wood); *HP* 5 9. 8: "Of woods

although the fruit drops of its own accord when ripened, yet before ripening it is stronger than the leaf and more firmly attached (which is why olive trees are cudgelled).

4.3 These occurrences and the like, then, have certain natural initiations, partly proceeding from the plants themselves and partly from the surrounding air.

The Unnatural [A1] : *Pieces of Wood that Sprout and the Like*

For that matter, even the pieces of wood that sprout of their own accord, as pieces of olive wood and the like,[1] and which are accounted as portents and signs, are not anything unreasonable. For the pieces are by nature tenacious of life and prone to sprout (owing to their close texture and possession of juice), and once they obtain some moisture from the outside, they quickly produce a shoot. For the most part they do so when buried in the ground and when they lie in a moist place, except for an occasional piece that sprouts shortly after being cut from the tree, when the piece contains a conflux of generative fluid already formed in itself, and its sprout-

prickly cedar and generally speaking those whose fluid is oily that give out exudations . . . It is mainly pieces of olive wood that sprout . . ."

στητικῆς[1] οὔσης).

4.4 τούτῳ[2] δὲ ὅμοιον τρόπον τινὰ[3] καὶ <τὸ>[4] ἐπὶ τῆς σκίλλης καὶ ἐπὶ τῶν ἄλλων τῶν ἐκβλαστανόντων·[5] ὡσαύτως δὲ καὶ τὰ ἐκ τῶν ξύλων ἐκφυόμενα, καὶ μάλιστα ἐκ τῶν ἐλατίνων, ἃ καλοῦσιν οἱ μάντεις εἰλειθυίας·[6] ἀνειμένου γὰρ ὄντος καὶ μαλακοῦ τοῦ ἀέρος ἐκφύεται μάλιστα, ὁπότε ἡ ἐνυπάρχουσα ὑγρότης συρρυεῖσα, καὶ ἡ ἔξωθεν προσπίπτουσα, συνεπάγη καὶ ἐποίησεν οἷον σφαιροειδές.

ὁμοίως δὲ καὶ ὅσα ἰδίει[7] τῶν ξύλων· καὶ γὰρ ταῦτα, νοτίου καὶ ὑγροῦ τοῦ ἀέρος ὄντος, τουτὶ πάσχει, καὶ οὐ πάντα, ἀλλ' ἐν οἷς ἐστι λίπος[8] (οἷον κέδρου κυπαρίττου ἐλάας), ἃ δὴ καὶ σημεῖα καὶ τέρατα νομίζουσιν.

4.5 ὅσα δὲ ἄλλα συμβαίνει καὶ προφέρεται τῶν

[1] u : βλαστικῆς U.

[2] τούτω u : τοῦτο U.

[3] Gaza (*simile quodammodo*), Schneider : ὁμοιό|τροπόν τινα U^{ar} : ὁμοιότροπόν τι U^{r} N aP. [4] ego.

[5] U (*suspensa germen emittunt* Gaza) : ἔξω βλαστανόντων Schneider.

[6] Heinsius : εἰλυθυίας U N aP : εἰληθυίας u.

[7] Gaza (*exudant* : ἰδίει Scaliger) : διει U.

[8] ego (*aliquid pinguedinis* Gaza : λιπαρότης Wimmer) : λεπτῆς U N : λεπὶς u : λεπτὴ aP.

ing season is close at hand.

The occurrence is also similar in a way in the case of squill[1] and the rest that send out a sprout. So too with the growths sent out of pieces of wood, especially from those of the silver fir (which the soothsayers call *Ilithyiai*[2]), for they mainly grow out when the air has lost its severity and is gentle, whenever the fluid present in the wood has formed a conflux and the fluid in the air comes in contact with it and the two coalesce and solidify to something resembling a ball. 4.4

Similarly too where pieces of wood sweat,[3] for this too occurs when the air is southerly and moist; and not all wood sweats in this way, but only the ones containing oiliness, as that of prickly cedar, cypress and olive, which are the cases that are regarded as signs and portents.[4]

All other such marvels as occur and are brought 4.5

[1] *Cf. CP* 1 7. 4.

[2] *Cf. HP* 5 9. 8: "The thing called 'menses of Ilithyia' [that is, of the birth-goddess], to avert the ominousness of which they perform a sacrifice, occurs on wood of the silver fir when a certain fluid forms, round in shape and more or less the size of a pear."

[3] *Cf. HP* 5 9. 8: "Of woods prickly cedar and generally speaking those whose fluid is oily send out exudations. This is why people assert that the statues of gods sometimes sweat."

[4] These were among the woods favoured for statues of the gods: *cf. HP* 5 3. 7.

τοιούτων, οἷον ὥς ποτέ φασιν ἐν πλατάνῳ φῦναι δάφνην (ἢ ἁπλῶς περὶ τὰς ἐμβλαστήσεις[1] τὰς ἐν ἀλλήλοις), ἐκ λανθανούσης ἀρχῆς ὑποληπτέον γίνεσθαι (καθάπερ ἐλέχθη)· τὸ γὰρ αὐτοῦ τοῦ δένδρου τοιαύτην τινὰ γίνεσθαι σῆψιν ἢ ἀλλοίωσιν, ἄλλως τε καὶ πολὺ διεστώσης, οὐκ εὔλογον, οἱ δὲ τρόποι τοιοῦτοι.

λέγω δὲ "λανθανούσης," εἰ ἐπιπέσοι σπέρμα καὶ ἔμβιον γένοιτο, σῆψιν ἔχοντος γεώδη τινὰ τοῦ δένδρου (βλάστοι[2] γὰρ ἂν οὕτως ἕτερον ἐν ἑτέρῳ). ἀλλ' ἡ τοιαύτη βλάστησις ὁμοία τῇ τῶν ἰξιῶν, ἢ ὡς διὰ πλείονος ἔτι,[3] τῇ πλατάνῳ ἐκ τοῦ χαλκοῦ τρίποδος· μᾶλλον δ' ἀμφότεραι τῇ ἐκ τῆς γῆς.

4.6 οὐ γὰρ οἷόν τε φύναι μὴ γεώδους τινὸς ἐνυπάρχοντος· ὥσπερ οὐκ[4] ἐκ τῶν τοίχων τῶν λιθίνων ἐὰν μὴ τοιαύτη τις συρροὴ γένηται καὶ σῆψις ἐξ ἧς

[1] a : ἐνβλαστησεις U (-ή- u N P).
[2] βλαστοι U : βλαστοὶ u[ac] : βλαστοίη u[c] N aP.
[3] ego : ἐπὶ U.
[4] U : οὐδ' Gaza (*neque*), Schneider.

---

[1] *Cf. CP* 2 17. 4.
[2] *CP* 1 5. 3.
[3] Unnoticed.
[4] *Cf. CP* 2 17. 4.
[5] *Cf. CP* 2 17. 5, 8 (it comes from bird-droppings that

forward, such as the story of a bay that grew in a plane tree[1] (or in general all instances of one plant sprouting in another) we must suppose are due (as we said)[2] to growth from an unnoticed origination, since it is not reasonable that "decomposition" or alteration of this productive sort should arise in the host tree by the host tree's own doing, especially when the shoot is of a kind very remote from the tree's own shoots (and the forms these marvels take involve such remoteness).

I mean by "unnoticed origination" the case where a seed[3] alights and germinates on a tree that has acquired some earthy decomposition,[4] for under these circumstances different kinds of plants might sprout in one another. But this sort of sprouting is like that of the mistletoe,[5] or, to take still remoter partners, like that of the plane tree that came up from the bronze tripod[6]; or rather both are like sprouting from the ground. For it is impossible for the seed to grow unless there is an earthy spot in the host, just as nothing can grow from stone walls unless this sort of collection of fluid has first arisen and the sort of decomposition[7] from which the wall-

4.6

contain the seed and bring about a certain change in the host).

[6] *Cf. HP* 3 1. 3 (where the case is cited to prove that the elm grows from seed). There was evidently also some earth in the tripod.

[7] It must be earthy.

πέφυκε βλαστάνειν, ἔτι δ' ὕστερον ἐπιρροήν τινα λαμβάνῃ τῆς τροφῆς (οὕτω γὰρ αὐτῶν ἡ αὔξησις).

ἀλλὰ τὰ μὲν τοιαῦτα καὶ αὐτομάτως (ἢ πάντα ἢ ἔνια) γίνεται (τάχα δὲ καὶ σπερμάτων τινῶν καταρρυέντων ἅμα τῇ σήψει καὶ συστάσει)· δάφνη δὲ (καὶ εἴ τι τοιοῦτον ἕτερον) ἀπὸ τῶν καρπῶν (εἰ δ' ἄρα καὶ τοῦτο ἀπὸ σήψεώς τινος, οὐδὲν διαφέρει πρὸς τὴν αἰτίαν).

4.7 εἰ δέ ποτε δένδρον ἐκπεσὸν ὑπὸ χειμῶνος ἤδη <κατέστη> πάλιν[1] αὐτόματον, ὥσπερ ἐν Φιλίπποις μὲν ἰτέα,[2] <ἐν> δὲ <'Αντά>νδρῳ[3] πλάτανος, καὶ τῆς μὲν οὐδὲν ἀφῃρέθη πλὴν ὅσοι τῶν

[1] ego (*restibilis . . . facta et vitae reddita est* Gaza : ἤδη πάλιν ἀνέστη Itali : ἀνέστη πάλιν Wimmer) : ἤδη πάλιν U.

[2] Gaza, Itali : εἴτε U.

[3] Wimmer (*in antandro* Gaza : ἐν ἀντάνδρῳ δὲ Itali [ut vid.], Basle ed. of 1541) : δενδρω U.

---

[1] The description covers both sprouting from seed and spontaneous generation.

[2] That grow on stone walls.

[3] The seed has apparently never been seen (like the "seeds" of Anaxagoras [*CP* 1 5. 1] that cannot be observed [*cf. HP* 3 1. 5 *init.*]).

plant is naturally fitted to sprout,[1] and unless the plant further obtains some subsequent supply of food (for it is under this condition that these plants grow larger).

But plants of this sort[2] (at least) are also produced (either all or some of them) spontaneously (as well as perhaps by seeds of a sort[3] that are carried down by the rain at the time when the decomposition and conflux of fluid are present).[4] But the bay[5] and the like[6] come on the other hand from their fruit[7]; and even if growth from the fruit in these instances involves some decomposition,[8] the circumstance makes no difference in the causation.

The Unnatural [A1] : *Trees that Righted Themselves*

4.7 If ever a tree blown down by a storm returned to its place of its own accord, as a willow did at Philippi and a plane at Antandrus, no wood being taken from

[4] Some sort of food is needed; no seed could otherwise grow from a stone wall. The conflux is needed to promote decomposition of the wall and provide food.

[5] Like the bay that grew in the plane tree (*CP* 5 4. 5 *init.*).

[6] "All instances of one plant sprouting in another" (*CP* 5 4. 5).

[7] That is, seeds.

[8] *CP* 5 4. 5, second paragraph.

ἀκρεμόνων κατεκλάσθησαν ἐν τῇ πτώσει, τῆς πλατάνου δὲ ἀφῃρέθη, καί τι παρεπεληκήθη—τὴν δ' αἰτίαν <τις ἂν>[1] ὑπελάμβανεν[2] ὅτι πεσοῦσα ἐπὶ θάτερον μέρος ἀνέσπασε πολλὴν γῆν, ἐπιγενομένου δ' εἰς νύκτα τότε πνεύματος ἐναντίου καὶ μεγάλου, κινήσαντος αὐτὸ[3] διὰ τὸ ἐμπίπτειν τοῖς ἀκρεμόσιν, ῥοπὴν ἐποίησεν ἐγκείμενον τὸ βάρος,[4] καὶ κατασπάσαν ὤρθωσεν· οὕτω γὰρ συνέβη τῷ[5] ἐν Φιλίπποις. τὸ δ' ἕτερον ἐκινήθη μὲν[6] ὁμοίως, καὶ τὴν ἀνάσπασιν εἶχε τῆς γης, διὰ δὲ τὴν περικοπὴν ἀνέστη ῥᾷον.

ἀλλὰ γὰρ ταῦτα μὲν ἴσως ἔξω φυσικῆς αἰτίας ἐστίν· ὑπὲρ δὲ τῶν ἐν αὐτοῖς τοῖς φυτοῖς ἐκ τῶν

[1] ego.
[2] U : ὑπελάμβανον Gaza, Scaliger.
[3] u : αὐτοῦ U.
[4] ego : μερος U.
[5] τῷ U N aP : τὸ u.
[6] u aP : ἐκινήθημεν U N.

---

[1] *Cf. HP* 4 16. 2–3: "Some trees endure being hewn with the axe both when standing and when blown down by the wind, to such an extent that they rise again and live and sprout, as willow and plane. This occurred both at Antandros and Philippi. When the plane had fallen and its branches had been cut off and its trunk hewn it rose up again in the night, relieved of the weight, and

the willow except the branches broken in the fall, whereas wood was removed from the plane and some hewn away at the side of the trunk,[1] one would have taken the causation to be this: the tree in falling to one side pulled up a quantity of earth on the other, next a strong wind arose in the night from the opposite direction and set the tree rocking by blowing on the branches, and the weight of the pulled up earth turned the scale by bearing down, pulling the tree back and righting it. For this is what happened to the tree at Philippi. The other tree was similarly set rocking by the wind and had a weight of similarly pulled up earth, but was more easily righted because wood had been removed from all sides of it.

Still these cases perhaps fall outside the realm of a natural cause.[2] In dealing however with occurrences in the plants themselves we must

recovered, and its bark grew round it once more. Wood had been hewn from two-thirds of its girth. The tree was tall, more than twelve cubits high, and so big around that four men could not easily have encompassed it. The willow at Philippi had had the branches on one side cut off (παρεκόπη ego : περιεκόπη U), but no wood had been hewn from the trunk. A certain soothsayer persuaded the people to hold a sacrifice and preserve the tree as having been a favourable portent."

[2] The cause was violent, and did not call into action the living processes of the tree.

εἰρημένων πειρατέον μετιέναι καὶ θεωρεῖν.

5.1 ἑπόμενα δέ πώς ἐστι τούτοις εἰπεῖν ὅσα διὰ τέχνης καὶ παρασκευῆς γίνεται τῶν[1] περιττῶν, ὑπὲρ ὧν φανερωτέρας ἄν τις ὑπολάβοι τὰς αἰτίας εἶναι, καθάπερ καί εἰσιν.

ἀγιγάρτους μὲν γὰρ ποιοῦσιν τοὺς βότρυς ἐξαιροῦντες τὴν μήτραν, ἀφ' ἧς γίνεται τὸ γίγαρτον· ἐκ τοῦ αὐτοῦ δὲ κλήματος[2] φέρειν λευκὸν καὶ μέλανα βότρυν, ἢ ἐν αὐτῷ τῷ βότρυϊ [τὰς][3] τοιαύτας, τὰς δὲ τοιαύτας, ὅταν διελόντες συνθῶσιν ἑκατέρου τὸ ἥμισυ, πλὴν [τὴν][4] τοῦ κάτω μέρους, καὶ συνδήσαντες καταπήξωσι· συμφύεται γὰρ

5.2 ἀλλήλοις. σύμφυτον μὲν γὰρ ἅπαν τὸ ζῶν τῷ ζῶντι (καὶ μάλιστα τὸ ὁμογενές) ὅταν ἀφελκωθῇ, καὶ γίνεται μία τις φύσις· ἑκάτερος δὲ καθ' ἑαυτὸν[5] καὶ τὴν τροφὴν διίησιν[6] ὥστε, μὴ ἐπιμιγνυ-

[1] u : γινετῶν U. [2] U[c] : κλήμακος U[ac].
[3] ego (ῥᾶγας τὰς μὲν Schneider). [4] Scaliger.
[5] U : ἑκάτερον δὲ καθ' ἑαυτὸ Schneider (*Sed uterque* [*sc. palmes*] *per se* Gaza).
[6] Schneider (*transmittunt* Gaza) : διείσιν U[c] (a miswritten εί superscribed) : διισιν U[t].

---

[1] *CP* 5 1. 2–5 4. 6. [2] *Cf. CP* 1 21. 2; 5 1. 1.

endeavour to investigate and understand them in the light of what has been said.[1]

Remarkable Effects of Art [B2]

5.1 Next in order (in a way) to these occurrences comes the discussion of the remarkable results of art and design. One would suppose that the causes here are more evident, as indeed they are.

(1) *In Trees*

Grape clusters without pits[2] are grown by removing the core, from which the pit is produced. The vine is made to bear from the same twig both white and black clusters, or the cluster itself both white and black grapes,[3] when two twigs are split and the halves of each (except for the lower part) put together and bound and the whole is then set in the ground, for the halves coalesce. 5.2 For anything alive can coalesce with what is alive (and especially if the source is a plant of the same kind) when a wound has been made, and the result is (in a sense) a single nature. But each of the two component shoots[4] also transmits its food separately (and since this is not

[3] *Cf. CP* 5 1. 1.

[4] The Greek leaves the noun to be understood; we supply βλαστός ("shoot") or καρπός ("fruiting shoot"). For this last *cf. CP* 1 12. 10.

μένης,[1] ἀποδιδόναι τὸν οἰκεῖον καρπόν, ὅπερ καὶ οἱ ποταμοὶ ποιοῦσιν οἱ συμβάλλοντες ἀλλήλοις, ὥσπερ ὅ τε Κηφισὸς ἐν τῇ Βοιωτίᾳ καὶ ὁ Μέλας, ἑκάτερος γὰρ ῥεῖ τὸν αὑτοῦ[2] πόρον. ἐνταῦθα δ' οὐδὲ συμβάλλουσιν, ἀλλὰ παρ' ἀλλήλας[3] ὀχετεύονται καὶ ῥέουσιν αἱ τροφαί.

5.3 ταὐτὸ δὲ καὶ παραπλήσιον τούτῳ[4] καὶ ὅταν τὸ αὐτὸ δένδρον παντοδαπὰς φέρῃ[5] ῥόας ἢ μῆλα· τῇ σφύρᾳ γὰρ οἷον μαλάξαντες τὰς ῥάβδους, ἵνα συμφυῶσιν [ἢ][6] διὰ τὴν ἀφέλκωσιν, συνδήσαντες ἐφύτευσαν, εἶτα γίνεται τὸ μὲν δένδρον ἓν τῇ συμφύσει, διατηρεῖ δ' ἕκαστον τὸ γένος, ἕλκον κατ' αὑτὸ[7] καὶ πέττον τὴν τροφήν, οὐδὲν δὲ ἄλλο ἢ τῆς συμφύσεως κοινωνοῦν.

σχεδὸν δὲ καὶ παρόμοιον τούτῳ καὶ ἐπὶ τῶν μειζόνων γίνεται, καὶ μάλιστα ἐπὶ τῶν ὑγρῶν[8] τῇ φύσει· περιπλακεῖσα γὰρ συκῇ καὶ εἴ τι ἄλλο τοιοῦτόν ἐστι συμφύεταί τε καὶ ἓν ποιεῖ τὸ στέλεχος.

5.4 καὶ τούτων τὰ μὲν ἐξεπίτηδες ποιοῦσιν, ἔνια δὲ καὶ αὐτομάτως[9] λαμβάνει τοιαύτην σύμφυσιν,

[1] Wimmer : επιμιγνυμένας U : ἐπιμιγνυμένως u.
[2] Scaliger (*suum* Gaza) : αὐτοῦ U.
[3] παραλληλας U : παρἀλλήλως u.
[4] u : τοῦτων U.
[5] φέρῃ u : φέρει U N aP.

intermingled, each brings forth its own fruit), which is what rivers do when they meet, like the Cephisus and Melas in Boeotia: each flows in a separate current. In the vine however the two currents do not even meet, but the food for each part flows in a separate and parallel channel.

5.3 The same or much the same occurs also when the same tree bears pomegranates or apples of all sorts. For growers first soften up (so to speak) the twigs with the mallet so that they may coalesce because of the bruising, and then bind them together and plant them. The resulting tree is to be sure a unity by reason of the coalescence, but each component preserves its character, drawing and concocting its food separately, and sharing with the others nothing but the coalescence.

Much the same (one might say) occurs also in larger trees and especially those of a fluid nature: thus the fig will entwine about another tree of this character and then coalesce with it and make a single trunk.

5.4 Some such unions are produced by design, but in others the trees come to coalesce in this way of their

---

[6] Schneider : *συμφυωσιν ἢ* U[cc] (from *σι-*) : *συμφύωσιν ἢ* u. Perhaps *η* once indicated a variant *συμφυῇ*.

[7] U : *καθ' αὑτὸ* Gaza (*per se*), Scaliger.

[8] ego : *ἀγριων* U.

[9] u : *αυτομάτας* U.

ὅσα προσφιλῆ τε καὶ μὴ ἐναντία ἀλλήλοις· ὅταν γὰρ ἅπαξ συμπλακῇ καὶ δέξηται, καθάπερ φύσις τις αὕτη μία γίνεται, διὸ κἂν ἀφαιρῇ κἂν λύῃ τις αὐαίνονται, καθάπερ καὶ τῶν μὴ ὁμογενῶν τὰ ὁμοβλαστῆ καὶ σύντροφα γενόμενα ἀλλήλοις, ὥσπερ ἐπὶ τῆς ἀναδενδράδος ἐλέχθη καὶ τῆς συκῆς· ἐπεὶ ὅσα γε[1] βλάπτει περιφυόμενα καὶ ἐμφυόμενα, καθάπερ ὁ κιττός, ἐκ τούτων γε[2] οὐ γίνεται μία φύσις, αὐαίνεται γὰρ θάτερον.

πολυφορεῖν[3] μὲν οὖν τοῦτο[4] διὰ ταύτας ὑποληπτέον τὰς αἰτίας, ὅμοια[5] γὰρ τρόπον τινὰ καὶ ὥσπερ εἴ τις ἐνοφθαλμίσειε δένδρον ἓν ἀπὸ πλειόνων καὶ διαφόρων· ἀρχὰς γὰρ πεποίηκεν καὶ φύσεις πλείους ἀπὸ μιᾶς οὐσίας, ἐκεῖνο δὲ ἐξ ἀρχῶν πλειόνων μίαν οὐσίαν τῇ φύσει.[6]

6.1 τὰ δὲ τῶν καρπῶν μεγέθη τῶν κατορυττομένων ἐν ταῖς χύτραις, ὅταν κατάγωσι[7] τοὺς ἀκρε-

§ 6.1: *Cf.* [Aristotle], *Problems*, xx. 9 (923 b 24–29).

[1] aP : τὲ U (τε N).
[2] γε Wimmer (Gaza omits) : γὰρ U.
[3] aP : -εῖ U N.
[4] U : *ab eodem arbore* Gaza : τὸ αὐτὸ δένδρον Schneider : ταὐτὸ Wimmer.
[5] U : ὅμοιον Schneider (*res . . . similis* Gaza).
[6] U : συμφύσει Schneider.

own accord, when they are friendly and not harmful to one another. For once they entwine and accept one another there results (as it were) a single nature. This is why if one of the partners is removed or the union broken up, both wither away, as also happens with trees that are not of the same kind when they sprout together and have been reared with one another, as we said[1] of the tree-climbing vine and the fig. (But such plants as injure a tree by growing round it and into it, like the ivy, give rise to no single nature, since the tree withers away.)

And so we must suppose that these are the causes that make the sort of tree we are discussing bear several sorts of fruit. For the case is in a way similar to that of grafting a single tree with buds from several trees of different kinds. For this last procedure takes a single entity and produces a plurality of starting-points and of natures from it, whereas the former takes several starting-points and produces from them an entity that is unitary in its nature.

6.1 The large size of fruit obtained by bending the branches of the tree down and burying the fruit in

[1] *CP* 3 10. 8.

---

[7] Gaza, Itali : καταγωσι U : κατεαγῶσι u (ε now in text; τεα, now erased, was once superscribed).

μόνας, οἷον ῥοῶν καὶ μήλων, εὐλόγως γίνεται· τό τε γὰρ ὑπὸ τοῦ ἡλίου καὶ τοῦ ἀέρος ἀφαιρούμενον[1] ἡ χύτρα κωλύει ἀποστέγουσα, καὶ ἅμα τὴν[2] ἐκ τῆς γῆς ἕλκει νοτίδα, δι' ἧς τρέφεται· τὴν γὰρ ἀπὸ τοῦ δένδρου ἐπιρροὴν οὐκ εὔλογον γίνεσθαι, πάρωρον οὖσαν, ἢ βραχεῖάν τινα πάμπαν, διὸ καὶ ὁ μὲν κόκκος οὐδὲν μείζων τῆς ῥόας γίνεται, τὸ δὲ σίδιον παχύτερον, ὡς οὐ δικνουμένης εἰς ἐκεῖνον τῆς τροφῆς. ἔοικεν γὰρ ὁ οἰκεῖος χυλὸς τῇ φυσικῇ δυνάμει πάντων γίνεσθαι καὶ πεπαίνεσθαι, διὸ καὶ τὰ μῆλα χείρω καὶ ἀχυλότερα γίνεται· τὸ δὲ σίδιον καὶ ἐκ τῶν ἔξωθεν λαμβάνει <τὴν>[3] αὔξησιν, ὡς ἀλλοτριώτερον τῆς φύσεως.

6.2 ἡ δ' ὁλκὴ[4] τῆς νοτίδος,[5] ἐξ ἧς ἡ τροφὴ καὶ

[1] u : -ρουμένων U.
[2] U^cc from της.
[3] Wimmer.
[4] u a : δολικῇ U : δ' ὁλικὴ N P.
[5] u : νοτίνος U.

---

[1] *Cf.* Varro, *On Farming*, i. 59. 3–4: "Pomegranates are also kept in sand when already picked and ripe, and even when unripe and still on the tree, if you lower them into a pot with no bottom and put it in the earth, tamping the earth round the branch . . . , you will find them on removal

pots, as with pomegranates[1] and apples, has its good reasons: the pot shuts in what is otherwise lost to the sun and air, and at the same time the fruit attracts the moisture in the earth and thus gets its nutriment (since it is unreasonable to suppose that any food, or any but very little, is supplied from the tree, since the supply would be out of season).[2] This is why the berry of the pomegranate gets no larger in this case, although the rind gets thicker: the food does not reach the berry. For it appears that in all trees the proper juice is produced and ripened by the natural power of the tree (which moreover is why the apples deteriorate and get less succulent under this treatment, whereas the rind of the pomegranate, as more foreign to the nature of the tree than the berry, gets its increase from external sources as well).

6.2 That the moisture,[3] which leads to the feeding and increase in size, is attracted is not unreasonable, just as it is not unreasonable in cuckoo-pint

not only entire but much larger than they ever were when hanging on the tree." *Cf.* also Palladius, *On Agriculture*, iv. 10. 5.

[2] The pomegranates were presumably kept in these pots long after their season, to be sold when the tree-ripened fruit was off the market.

[3] That is, moisture in the ground, attracted directly and not by way of the roots.

ἐπίδοσις, οὐκ ἄλογος, ὥσπερ τοῦ ἄρου καὶ ἑτέρου (περὶ ὧν[1] πρότερον εἴπομεν).

ὁμοίως δὲ τοῦτο συμβαίνει καὶ ἐν τοῖς λαχάνοις ἐφ' ὧν παχύνουσιν[2] τὰς ῥίζας, τῶν μὲν ἀφαιροῦντες τὰ φύλλα, καθάπερ τῆς ῥαφανῖδος, ὅταν μάλιστα ἀκμάζωσιν[3] τοῦ χειμῶνος, καὶ κατασάττοντες τὴν γῆν, ὥστε καὶ τὸ ὕδωρ ἀποστέγειν· ἐν 6.3 γὰρ τῷ θέρει γίνονται θαυμασταὶ τῷ πάχει· τοῦ δὲ σελίνου, περιορύξαντες κάτω μέχρι τῶν ῥιζῶν, καχρύδιον[4] περιβάλλοντες[5] καὶ ἄνωθεν τὴν γῆν.

αἴτιον δέ, ὅτι τὴν τροφὴν ἅπασαν αὐταὶ[6] λαμβάνουσιν καὶ οὐ διδόασιν[7] εἰς τοὺς βλαστούς·

§ 6.2: [Aristotle], *Problems*, xx. 13 (924 a 24–27).
§ 6.3: *Cf.* [Aristotle], *Problems*, xx. 8 (923 b 10–15); *ibid.* xx. 13 (924 a 27–35).

[1] περὶ ὧν ego : ὥσπερ U.
[2] u : ταχύνουσιν U.
[3] u : ἀκμάζουσιν U[c] (from ἀμά-) : ἰκμάζωσι N : ἐκμάζωσι P : ἐκμάξωσι a.
[4] H. Stephanus : καὶ χρύλιον (-υ- U) N aP : καχρύλιον u.
[5] Scaliger : περιλαμβάνοντες U.
[6] ego : αὗται U.
[7] διαδιδόασιν Gaza (*transmittunt*), Schneider.

---

[1] *HP* 7 12. 2 (of cuckoo-pint): "To make the root larger they dig it up and turn it upside down after stripping off the leaves (which are very large), to keep it from sprout-

and another bulbous plant (which we mentioned before).[1]

(2) *In Lesser Plants*

It occurs equally in the vegetables whose roots are made thick by growers, who with some strip off the leaves, as with radish,[2] when the roots are at their best in winter, and tamp down the ground to keep the water out as well[3]; for then the roots become remarkably thick in summer. Growers do this with celery by digging round the plant as far as the roots and then putting parched barley in the hole and covering it with earth.

6.3

The cause is this: the roots then take all the food themselves and do not pass it on to the shoots; and

ing and make it draw all the food to itself. Some gardeners also do this with purse-tassels, putting several roots together"; *HP* 1 6. 10 (perhaps the bulb of bulbous plants is a root, and the plants have two kinds of root, the upper one fleshy and fed by the lower): "Yet the fleshy roots too appear to attract food by themselves. Thus gardeners turn the roots of cuckoo-pint upside down before they sprout, and this makes them larger, since they are then prevented from passing the food to the shoot."

[2] *Cf. HP* 7 2. 5 (of vegetables): "The roots of most persist, but some sprout again, others do not. Thus radish and turnip last till summer if earth is thrown on them, and grow larger, and some gardeners do this by design . . ."

[3] That is, to keep the rain of winter out as well as to retain the ground moisture in summer.

<μὴ>[1] μεριζομένης δὲ πλείων ἡ αὔξησις. ὅσα μὲν οὖν παραβλαστητικά, οἷον κρόμμυα, ῥίζας ἑτέρας ἀφίησιν, καὶ ἄλλα δὲ τῶν ὑγρῶν·[2] ἡ δὲ ῥαφανίς, μὴ οὖσα παραβλαστητική, τροφὴν δὲ λαμβάνουσα καὶ οὐ διαπέμπουσα εἰς τὸ[3] ἄνω, παχύνεται καὶ μείζων γίνεται. τοῦ δὲ μὴ συμβαίνειν φθορὰν καὶ σῆψιν ἡ ἐπίσαξις[4] αἰτία τῆς γῆς, ἀποστέγουσα τὸ ὕδωρ καὶ ὅλως πᾶν τὸ ἀλλότριον.

τοῖς δὲ σελίνοις τὸ καχρύδιον,[5] θερμὸν καὶ μανὸν[6] ὄν, συνανέλκει μὲν τῇ μανότητι[7] τὴν τροφήν, κατέχει δὲ καὶ οὐ διαδίδωσιν εἰς τὸ ἄνω, καὶ ἅμα τῇ θερμότητι πέττει· πολλῆς οὖν τροφῆς γινομένης καὶ πεττομένης, πολλὴ καὶ αὔξησις.

6.4 παραπλήσια δὲ τούτοις καὶ τὰ περὶ[8] τοὺς σικύους ἐστὶ καὶ τὰς κολοκύντας γινόμενα κατά τε τὴν ἁπαλότητα καὶ τὴν αὔξησιν, οἷον ἐάν τις μικρὰς οὔσας κρύψῃ καὶ μικρούς· οὐδὲν γὰρ ἀφαιρεῖται τῆς τροφῆς (ὁ δ' ἥλιος καὶ τὰ πνεύματα

§ 6.4: *Cf.* [Aristotle], *Problems*, xx. 9 (923 b 16–29).

1 Schneider.
2 ego : ἀγρίων U.
3 U^{r} : τὸν U^{ar}.
4 aP^{c} : ἐπισταξις U (-πί- u N P^{ac} [?]).

when the food is not divided up the growth of the root is greater. Now when such plants are capable of sending out side-growths, as onion, they produce other roots in this case, and so do some other fluid plants; but the radish, since it lacks the capacity to do so, and gets food which it does not transmit upward, grows thick and gets longer. The root remains sound and no decomposition occurs because of the heaping up of earth which keeps out the water and in general everything unfavourable.

In celery it is the parched barley that does this. The barley, being warm and of open texture, by its openness of texture helps the roots to draw the food, but keeps it there and does not let it pass upward; and at the same time, by its heat, concocts the food. In consequence, since a great deal of food is attracted and this gets concocted, the growth is also great.

6.4 Similar to these procedures are those used with cucumbers and gourds to improve their tenderness and size, such as covering them when they are too small.[1] For then no food is lost; whereas exposure to

[1] *Cf. CP* 2 9. 1.

[5] H. Stephanus : -υλιον U.
[6] ego (σομφὸν Schneider) : πυκνον U.
[7] ego (σομφότητι Schneider) : πυκνότητι U.
[8] u : πε U.

ἀναξηραίνει καὶ τοὺς ὄγκους ἐλάττους ποιοῦσιν, ὥσπερ καὶ τῶν δένδρων τῶν ἐν τοῖς προσηνέμοις καὶ εὐείλοις). ὡσαύτως δὲ καὶ οἱ ἐν[1] τοῖς ἀγγείοις[2] τιθέμενοι, καθάπερ ἐν νάρθηκι καὶ καλυπτῆρσιν· ἡ μὲν γὰρ τροφὴ πλείων, διὰ τὸ μήτ' ἀποπνεῖσθαι μήτε ἀποξηραίνεσθαι μηδέν, ἡ δ' αὔξησις <εἰς>[3] μῆκος, καὶ διὰ τὸ εὐροεῖν[4] τὴν τροφὴν εὐθυποροῦσαν, καὶ διὰ τὸ μηδὲν ἀντισπᾶν μηδ' ἀντιπίπτειν.

6.5 ὁμοία δ' αἰτία καὶ τοῦ διαμένειν χλωροὺς ἐάν τις φυτεύσας περὶ φρέαρ, ὅταν ὦσιν ὡραῖοι καθεὶς ἀποστεγάσῃ· ἡ μὲν γὰρ ἀπὸ τοῦ ἡλίου καὶ τοῦ πνεύματος οὐ γίνεται ξηρότης, ἅμα δὲ καὶ ἡ ἀπὸ τοῦ ὕδατος ἀτμὶς οἷον θάλλοντά[5] τε παρέχει καὶ κωλύει ξηραίνεσθαι, τροφὴν δὲ λαμβάνοντα[6] διαμένουσιν ἑωμένων[7] τῶν ῥιζῶν.

§ 6.5–6: *Cf.* [Aristotle], *Problems*, xx. 14 (924 a 36–b 14).

[1] οἱ ἐν u : οἷον U.
[2] u : αἰτίοις U : αἰ(blank of 3–5 letters)ιοις N aP.
[3] aP.
[4] Schneider : εὐρεῖν U N : εὖ ρεῖν u : εὑρεῖν aP.
[5] U : θάλλοντάς Schneider.
[6] U : λαμβάνοντες Schneider.
[7] -σιν ἑωμένων ego (from [Aristotle] *Probl.* 924 b 2) : -σι δὲ σωζομένων U N : -σι διασωζομένων u : -σι σωζομένων aP.

sun and wind makes them dry and smaller in size, as it reduces the size of trees too in windy and sunny positions.[1] So too with the cucumbers that are grown in containers,[2] such as a fennel stalk and tiles. For this enclosure increases the amount of food, none of it being carried off by the wind or dried up by the sun; and the growth of the cucumber in length is due not only to the rapid flow of the incoming food, which moves in a straight line, but also to absence of anything that draws the food aside or obstructs it.

6.5 A similar cause makes them stay fresh if you grow them round a well and lower them into it when they are in season and cover it; for then there is no dryness from sun and wind, and at the same time the vapour keeps the plants flourishing (as it were) and prevents their getting dry; and they continue to live and receive food so long as the roots are left alone.

[1] *Cf.* *HP* 2 7. 5: "At Megara this is done also with cucumbers and gourds. When the Etesian winds have begun to blow [after the summer solstice and the rising of Sirius: *cf.* Aristotle, *Meteorologica*, ii. 5 (391 b 35–36); cucumber and gourd were sown in Munychion or April: *HP* 7 1. 2] the gardeners hoe up the ground and cover them with the dust, thus making them sweeter and more tender without watering."

[2] *Cf.* Pseudo-Hippocrates, *On Generation*, chap. ix (vol. vii, p. 482. 14–19 Littré), for a cucumber assuming the size of the vessel in which it is grown.

ἔμβιοι δὲ γίνονται καὶ θεραπευόμεναι, πλείω χρόνον· διὸ καὶ ἄν τις περιτεμὼν τὴν βλάστησιν, ὅταν καρποτοκήσωσιν, περισάξῃ τὰς ῥίζας εὖ μάλα τῇ γῇ, καὶ καταπατήσῃ, γίνονται πάλιν ἐκ τῶν ῥιζῶν σίκυοι, καὶ πρωΐτεροι πολὺ τῶν σπειρομένων, ὅτι προϋπάρχει τὸ τῶν ῥιζῶν. ἔτι δ'[1] ἡ περίσαξις, ἀλέαν παρέχουσα, θᾶττον ἀνιέναι ποιεῖ βλαστόν· μέγα γὰρ καὶ ἡ ἀλέα πρὸς τὸ πρωϊβλαστεῖν. σημεῖον δὲ καὶ τούτου[2] φανερόν· ἐὰν γάρ τις χειμῶνος ἐν ταλάροις φυτεύσῃ σικύου σπέρμα,[3] καὶ ἄρδῃ τε[4] θερμῷ καὶ πρὸς τὸν ἥλιον ἐκφέρῃ καὶ πρὸς τὸ πῦρ τιθῇ, καὶ ὅταν ἡ ὥρα τοῦ σπείρειν καθήκῃ, σὺν αὐτοῖς ταλάροις φυτεύσῃ, πρώϊοι σφόδρα γίνονται.

6.6

ταῦτα μὲν οὖν διὰ τὰς εἰρημένας αἰτίας συμβαίνει.

6.7 τῷ δὲ σχήματι καὶ τῇ μορφῇ μεταβάλλει τὸ σέλινον ἐὰν σπαρὲν καταπατηθῇ καὶ ἐπικυλινδρωθῇ· γίνεται γὰρ οὖλον διὰ τὸ μὴ διιέναι τὴν βλάστησιν.

[1] Itali (*et* Gaza) : ὅτι δ' U.

[2] U : τοῦτο Wimmer.

[3] [Aristotle] *Probl.* 924 b 11 : σπέρματος U N : σπέρματα aP.

[4] U : ἄρδηται [Aristotle] *Probl.* 924 b 11.

The roots also stay alive longer when tended. This is why, if you cut off the upper part when the plants have borne their fruit and heap a generous amount of earth over the roots and tread it down, new cucumbers will come from the roots,[1] and the new ones are much earlier than those grown from seed, since the roots are already present. Furthermore packing the ground around the roots keeps them warm and so makes them send out shoots earlier; for warmth too is important for early sprouting. For this importance we also have clear proof: if you plant cucumber seed in winter in baskets, and then water it with warm water and carry the baskets out into the sun and put them by the fire, and then, when the sowing season comes around, plant them in the ground, baskets and all, they come out very early. 6.6

These results, then, are due to the causes mentioned.

Celery changes in shape and form when the ground is trodden and rolled after sowing, for the plants then become curly because the ground does not let the shoot pass through.[2] 6.7

[1] *Cf. HP* 7 3. 1: "... for cucumber can have a second growth."

[2] *Cf. HP* 2 4. 3: "Thus with celery: if after sowing the soil is trodden down and a roller passed over it the celery is said to come up with curly leaves."

ἄλλα δ' ἐξομοιούμενα· διεξομοιοῦται γὰρ[1] ἐν ᾧ ἂν τεθῇ [ἀγγείῳ]·[2] τοῦτο δ' ὅτι ἡ τροφή, κωλυομένη καὶ ἀποστεγομένη τῷ[3] πέριξ, φέρεται πρὸς τὸ ἐφελκόμενον καὶ εὐοδοῦν (οἷον γὰρ ὀχετεία τίς ἐστιν), ὥστε λαμβάνειν τὴν ὁμοιότητα τῷ περιέχοντι. (συμβαίνει δὲ τρόπον τινὰ καὶ ἐπὶ τῶν ζῴων τοῦτο, κατὰ <δὲ>[2] μικρότητα καὶ μέγεθος καὶ βραχύτητα καὶ μῆκος, μεμορφωμένα γὰρ εὐθὺς ἐκεῖνα, ταῦτα δ' ἅμα τῇ γενέσει μορφοῦται.) διὸ καὶ τὸ σέλινον, ὅταν μεταφυτεύηται, κελεύουσιν ὁπόσον[4] ἄν τις βούληται ποιῆσαι, τηλικοῦτον πάτταλον κατορύττειν,[5] ὡς ἐκπληροῦν πάντα[6] τὸν τόπον τὴν ῥίζαν.

6.8 ὅμοιον δὲ τρόπον τινὰ τούτῳ καὶ ἡ τῶν ῥιζῶν

[1] ἄλλα—γὰρ ego (ἄλλα δὲ ἐξομοιοῦται Schneider : ἀλλὰ—ἐξομοιοῦται γὰρ Wimmer): ἀλλα διεξομοιοῦσθαι· ἐξομοιοῦται γὰρ U.

[2] ego.

[3] τῷ u : το U.

[4] u : ὁποσ' U.

[5] U : κατακρούειν Schneider.

[6] u : παν U.

Other plants change through assimilation of their shape to that of whatever they are placed in.[1] This happens because the food is checked and shut in by the wall of the container and so moves to what attracts it and provides a passage, the process resembling the directing of the flow of water by opening and closing irrigation channels, with the consequence that the plant acquires its conformity to the shape of the container. (In a way the change occurs also in animals, but in smallness and bigness and shortness and length, since animals have their shapes from birth,[2] whereas these plants acquire their shape in the course of production.) This is why we are told when transplanting celery to put in the ground a peg of the size desired for the celery, to let the root fill out the whole space.[3]

6.8 Similar in a way to this is the size to which roots

[1] *Cf. HP* 7 3. 5: "Some plants even change in their shape to fit the surrounding space, for bottle-gourd assumes the shape of what it is put in."

[2] *Cf.* Aristotle, *History of Animals*, ii. 1 (500 b 26–501 a 7) [A full-grown man has the upper part of his body smaller than the lower; in the other blooded animals the reverse is true. As man grows the comparative size of upper and lower part is reversed; some of the animals keep the same relation, in others the upper part becomes larger.]

[3] *Cf. HP* 7 3. 5: "The surrounding space too contributes to growth in size. Thus we are told when transplanting celery to hammer a peg into the ground of the size desired for the celery . . ."

αὔξησις ἐν τοῖς ἡμερώμασιν· εὐοδοῦσαι γάρ, καὶ ἔχουσαι τροφήν, αὔξονται μᾶλλον καὶ εἰς μῆκος καὶ εἰς πάχος.

6.9 ποιεῖ δὲ μεγάλας καὶ τὰς ῥίζας καὶ τὰς βλάστας καὶ ἐὰν πλείω τις εἰς ταὐτὸ σπέρματα[1] ξυνδήσας εἰς ὀθόνιον φυτεύσῃ, διὸ καὶ ἐπί τινων τοῦτο[2] ποιοῦσι, καθάπερ ἐπὶ τοῦ πράσου καὶ σελίνου καὶ ἑτέρων· ἰσχύει[3] γὰρ τὰ πλείω δῆλον ὅτι μᾶλλον, καὶ ἐξ ἁπάντων γίνεται μία τις φύσις.

ἔνια δὲ κατὰ τὴν[4] τῆς σπορᾶς ὥραν λαμβάνει μορφὴν ἀλλοίαν, οἷον ἡ γογγυλίς, ἂν εὐθύς τις ἀπὸ τῆς ἅλω[5] φυτεύσῃ, πλατεῖα γίνεται· τοῦτο δέ, ὅτι ῥιζοῦται καὶ διευρύνεται μᾶλλον.

[1] u : σπέρμα U.
[2] u : του U.
[3] u : ἴσχύϊ U.
[4] κατὰ τὴν u : καὶ η U.
[5] u aP (ἅλο N) : ἄλλω U.

---

[1] *Cf. HP* 1 7. 1: ". . . but no root goes down farther than the sun reaches, for it is heat that generates. Nevertheless these other characters contribute greatly to the depth and still more to the length of roots: the light, open, and therefore penetrable nature of the land, for in ground such as this the roots grow farther and get larger. This is evident in new land; for once the trees have water

grow in new land[1]; for having an easy passage and getting food, they increase more both in length and thickness.

6.9 It also makes both the roots and the shoots grow large to plant by tying several seeds together in a bag. Hence farmers do this with some plants, as leek, celery[2] and others, since in greater numbers the seeds evidently have greater strength, and from the combination arises a single nature.

Some plants get a different kind of shape, depending on the season of sowing. For instance if one sows turnip seed immediately after threshing, the turnip produced is of the flat variety.[3] This happens because it then turns to root and so gets broader.

they penetrate the ground practically everywhere with their roots, when the place is empty of other trees and offers no obstacle. Thus the plane tree in the Lyceum by the new irrigation ditch, while still young, sent roots to the distance of thirty-three cubits, since it both had room for expansion and got food."

[2] For leek and celery *cf.* *HP* 7 3. 4; for celery *HP* 7 3. 5.

[3] This is the "female" turnip of *HP* 7 4. 3. Turnip is usually sown in Metagitnion (July), after the summer solstice (*HP* 7 1. 2) and matures in summer (*HP* 7 1. 6). So if it is sown immediately after maturing it will live through the winter. *Cf.* *HP* 7 2. 5 (translated in n. 2 on *CP* 5 6. 2) and *HP* 7 4. 3: "Both turnip and radish enjoy cold weather, and it is supposed that they then get sweeter and grow in root rather than in leaf."

6.10 ὅσα δ' ἐν σχίνῳ φυτεύουσιν ἢ σκίλλῃ, πάντα τῆς εὐβλαστείας[1] ἕνεκα καὶ εὐτροφίας φυτεύουσιν· ἔχει γάρ τινα ἄμφω θερμότητα καὶ ὑγρότητα, καὶ γίνεται καθάπερ ἐμφυτεία τις.

ὁμοίως δὲ καὶ εἴ τι ἕτερον ἑτέρῳ,[2] καθάπερ τὸ πήγανον ἐν συκῇ, δοκεῖ γὰρ δὴ κάλλιστον γίνεσθαι· φυτεύεται δὲ παρὰ τὸν φλοιὸν παραπηγνύμενον, καὶ τῇ γῇ κατακρύπτεται. καὶ ξυμβαίνει δὲ τὸν ὀπόν, ἅμα τῇ τροφῇ, διὰ θερμότητα καὶ βοήθειάν τινα ἔχειν εὔκαιρον (ὥσπερ καὶ τὴν τέφραν παραπαττομένην, εἴτ' οὖν πρὸς τὸ μὴ σκωληκοῦσθαι τὰς ῥίζας, εἴτε καὶ πρὸς τὸ τρέφεσθαι <τῇ>[3] ἅλμῃ· ἔχει γάρ τιν' ὁμοίαν θερμότητα).

6.11 μέγεθος δὲ γίνεται φακῶν καὶ ἐρεβίνθων, τῶν

§ 6.10: *Cf.* [Arist.], *Problems*, xx. 18 (924 b 35–925 a 5).

[1] ego (-τίας u) : ευβλαστήας U.
[2] ἐν ἑτέρῳ Γαζα, Σξηνειδερ.
[3] Schneider.

---

[1] *Cf. CP* 5 9. 5.
[2] *Cf. HP* 2 5. 5: "A fig cutting stuck in a squill and planted comes up faster and is not infested to the same degree by grubs; and in general anything planted in a squill sprouts well and grows faster"; *HP* 7 13. 4 (of

Special Treatment of Cuttings

All cuttings that are planted in a pine-thistle[1] or a squill[2] are so planted with a view to their sprouting and feeding well. For both pine-thistle and squill possess a certain warmth and fluid and the result is (as it were) a kind of twig-grafting. 6.10

This is equally the case moreover when one plant is grown in another, as rue in a fig-tree, since the best rue is held to be so produced. The planting is done by inserting the cutting alongside the bark and covering with earth. The result is that the fig-juice, besides feeding the rue, furthermore, owing to its heat, provides a certain remedy when it is needed (just as ashes do when scattered around rue,[3] whether they keep the roots from getting grubs or feed the plant by means of their brine, which has a heat resembling that of fig-juice).

Similar Treatment of Seeds

Size is obtained in lentil and chickpea[4] if the 6.11

squill): ". . . again some cuttings when planted in it sprout faster . . ."

[3] *Cf. CP* 3 17. 1.

[4] *Cf. HP* 2 4. 2: "To produce vigorous lentils farmers plant them in cow-dung; to produce large chickpeas it is recommended to soak them first and then sow them pod and all."

μέν, ἂν ἐν βολίτῳ[1] φυτεύηται τὸ[2] σπέρμα, συνεκτρέχει γὰρ τῇ θερμότητι καὶ ξηρότητι· τῶν <δ'>[3] ἐρεβίνθων, ἐὰν μετὰ τῶν κελυφῶν βρεχθέντες, ἔλαττον γὰρ τὸ ἀποσηπόμενον, καὶ ἡ τροφὴ πλείων ἡ πρώτη, πρώϊοι δ' ἐὰν ἅμα τοῖς ἄλλοις σπαρῶσιν.

καὶ ταῦτα μὲν δὴ τὸ θαυμαστὸν ἔχει, καὶ ἔνια δοκεῖ καὶ[4] παρὰ φύσιν.

6.12 αἱ δὲ τοιαῦται μεταβολαὶ καὶ αὐτόματοι γίνονται καὶ τεχνηταὶ[5] κατὰ φύσιν.

οἷον ἐπὶ τίφης καὶ ζειᾶς περιπαλαττομένης[6] <εἰς> πυρούς,[7] ὥσπερ καὶ ὅσα τῶν σπερμάτων ἢ

[1] ἐν βολίτῳ Gaza (*stercore bubulco involutum*), Scaliger : εμβολίτω U : ἐμβολητῶ u.

[2] Itali (Gaza omits, Scaliger deletes) : εἰ U N P : εἰς a.

[3] Gaza (*autem*), Schneider.

[4] Schneider : δοκεῖται U N : δοκεῖ γε aP.

[5] ego : τεχνήται U : τέχνη τὰ u : τέχνη N aP.

[6] ego : περιπλαττομένης U. [7] εἰς πυρούς ego : πυροῖς U.

---

[1] The greater heat promotes the digestion and transmission of the food; the greater dryness keeps the plant from receiving more food than it can master.

[2] *Cf. Geoponica*, ii. 36: "On Chickpeas. From Florentinus. If you soak the chickpeas in warm water one day before sowing they grow larger. Some follow a more elaborate treatment, desiring much larger chickpeas, and sow them with the pods after soaking them in the same

seed of the first is planted in cow-dung, for the heat and dryness[1] of the dung make the plant run up fast; in chickpeas if they are soaked and planted with the pod. For a smaller proportion decomposes and fails to grow, and there is more initial food; and the chickpeas come out earlier when sown with the rest.[2]

Now all these cases have the character of being remarkable, and a few are even regarded as unnatural.

Changes of the Fruit Through the Producer

6.12 But such changes as the following occur naturally as a combination of spontaneity and contrivance.[3]

So with the change of single-seeded and double-seeded wheat to wheat when they have been bruised in a mortar,[4] as also with the improvement in

way with water containing carbonate of soda. If you wish to make them early, sow them at the time of barley." The last sentence looks like a misinterpretation of Theophrastus' "with the rest (*sc.* of the chickpeas)."

[3] The change in the seed (or twig) is due to art, the change in the fruit is the plant's own doing.

[4] *Cf. HP* 2 4. 1: "The changes in annuals are due to the operation of art: thus single-seeded and double-seeded wheat change to wheat if they are bruised before sowing, and they do so not at once but two years later [*i.e.*, they must be bruised for three successive generations]."

ἐν λίτρῳ προβρεχόμενα τεραμονέστερα, ἢ ἐν μέλιτι καὶ γάλακτι γλυκύτερα γίνεται. πέφυκεν γὰρ οἷον ἂν[1] σπαρῇ, τοιοῦτον καὶ γεννᾶν·[2] σπείρεται δὲ διηλλοιωμένον[3] καὶ μεταβεβηκός·[4] ἐν ἀμφοῖν δέ πως, ἢ ἐν ἅπασιν συμβαίνει μετακινεῖσθαι τὴν ἀρχήν, ἔνθα μὲν κατὰ τὸ ποιόν, ἔνθα δέ, καθάπερ ταῖς τίφαις καὶ ζειαῖς, καὶ τῷ μόριόν[5] τι μὴ ἔχειν, ὅπερ ἐξ ἀρχῆς μὲν περιαιρεθέν, οὐκ ἀδυνατεῖ γεννᾶν, πλεονάκις δὲ τοῦτο παθοῦσα καὶ ὥσπερ τελέως γυμνωθεῖσα καὶ παθητικωτέρα γινομένη, τῷ τε ποιῷ μεταβάλλει καὶ τὸ πλῆθος οὐκ ἴσχει[6] τοῦ ἀχύρου.

6.13 συμβαίνει δὲ καὶ ἐν ἄλλοις, μορίων τινῶν ἀφαιρουμένων, ποιεῖν τινα διαφοράν, ὥσπερ τὰς ἀμπέλους,[7] ὅταν ἡ μήτρα τοῦ κλήματος[8] ξυσθῇ, τοὺς βότρυς ἀγιγάρτους·[9] ᾗ[10] καὶ πίστιν ταῦτα παρά-

[1] U^c : U^ac omits. [2] u : γεννᾶ U.
[3] U^cc (η from α). [4] U : -βληκὸς u.
[5] τῷ μόριόν Gaza (*quod pars*), Schneider : των μωρίων U.
[6] u : ἴσχύει U.
[7] U : ταῖς ἀμπέλοις (*viti* Gaza) Schneider.
[8] U^c (τ from κ). [9] aP : ἀγιγαρτος U (-γί- u N).
[10] ᾗ U^ac : ᾗ U^c N : εἶναι aP.

---

[1] *Cf. HP* 2 4. 2: "Some changes are brought about . . . by husbandry alone: for example, to keep pulses from pro-

amenability when the seed is previously soaked in a solution of soda,[1] or in sweetness when it is soaked in honey and milk.[2] For it is natural that the character of the seed when it is sown should be the character of the seed that it produces; but it is sown in an altered state and when it has passed to something else, and in both cases or in all three[3] a certain shift occurs in the starting-point, a shift in quality in the last two, a shift that also involves the absence of a part in the first, as in single-seeded and double-seeded wheat. When this part is first removed, the seed is not incapable of generating the seed that it generated before; but after this type of wheat has undergone the removal more than once and has been completely (as it were) laid bare and gains in responsiveness, it changes in quality and no longer produces the same abundance of bran.

6.13 In other plants too when certain parts are removed the result is that the plant makes a certain difference in its product: so the vine when the core of the twig is scraped away produces clusters with no stones.[4] And so these cases might perhaps provide a

ducing stubborn crops we are told to soak the seeds in soda at night and sow them the following day on dry ground."

[2] *Cf. CP* 2 9. 4; *HP* 1 7. 6; 7 5. 3.

[3] The three cases are: bruising the seed; soaking it in soda; soaking it in honey or milk.

[4] *Cf. CP* 5 5. 1.

σχοιτ' ἂν ἴσως τοῖς ἀφ' ἑκάστου τῶν μερῶν λέγουσιν ἀπιέναι σπέρμα· λύσις δ' ἥπερ εἴρηται καὶ ἐπὶ τῶν ζῴων.

αἱ μὲν οὖν τούτων μεταβολαὶ διὰ τὰς εἰρημένας αἰτίας.

7.1 ἡ δὲ τοῦ σισυμβρίου εἰς μίνθαν ὥσπερ ἐναντία, δι' ἀργίαν γεγενημένη· συμβαίνει γὰρ ὅταν μή τις ἐξεργάζηται, μηδ' ἀποδιδῷ τὴν οἰκείαν θεραπείαν, ῥιζοῦσθαι μᾶλλον εἰς τὸ κάτω, ῥιζούμενον δὲ καὶ τὴν δύναμιν ἐκεῖσε τρέπον πᾶσαν, ἀσθενέστερον ἄνωθεν γίνεσθαι καὶ τὴν δριμύτητα ἀποβάλλειν τῆς ὀσμῆς, ὥστε[1] ἐξ ἀμφοτέρων ἡ ὁμοιότης, τῆς τε βλάστης καὶ τῆς ὀσμῆς. τῆς γὰρ δριμύτητος ἀφαιρουμένης, ἡ κατάλοιπος ὀσμή, μαλακή τις οὖσα καὶ ἀνειμένη, προσεμφερὴς τῇ[2] μίνθῃ γίνεται, διὸ μεταφυτεύειν κελεύουσιν πολ-

[1] Wimmer (ὧνπερ Schneider) : ὥσπερ U.
[2] u : της U.

---

[1] The proponents of the view that the seed comes from every part of the parent had argued that crippled parents produce crippled offspring, since no seed comes from the

piece of plausible evidence[1] to those who say that the seed comes from every part of the parent.[2] But the solution of that difficulty is the same as the one given[3] for the seed of animals.

The changes of these plants, then, are due to the causes given.

Mutations Occurring of their own Accord

7.1 The change of bergamot-mint to mint is an opposite one (so to speak), arising from neglect of cultivation. For when the plant is not carefully cultivated and does not receive the kind of tendance that it requires, the result is that it pushes its roots deeper, and when it is pushing its roots and turning its whole power in that direction it becomes weaker in the part above ground and loses the pungency of its odour (so that the similarity to mint has the two sources, the growth above ground and the odour). For when pungency is removed the odour that remains is of a soft and languid sort and so comes to resemble that of mint. This is why we are told to transplant bergamot-mint often to prevent the

missing part (Aristotle, *On the Generation of Animals*, i. 17 [721 b 17–20]).

[2] *Cf.* Aristotle, *On the Generation of Animals*, i. 17 (721 b 8–722 a 21); iv. 3 (769 a 11–12).

[3] Aristotle, *On the Generation of Animals*, i. 18 (722 a 2–726 a 28) [722 a 11–16 refers to plant seeds].

λάκις, ὅπως τοῦτο μὴ συμβαίνῃ.

καὶ τοῦ μὲν σισυμβρίου τοιαύτην τὴν αἰτίαν ὑποληπτέον.

7.2 τὸ δ' ὤκιμον [τὸ][1] ἐν εὐηλίῳ[2] πολλάκις [κείμενον][1] ἀφερπυλλοῦται διὰ τὸ καταξηραίνεσθαι μᾶλλον· καὶ γὰρ τὸ φύλλον ἔλαττον γίνεται, καὶ ἡ ὀσμὴ δριμυτέρα τῶν ξηρῶν, ἐλάττων γὰρ καὶ ἡ τροφή. (δεῖ δὲ τὰς μεταβολὰς τοιαύτας ὑπολαμβάνειν, ὡς ἂν ὁμοιότητά τινα ἐχούσας, οὐχ ὡς τελέας.)

ἡ δὲ λεύκη πλατυφυλλότερόν τε τῆς αἰγείρου καὶ λειοφλοιότερον, καὶ τὸ ὅλον εὐτροφώτερον· ἀπογηράσκουσαν δὲ καὶ ἐλάττονι τροφῇ χρωμένην, οὐκ ἄλογον καὶ τὸ φύλλον στενότερον καὶ τὸν φλοιὸν τραχύτερον ἔχειν καὶ εἴ τι ἄλλο συνακολουθεῖ τοῖς μὴ ὁμοίως εὐτρόφοις.

7.3 τὸ δ' ὅλον οὐκ ἄγαν ἴσως τὸ συμβαῖνον θαυμαστόν, τῷ τε εἰς τὸ σύνεγγυς[3] καὶ εἰς τὸ ὅμοιόν

§ 7.2: *Cf.* Pliny, *N. H.* 19. 176.

[1] ego (τὸ κείμενον was once a note stating that τυλλωι—see below—stood in the text of an exemplar).

[2] Gaza (*loco soli exposito*), Basle ed. of 1541 (ἡλίῳ Heinsius) : τύλλωι U.

[3] U^{c} : συγγυς U^{ac}.

occurrence.[1]

Such, then, we must suppose to be the cause of the change in bergamot-mint.

7.2 Basil in a sunny place often changes in the direction of tufted thyme,[2] because it then gets too dry, for the leaf in the dry plant gets smaller and the odour more pungent, since here there is also less food. We must take these changes to be such as involve a certain resemblance and not a complete new identity.

White poplar is a tree with broader leaves and smoother bark than black poplar,[3] and is in general better nourished. But it is not unreasonable that as it ages and uses less food it should get both a narrower leaf and a rougher bark and whatever other characters go with plants that are less well nourished than they were before.

7.3 But in general the occurrence is perhaps not very astonishing, since the change is to a thing in a way

[1] *Cf.* *CP* 2 16. 2; *CP* 4 5. 6; *HP* 2 4. 1: "... bergamot-mint is held to change to mint if not restrained by husbandry. This is why gardeners transplant it frequently." Transplanting evidently checks the growth of the root, thereby strengthening the upper growth and preserving the odour.

[2] Not elsewhere mentioned in the *CP* or *HP*; *cf.* Pliny, *N.H.* 19. 176; Palladius, v. 3. 4.

[3] For the change of white poplar to black *cf.* *CP* 2 16. 2; *CP* 4 5. 7.

πως μεταβάλλειν, καὶ ἔτι[1] τῷ ὁρᾶν καὶ ἐπί γε τῶν ζῴων γινομένας τοιαύτας τινὰς μεταβολάς, τὰς μὲν κατὰ τὴν γέννησιν[2] (οἷον ἐπ' ἄλλων καὶ ἐπὶ τῶν καλουμένων ψυχῶν· ἐκ κάμπης γὰρ χρυσαλλίς, εἶτα ἐκ ταύτης ἡ ψυχή· τοῦτο γὰρ ἐπὶ τῶν φυτῶν[3] οὐδενὸς συμβαίνει), τῶν δὲ καὶ τετελειωμένων ἤδη κατὰ <τὰ>[4] πάθη καὶ τὰς διαθέσεις τοῦ σώματος, αἳ γίνονται διὰ τὰς ἐπετείους ὥρας (ὥσπερ τοῖς[5] ὄρνισιν, ἐπὶ τούτων γὰρ μάλιστα καὶ λέγεται καὶ ἔνδηλος[6] τῶν χρωμάτων μεταβολὴ καὶ τῶν ὅλων σωμάτων, ὥστε δοκεῖν ἑτέρους εἶναι).

[1] Wimmer : ἐπὶ U.
[2] N aP : γένησιν U.
[3] ego (*aliis* Gaza) : ψυχῶν U : ψύχων u.
[4] Schneider : κατὰ U : καὶ τὰ N aP.
[5] v[t], Schneider : ταῖς U N aP.
[6] ἔνδηλος ἡ Schneider.

---

[1] *Cf. HP* 2 4. 4: "It might appear odder if such changes are in animals natural and more numerous; for some animals are held to change according to the seasons, as the hawk and hoopoe and other similar birds, and other animals change with alterations of their habitat, as water-snake to viper when the marshes are dried. Some animals also change most obviously with each birth, as chrysalis from caterpillar and butterfly from chrysalis, and this change occurs in a number of other animals as well."

proximate and similar, and moreover since one observes certain similar changes taking place in animals too,[1] some of these changes occurring in the process of generating the animals, as among other animals in butterflies, the caterpillar changing to the chrysalis and this to the butterfly[2] (for such a change is found in no plant), and others in animals already adult, affecting the qualities and states of the body, these changes being due to the yearly seasons, as in birds; for in birds change of colour and of the whole body is most spoken of and most evident, to the point that the bird is taken to be of another kind.[3]

[2] *Cf.* Aristotle, *History of Animals*, v. 19 (551 a 13–24): "Butterflies come from the caterpillars found on green leaves, especially on cabbage; first they are smaller than a millet seed, then they grow to small grubs, then in three days to small caterpillars; next, when grown, they remain motionless and change their form and are called chrysalises, and have a hard case . . . Not long afterwards the case breaks open and winged animal, the butterfly, flies out." Aristotle (*On the Generation of Animals*, iii. 9 [959 a 3]) calls such animals "thrice-born" (*trigenē*).

[3] *Cf.* Aristotle, *History of Animals*, iii. 12 (519 a 7–9): "Again most birds also change colour depending on the season, to the point that one ignorant of the birds would not recognise the change"; *ibid.*, ix. 14 (616 b 1–2); ix. 49 (632 b 14–633 a 28) [birds change in colour and song; cases where the birds are thought to be different]; *On the Generation of Animals*, v. 6 (786 a 29–34).

ταῦτα μὲν οὖν (ὥσπερ εἴρηται) φυσικῶς ὑποληπτέον· τὰ δ' ἐξ ἀρχῆς λελεγμένα μᾶλλον κατὰ φύσιν.

8.1 *ἀκόλουθα δέ πως τοῖς εἰρημένοις, καὶ ὥσπερ ἐσχάτης θεωρίας, περὶ νοσημάτων καὶ φθορᾶς εἰπεῖν, ἀμφοτέρων μετέχοντα καὶ τῶν κατὰ φύσιν καὶ τῶν παρὰ φύσιν. φθοραὶ μὲν γὰρ εὐθὺς αἱ μὲν οὕτως, αἱ δ' ἐκείνως*[1] *λέγονται· νόσοι δὲ τῇ μὲν ὅλως δόξαιεν ἂν εἶναι παρὰ φύσιν (ἔκβασις γάρ τις αἰεὶ καὶ σύγχυσις τοῦ κατὰ φύσιν ἡ νόσος), τῷ δ' εἰωθέναι καὶ πολλάκις συμβαίνειν, κατὰ φύσιν λέγομεν ὁμοίως ἔν τε ζῴοις καὶ φυτοῖς, αὐτὰ ταῦτα διαιροῦντες, τὰ βίαια πάθη καὶ φανερῶς ἀπὸ τῶν ἔξωθεν αἰτιῶν, οἷον τραυμάτων καὶ πληγῶν. τὰ γὰρ ὑπὸ ψύχους ἢ καύματος ἤ τινος ἑτέρου συμβαίνοντα τῶν ἐν τῷ ἀέρι παθημάτων οὐ λέγομεν παρὰ φύσιν· καίτοι βίᾳ*[2] *γέ πως καὶ*

[1] U^c : *ἐκεῖνος* U^ac.
[2] *βίᾳ* U : *βίαιά* Dalecampius.

---

[1] The remarkable effects of art [B2] of *CP* 5 5. 1–5 7. 3, some of which also are produced spontaneously (*CP* 5 5. 4; 5 6. 12). [2] *CP* 5 6. 12 *init.*

[3] The purportedly unnatural phenomena of *CP* 5 1. 2–5 4. 6.

These phenomena,[1] then, we must take to occur by natural causation (as we said)[2]; whereas those discussed initially[3] we must take rather to be natural.[4]

Diseases and Death, Natural and Unnatural

Next in order (in a way) to the preceding, and belonging (so to speak) to the end of the investigation, is the discussion of diseases and death, partaking as they do of both the natural and unnatural. Thus among the forms of death some are with no further ado called unnatural, others natural.[5] Diseases on the other hand from one point of view would appear to be unnatural, since disease is always a departure from the natural and a disturbance of it; but because they are customary and occur frequently, we call them natural in animals and plants alike, setting off however from the rest the special effects that are due to violence and manifestly proceed from external causes such as wounds and blows. As for diseases arising from cold or hot weather or some other character in the air, we do not call them unnatural, although these too are in a 8.1

[4] That is, rather to be natural than to be (as is supposed) "unnatural"; *cf. CP* 5 1. 2 "the occurrences that are either unnatural or regarded as such."

[5] *Cf. CP* 5 11. 1.

ταῦτα, καὶ ἀπὸ τῶν ἔξωθέν ἐστιν· ἀλλὰ γὰρ τοῦτο μὲν οὔτ' ἴδιόν ἐστιν ἐπὶ τῶν φυτῶν, ἀλλὰ καὶ ἐπὶ τῶν ζῴων, οὔτε διάφορον[1] πρὸς ὃ νῦν ζητοῦμεν.

8.2 τῶν δὲ νόσων ἀρχαί, καθάπερ τοῖς ζῴοις, ἢ ἀπ' αὐτῶν ἢ ἀπὸ τῶν ἔξωθεν, καὶ ἡ φθορὰ ὅλως ἢ εἰς καρπογονίαν. ἀπ' αὐτῶν μὲν ὅταν ἢ πλῆθος ἢ ἔνδεια τῆς τροφῆς, ἢ ποιότης·[2] ἀπὸ δὲ τῶν ἔξωθεν ὅταν ἢ χειμῶνες ὑπερβάλλοντες ἢ καύματα, ἢ ἐπομβρίαι[3] ἢ αὐχμοί, ἢ ἄλλη τις δυσκρασία τοῦ ἀέρος, ἔτι δὲ ὅσα <διὰ>[4] πληγὴν ἢ ἕλκωσιν ἐκ σκαπάνης ἢ τομῆς[5] ἢ διακαθάρσεως[6] (ἢ ἐξ ἄλλης [ἢ][7] τοιαύτης αἰτίας, ὥς[8] γ' ἔνιαι καὶ ἀπὸ τοῦ δαίμονος συμβαίνουσιν, καθάπερ ἡ χαλαζοκοπία)· εἰ δὲ καὶ ἡ ἔνδεια καὶ ἡ ὑπερβολὴ τῆς τροφῆς ἀπὸ τῶν ἔξωθεν, ὥς τινές φασιν, οὐδὲν <ἂν>[9] διαφέροι.

[1] U : διαφέρον Schneider. [2] M : ποιότητος U N aP.
[3] u N aP : ὑπομβρία U. [4] Heinsius.
[5] ἐκ . . . τομῆς Schneider : ἢ σκαπάνην ἧς (ῆς U^c) ἡ τομὴ U. [6] U N : διὰ καθάρσεως u aP.
[7] ἐξ ἄλλης Schneider (*aliqua* Gaza) : ἐξαλλαγῆς · ἢ U^cc (γ from λ).
[8] U : ὧν Schneider.
[9] Schneider.

way also due to violence and proceed from the outside. But we do not insist; the point is neither exclusively confined to plants, but applies to animals as well, nor does it make any difference for our present investigation. [1]

Diseases: Origination

8.2 Diseases (as in animals) have their origins either in the individual itself or outside it, and the destruction is either total or limited to the production of fruit. [2] Diseases arise from the tree itself when there is too much or too little food or food of the wrong quality. They arise from the outside if the spells of cold or heat or rain or drought are excessive or there is some other unfavourable tempering of the air. There are moreover the results of a blow or wound inflicted by hoeing or pruning or thinning or some other causation of the sort (indeed some causes, such as hail-stroke, are an act of God). If moreover deficiency and excess of food come from the outside, as some assert, this would make no difference. [3]

[1] Because all are discussed in terms of internal affections.

[2] Diseases affecting the trees themselves are discussed in *CP* 5 8. 3–5 9. 13; those affecting the fruit are discussed in *CP* 5 10. 1–5 10. 5.

[3] Outside origination may be due (1) to the weather or (2) to acts of man (or of God).

8.3 πάντα δὲ ἰσχύει[1] μᾶλλον τὰ ἀπὸ τοῦ ἀέρος ἐν τοῖς ἀσθενῶς διακειμένοις· τὰς γὰρ ὑπερβολὰς ἧττον δύνανται φέρειν. ἀσθενεστάτη δὲ διάθεσις μελλόντων τε καὶ ἀρχομένων βλαστάνειν, καὶ πάλιν μετὰ τὴν καρποτοκίαν ὥσπερ ἐξηραμμένων· τότε γὰρ ἐν μεγίστῃ μεταβολῇ. διὸ καὶ τὰ ἄγρια μάλιστα πονεῖ πρὸ τῆς βλαστήσεως, ἢ ὑπ' αὐτὴν τὴν βλάστησιν, ὅταν χαλαζοκοπηθῇ, ἢ πνεύματ' ἐπιγένηται ψυχρὰ σφόδρα ἢ θερμά· κρατεῖται γὰρ ταῖς ὑπερβολαῖς. οἱ δ' ὡραῖοι χειμῶνες ὠφελοῦσι καὶ τὰ ἄγρια καὶ τὰ ἥμερα· καὶ γὰρ <καὶ>[2] ἰσχὺν ἐμποιοῦσιν [καὶ][2] τῇ πιλήσει[3] καὶ πυκνώσει καὶ εὐβλαστίαν.

ταῦτα μὲν οὖν ἀμφοῖν κοινά.

9.1 νοσήματα δὲ τῶν ἀγρίων οὐ λέγεται· τῶν δ' ἡμέρων λέγεται πλείω, τάχα δὲ καὶ ἔστιν, διὰ τὴν

[1] u : ἴσχύοι U.
[2] ego.
[3] u (η ss.) : πιλώσει U.

---

[1] *Cf. CP* 2 1. 4.

[2] *Cf. CP* 2 1. 7 and *HP* 4 14. 1: "It is asserted that killing diseases do not occur in wild trees, but that the trees get into a poor condition, and most noticeably when struck by hail as they are about to sprout or are sprouting

Effects Common to Wild and Cultivated Trees

8.3 All effects arising from the air are stronger on trees that are in a weak state, since they are then less able to bear excess. The weakest state is when they are about to sprout or beginning to do so, and again when they are dried out (as it were) after bearing their fruit, for at these times they are involved in the greatest change. This is why wild trees too suffer most before or during sprouting when they are struck by hail or the ensuing winds are extremely hot or cold, for they are mastered by the excess. Seasonable winters on the other hand benefit both wild and cultivated trees, not only giving them strength by compressing and condensing them,[1] but good sprouting as well.[2]

Now these effects are found in both cultivated and wild trees.

Cultivated Trees: Diseases of the Whole Tree

9.1 As for diseases, people do not speak of any in wild trees. In cultivated trees on the other hand we hear of several, and several perhaps in fact exist, owing

or flowering, and when a very cold or hot wind comes up on these occasions; that however they are not affected by seasonable wintry weather, not even when it is excessive, but all profit from exposure to winter, since when they are not so exposed their sprouting is poorer."

ἀσθένειαν, ὧν τὰ μὲν ἀνώνυμα, τὰ δ' ὠνομασμένα, καθάπερ ἡ ψώρα, καὶ ὁ σφακελισμός, καὶ ἀστρόβλητα καὶ σκωληκόβρωτα γίνεσθαι.

ἀστρόβλητα μὲν οὖν μάλιστα γίνεται τὰ φυτὰ καὶ οἱ μόσχοι,[1] διὰ τὴν ἀσθένειαν· συμβαίνει δὲ τὸ πάθος ὅταν ἡ γῆ ξηρανθῇ, καὶ μὴ ἔχωσιν ἕλκειν ὑγρότητα, διὸ καὶ ὑπὸ <τὸ>[2] ἄστρον πλεῖστα.

9.2 τὰ δὲ πρεσβύτερα[3] διὰ τὴν ἰσχὺν ἀντέχει τε μᾶλλον καὶ ἕλκει πόρρωθεν· ἔτι δ' ἡ ὑγρότης ἡ οἰκεία πιοτέρα[4] καὶ λιπαρωτέρα·[5] καὶ πρωϊκαρπότερα[6] τῶν νέων, καὶ ὅσα ἂν προδείξωσιν, πάντ' ἐκφέρουσιν· ὅλως δὲ πᾶσαν μεταβολὴν ἧττον δύναται τὰ ἀσθενέστερα φέρειν. ὑπὸ δὲ τὸ ἄστρον καὶ διυγραίνεταί πως μᾶλλον τὰ δένδρα

[1] Gaza (*surculi*), Dalecampius : μίσχοι U.
[2] Schneider. [3] N aP : πρεσβύτευ|ρα U.
[4] P (ποιοτέρα a : πιωτέρα u N) : πιότερα U.
[5] u : -ώτερα U. [6] U : -τέρα u.

---

[1] Wild trees are stronger: *CP* 1 15. 3; *cf. HP* 3 2. 1.

[2] Contrast Aristotle, *On Life and Death*, chap. vi (470 a 27–32): "But if . . . spells of intense heat occur in summer

to the weakness of the kind.[1] Some of the diseases have no special names; others do, as "scab," "necrosis," getting "sun-scorched" and getting "grubby."

Sunscorch

Cuttings and layerings are the main victims of sunscorch because of their weakness. The ailment occurs when the ground is dried up and the plants are unable to attract fluid[2] (which is why most cases occur in the dog days).[3]

9.2 Older trees on the other hand by reason of their strength hold out better and draw food from a distance; again, the fluid proper to them is fatter and oilier,[4] and they bear earlier than the young ones and mature all of their fruit[5]; and in general weaker plants are less able to endure any change at all. In the dog days trees[6] also come in a way to get

and they (*i.e.* the plants) are unable to cool the fluid attracted from the ground, their heat dies down and perishes, and trees are said to suffer on these occasions from necrosis and to get sun-scorched."

[3] Literally "under the Star," the star (to which the word for "sun-scorched," literally "star-struck," refers) being Sirius.

[4] And so more resistant to evaporation.

[5] Their fluid is therefore at this time well on the way to its oily character and there is more of it.

[6] That is, maturer trees.

(καθάπερ πρότερον ἐλέχθη), διὸ καὶ ἡ ἐπιβλάστησις.

ἡ μὲν οὖν ἀστροβολία διὰ ταῦτα συμβαίνει, καὶ ἔτι διὰ τὰς πληγὰς τὰς ἔξωθεν καὶ τὰς ἑλκώσεις· ἀσθενέστερα γὰρ καὶ ταῦτα γίνεται, καὶ ἐπὶ πλέον αὐτῶν αἱ δυνάμεις διικνοῦνται.

9.3 σκωληκοῦται δὲ μᾶλλον τὰ πρεσβύτερα. συμβαίνειν δὲ[1] δοκεῖ καὶ τοῦτο τὸ πάθος ἢ διὰ τὰς ἑλκώσεις τῶν περισκαπτομένων, ἢ ὅταν ἐκδιψήσῃ διὰ τοὺς αὐχμούς. ἐκ μὲν γὰρ τῆς πληγῆς σήπεται διελκούμενα, τῇ δὲ[2] σήψει ζῳογονεῖ, καθάπερ καὶ τἆλλα· ἐκ δὲ τοῦ διψῆσαι, διὰ τὸ ἔλαττον ἔχειν τοῦ συμμέτρου τὸ ὑγρόν· οἷον γὰρ ἔκστασις γίνεταί τις ἐκ φύσεως,[3] ἐν δὲ τῇ ἐκστάσει, μεταβολὴ καὶ ἀλλοίωσις, ἐν δὲ τῇ μεταβολῇ, διὰ τὴν

[1] συμβαίνειν δὲ U^c in an illegible erasure.
[2] τῇ δὲ ego : δὲ τῇ U.
[3] τὶς ἐκ φύσεως Itali (*ex natura* Gaza : ἐκ τῆς φύσεως Schneider) : τῆς ἐκφυσεως U.

---

[1] *CP* 1 13. 5.
[2] *Cf. CP* 1 13. 3 (more trees have their second sprouting after the rising of Arcturus and in the dog days).

fuller of fluid, as we said[1] (which is why they have their second sprouting then).[2]

Sunscorch, then, is due to these circumstances; it is also due to blows from the outside and the consequent wounding,[3] since wounded trees too become weaker, and forces penetrate them farther.

Grubs

9.3 Older trees on the other hand are more apt to get grubby. This ailment too is considered to be due either to wounds received by the trees when hoed[4] or to their getting parched in time of drought.[5] For from the blow comes wounding and decomposition, and the tree, like other decomposing things, breeds animals. When a tree gets parched on the other hand grubs result because it has then less than the right amount of fluid, since a departure from nature arises (as it were),[6] and the departure involves change and alteration, and the change involves the breeding of grubs because of the decom-

[3] *Cf. HP* 4 14. 7: "And, as some suppose, most ailments (so to speak) arise from a blow; for the cases of so-called 'sunscorch' and of necrosis are (they say) due to the roots suffering from a blow."

[4] Like sunscorch: *CP* 5 9. 2.

[5] Like sunscorch: *CP* 5 9. 1.

[6] *Cf.* Plato, *Philebus*, 31 D 4–E 10 (of thirst as a departure from "nature").

σῆψιν ἡ ζῳογονία. (συμβαίνει δὲ καὶ[1] τοῦτο καὶ ἐν ταῖς ἀφαυαινομέναις· ἐνισχύει γὰρ τότε μᾶλλον τὸ θερμόν.) σκωληκοῦνται δὲ μάλισθ' αἱ μηλέαι, διὸ κἂν ταῖς ἐμπύροις φαῦλαι, ταχὺ γὰρ αἱ ῥίζαι ξηραίνονται.

9.4 κοινότατα δὲ φαίνεται τῶν νοσημάτων εἶναι τοῦτό τε καὶ ἡ ἀστροβλησία·[2] πάντα γὰρ (ὡς εἰπεῖν) σκώληκας ἴσχει, πλὴν τὰ μὲν πλείους, καὶ θᾶττον ἀπόλλυται, καθάπερ μηλέα συκῆ ἄπιος, τὰ δ' ἐλάττους, καὶ βραδύτερον. ἥκιστα δὲ σκωληκοῦται τὰ δριμέα, οὐχ ὅτι ἄσηπτα μόνον, ἀλλ' ὅτι καὶ ἡ δριμύτης κωλύει ζῳογονεῖν· σημεῖον δέ, τὸ τῆς δάφνης· αὕτη μὲν γὰρ σήπεται ταχέως, σκωληκοῦται δ' οὐχ ὁμοίως (ἐπεὶ διὰ τοῦτο οὐκ[3] ἐρινεὸς ὁμοίως τῇ συκῇ· δριμύτερος γὰρ ὁ ὀπός).

9.5 ὅλως[4] γὰρ τὰ γλυκέα σήπεται θᾶττον, εὐμετάβλητος γὰρ ὁ χυμός,[5] ἀσθενέστερος ὤν, διὸ καὶ αἱ μηλέαι καὶ αἱ ῥόαι μᾶλλον αἱ γλυκεῖαι τῶν ὀξειῶν·[6]

[1] [καὶ] Schneider.

[2] Liddell-Scott-Jones : ἀστροβολησία (*cf.* ἀστροβολία *CP* 5 9. 2).

[3] u aP : οὐχ U N : οὐδ' Schneider.

[4] Gaza (*omnino*), Itali : ὅπως U.

[5] U^{cc} (χ from μ) : χυλὸς Schneider.

[6] Schneider : ὀξέων U.

position. (Grubs also arise in roots that wither, since at that time the heat has greater power to affect them.) Apple trees are the most apt to get grubby, which is why the trees are poor in hot countries, since the roots soon get dry.

9.4 Grubs and sunscorch appear to be the diseases affecting the greatest number of different kinds of trees, for practically all get grubs, only some get more of them and perish sooner, as apple, fig and pear, whereas the rest get fewer and perish more slowly. Pungent trees are the least liable to get grubby,[1] not only because they resist decomposition, but also because their pungency prevents the breeding of animals when decomposition occurs. Proof of this is the case of the bay: it is quick to decompose but not quick to the same degree to get grubby. Indeed this is why the wild fig suffers less from them than the cultivated fig[2]: its juice is more pungent. 9.5 For in general the sweet fruit trees decompose faster, since the savour, being weaker, is more subject to change. This is why sweet apple and pomegranate decompose faster than acid, and the

[1] *Cf. HP* 4 14. 2: "For all trees (practically speaking) also get grubs; only some get fewer, some more, as fig, apple and pear. Broadly speaking pungent trees and those with a juice like that of the fig are the least liable to grubs . . ."

[2] *Cf. HP* 4 14. 4: "But the wild fig suffers neither from black twig or necrosis or scab or grubs in its roots to the same extent [*sc.* as the cultivated fig] . . ."

πάντων δ' αἱ ἠριναὶ μᾶλλον, διὰ τὴν ἀσθένειαν τοῦ τε χυλοῦ καὶ τῆς ὅλης φύσεως. ἐν σχίνῳ δὲ φυτευόμενα πάνθ' ἧττον σκωληκόβρωτα, διά τε τὴν θερμότητα καὶ τὴν ὀσμήν.

ταῦτα μὲν οὖν τὰ νοσήματα συμβαίνει διὰ τὰς ῥηθείσας αἰτίας.

9.6 οἴονται δέ τινες καὶ τὸν σφακελισμὸν ἀπὸ τῶν πληγῶν καὶ τῶν ἔξωθεν γίνεσθαι τραυμάτων, καθάπερ καὶ ἐπὶ τῶν ζῴων, μετενεχθῆναι γὰρ τοὔνομα κατὰ[1] τὴν ὁμοιότητα τοῦ πάθους. οὐ μὴν ἔοικεν ἀληθὲς οὐδὲ τοῦτ' ἐπὶ πάντων, εἴπερ ὁ ἐρέβινθος ἀπόλλυται σφακελίσας ὅταν ἀνθοῦσιν ὕδωρ ἐπιγινόμενον ἀποκλύσῃ[2] τὴν ἅλμην (τὸ γὰρ ἀφαλμᾶν[3] οἷον σφακελίζειν ἐστίν)· φαίνεται δὲ καὶ ἄλλα[4] χωρὶς πληγῆς πάσχειν τοῦτο τὸ πάθος.

[1] Gaza (*a*), Itali : καὶ U.
[2] Schneider : ἐπικλύσῃ U.
[3] ego : ἔφαλμα U.
[4] ἄλλα U^{c} : ἄλλα καὶ U^{ac} N aP.

---

[1] *Cf. CP* 2 11. 6 and *HP* 4 14. 7: "The weakest of all trees is the spring apple, and of these the sweet spring apple."

[2] The term is applied mainly to the roots: *cf. HP* 4 14. 4 (of the fig-tree): "The term necrosis is used when the roots turn black, and the term *krádos* when this happens to the

sweet spring apple[1] and pomegranate more than the rest by reason of their juice and their whole nature as well. When planted in a pine-thistle all trees are less liable to grubs because of its heat and its odour.

So these diseases occur owing to the causes mentioned.

Necrosis[2]

9.6 Some persons suppose that necrosis too comes from blows and wounds inflicted from the outside,[3] as it does in animals, the name being transferred to plants in virtue of the similar result. Still this explanation does not apply to all the plants affected any more than it did before,[4] inasmuch as chickpea gets necrosis and perishes when rain falls when the plant is in flower and washes the brine away,[5] the "brine-wash" being so to speak a suffering from necrosis[6]; further other plants as well are observed to suffer in this way without receiving a blow.

twigs . . ."

[3] *Cf. HP* 4 14. 7: "And, as some suppose, most diseases (so to speak) arise from a blow; thus both the case of so-called 'sunscorch' and of necrosis are due to the suffering of the roots that comes from this."

[4] In the case of sunscorch: *cf. CP* 5 9. 2.

[5] *Cf. CP* 3 23. 3; *CP* 3 24. 3.

[6] *Cf. CP* 3 24. 3.

9.7 ἔνιοι δὲ καὶ τὸ ἀστροβολεῖσθαι σφακελίζειν καλοῦσιν. τοῦτο μὲν οὖν τάχ᾽ ἂν ὀνόματος εἴη διαφορά· δι᾽ ὑπερβολὴν γὰρ καὶ ἔνδειαν τροφῆς καὶ οὕτως ἀπόλλυται τὰ δένδρα (τάχα δὲ μᾶλλον δι᾽ ἀπεψίαν), καὶ διὰ τὴν τῶν ἔξωθεν δύναμιν (οἷον ὅταν μετὰ τὴν καρποτοκίαν,[1] λεπτῆς οὔσης ἔτι τῆς ὑγρότητος διὰ τὴν ἀσθένειαν, ἐπιγένηται πάγος ἰσχυρός· ἀποθνήσκει γὰρ καταψυχόμενα, τὸ γὰρ ὀρρῶδες <καὶ>[2] λεπτόν, ψυχρόν, τὸ δὲ παχύ, καὶ ὥσπερ πῖον, θερμόν).

αὕτη μὲν οὖν εἴτε διὰ πλῆθος, εἴτε δι᾽ ἀπεψίαν, εἴτε δι᾽ ἄμφω καὶ ἔτι ἀσθένειαν, ἡ φθορά.

9.8 δι᾽ ἔνδειαν δὲ κατ᾽ ἄλλον τε τρόπον, καὶ ὅταν αἱ ῥίζαι μετέωροι γένωνται καὶ ἐπιφανεῖς· οὐ γὰρ δύνανται παρέχειν ἰκμάδα διὰ τὸν ἥλιον, ὥστε συγκάεται καὶ παχύνεται μᾶλλον τοῦ δέοντος ὁ ὀπός, ἐκ τούτου δὲ νόσος καὶ φθορά.

[1] N aP : -τομίαν U.
[2] Gaza (*et*), Schneider.

---

[1] Aristotle: *cf. On Life and Death*, chap. vi (470 a 31–32), cited in note 2 on *CP* 5 9. 1.

[2] To call sunscorch "necrosis" is merely to give it a new name. That necrosis is not the same as sunscorch is

Some[1] also call sunscorch "necrosis." Now this would perhaps be no more than a different name, since trees perish in this different way too[2] from excess and deficiency of food (or rather perhaps from failure to concoct)[3] and from the power of things outside them (as when a heavy frost follows the bearing of fruit, when they are so weak that their fluid is still thin, for then they are killed by the chilling, since serous and thin fluid is cold, whereas thick and as it were "fat" fluid is hot). 9.7

Here then is a form of death that is due either to the great amount of food or to failure to concoct or to both of them with weakness added.

Diseases from Deficiency of Food

Trees perish by deficiency of food in two ways. One is when the roots come to be at the surface and are exposed to view, since the sun renders them incapable of providing moisture. In consequence the sap is burnt by overheating and thickened more than is proper,[4] and this leads to disease and death. 9.8

shown by the inclusion of "excess" and "frost" in the following description.

[3] Such failure results from both deficiency and excess of food: there is either not enough to concoct, or too much for the plant to master.

[4] So that it cannot be transmitted upward.

παραπλήσιον δὲ τὸ συμβαῖνον καὶ ὅταν ἐξαυχμῶσι[1] δι' ἀνυδρίαν· οὐδὲ γὰρ τότε διαδιδόασι.

βοήθεια δὲ καὶ φυλακὴ πρὸς τὰς ἐνδείας τὸ κινεῖν ἀεὶ καὶ ὁμαλίζειν τὴν γῆν· οὕτω γὰρ τοῦ θέρους ἕλξει τινὰ ἰκμάδα, καὶ τοῦ χειμῶνος ἐνυπάρχει[2] μᾶλλον τὸ θερμόν, ὃ[3] διαδίδωσι τὴν τροφήν.

9.9 ἐξ ὑπερβολῆς <δὲ>[4] διὰ τὴν ἐπομβρίαν· τότε γὰρ διὰ τὴν ἀτροφίαν, ἀτροφεῖ γὰρ τὸ μὴ κρατοῦν μηδὲ πέττον (ἐκλευκαίνεται γὰρ[5] καὶ ἀπόλλυται τὰ δένδρα καθάπερ ὁ σῖτος)· ἐνίοτε δὲ οὐκ ἀπόλλυται μέν, εἰς δὲ τὴν καρπογονίαν νοσοῦσιν, ἀκαρπεῖ γὰρ τὸ μὴ πέττον μηδὲ κρατοῦν.

τῇ δὲ συκῇ καὶ νόσημά τι συμβαίνει περὶ τὰς ῥίζας, ὃ καλοῦσι λοπᾶν· τοῦτο δ' οἷον μάδησίς τίς ἐστιν τῶν ῥιζῶν καὶ[6] μικρὸν ἐπάνω διὰ τὴν

[1] Gaza, Schneider : ἐξαυχμώσῃ U.
[2] U : ἐνυπάρξει Schneider after Gaza.
[3] u : ὧ U.
[4] Schneider.
[5] U : Schneider deletes : τε Wimmer.
[6] καὶ τῶν Wimmer.

Close to this is the case when the roots go dry through lack of rain, since here too they fail to distribute food.

The remedy and prevention of cases of deficiency is to keep turning up the soil[1] and levelling it off. For when this is done the tree will attract some moisture in summer and in winter more heat is present,[2] and it is heat that transmits the food.

Diseases from Over-Supply of Food

9.9 Trees perish from excess because of heavy rains, for then they perish from undernourishment, since a tree is undernourished that does not master and concoct its food (for the trees turn white and die, just as cereal does).[3] But sometimes they do not die but ail in their production of fruit, since a tree that does not concoct and does not master its food bears no fruit.

The fig-tree is moreover liable to a certain disease called "peeling," that affects its roots. This is a kind of soddenness of the roots and of the parts

[1] *Cf. HP* 2 7. 5: "Some also dig up the soil round the fig-tree when this is required."

[2] Presumably the deeper earth is warmer; the cold has driven the warmth under the surface of the ground.

[3] *Cf. CP* 4 10. 1–2 and *HP* 8 6. 5 (of rain): ". . . but it is harmful to wheat, barley and cereals when they are in flower, for it destroys them."

πολυυδρίαν.

9.10 *ἐξ ὑπερβολῆς δὲ καὶ τὸ τραγᾶν τῆς ἀμπέλου, καὶ ὅσοις ἄλλοις ἀκαρπεῖν συμβαίνει διὰ τὴν εὐβλάστειαν·*[1] *οὐ δύνανται γὰρ οὐδὲ ταῦτ' ἐκπέττειν, ἀλλ' εἰς τὴν βλάστησιν ἡ ὁρμὴ τρέπεται, καθάπερ ἐπισπωμένη διὰ τὸ πλῆθος. ὡς ἐπὶ τὸ πολὺ δ' ἐκ τῶν τοιούτων συμβαίνει συκῇ μὲν ψωριᾶν, ἐλαίᾳ δὲ λειχηνιᾶν, ἀμπέλῳ*[2] *δὲ ῥεῖν, ὥσπερ καὶ Κλείδημός φησιν· λεπτὸς γὰρ ὁ καρπός, ἄπεπτος ὤν, καὶ ῥοώδης.*

9.11 *ἡ βοήθεια δ' ἐν δυεῖν·*[3] *ἔν τε τῷ παραιρεῖσθαι*

§9.10: Clidemus, Frag. 6 Diels-Kranz, *Die Fragmente der Vorsokratiker*, vol. ii^8, p. 50.

[1] Wimmer : *εὐβλαστείαν* U.
[2] u (*-ω* aP) : *ἄμπελων* U : *ἄμπελον* N.
[3] U P : *δυοῖν* N a.

---

[1] *Cf. HP* 4 14. 5: "The fig-tree gets diseased also if there are heavy rains, the parts close to the root and the root itself getting as it were sodden, and this is called 'peeling.'"

[2] *Cf. HP* 4 14. 6: "The vine gets the 'he-goat' (this with sunscorch being the disease to which it is most prone) either when the shoots are struck by winds or when it has been hurt in the course of cultivation, or third when it is pruned so that the cut faces up." (When the cut faces

just above them[1] due to excess of rain.

9.10 Also due to excess are the "he-goat" in the vine[2] and all other cases where trees sprout so well that they fail to bear. For these others too, like the vine, are unable to concoct their fruit properly, the drive turning in the direction of vegetative growth instead, pulled (as it were) in that direction by the great amount of such growth present. It is from such circumstances as these, for the most part, that the fig-tree gets scab,[3] the olive lichen,[4] and the vine its grape-drop,[5] as Clidemus says, for the fruit is thin, since it is not concocted, and tends to drop.

The Remedy

9.11 The remedy has two aspects: it reduces the good

down there is more drainage of sap.)

[3] *Cf. HP* 4 14. 5 (of the fig-tree): "Scab occurs principally when light rain falls after the rising of the Pleiades (if the rain is heavy the scab is washed off), and at that time both the *eriná* (*sc.* the *prodromi*) and the *ólynthoi* drop off."

[4] *Cf. HP* 4 14. 3: "The olive-tree . . . also grows a 'stud' (some call this a 'mushroom,' some a 'dish'); it resembles a patch burnt by the sun."

[5] *Cf. HP* 4 14. 6 (of the vine): "Shedding, which some speak of as *psínesthai* (*sc.* wasting away), occurs when the sky gets cloudy as the vine sheds its flower or when the vine gets 'lusty,' and the ailment consists in the grapes dropping and the remaining ones being small."

τὴν εὐτροφίαν (ὥσπερ καὶ τὰς ἀμυγδαλᾶς καὶ τὰς ῥόας τινὲς κολάζουσιν), καὶ ἐν τῷ δύναμίν τινα προστιθέναι τῷ δένδρῳ καὶ ἰσχύν. τοῦτο γὰρ αἵ τε διακόψεις ποιοῦσι τῆς συκῆς ὅταν ἄρχηται βλαστάνειν, ὅπως μή, παλίσκιος οὖσα, παχύνῃ τὸν ὀπόν, καὶ αἱ ἀποψιλώσεις τῶν ἀμπέλων καὶ ἀποκνίσεις τῶν οἰνάρων τῶν μεγίστων παρ' ὅλον τὸ θέρος, ὥς τινες κελεύουσιν,[1] καὶ αἱ περικαθάρσεις δὲ τῶν ἀνωτάτων[2] ῥιζῶν, καὶ τὰ κλήματα ἐπιβαλλόμενα κἀποκατώρυχες[3] καθιέμεναι, καὶ ἡ σκαπάνη βοήθεια γενομένη, καὶ γυμνοῦσα τὰς ῥί-

[1] κελεύουσι u : καιλύουσιν U.
[2] U N : -τω aP.
[3] καποκατωρυχες U : ἢ ἀποκατώρυγες u : κἀποκατώρυγε N : καὶ ὑπο(ὑπὸ P)κατώρυχα aP. LSJ (suppl.) suggest that the reading is due to confusion of ἀπῶρυξ and κατῶρυξ.

---

[1] *Cf. HP* 2 7. 6: "With the almond they even hammer in an iron peg and then replace it with one of oak and cover the spot with earth, a procedure that some call 'castigation,' with the implication that the tree is getting out of hand." *Cf.* also *CP* 1 17. 9; 2 14. 1; 3 18. 2; 5 17. 3.

[2] *Cf. CP* 2 14. 1.

[3] *Cf. HP* 2 7. 6: "If a tree bears no fruit but turns to vegetative growth they slit the part of the stem that is underground and insert a stone to split it and assert that

feeding (as some persons "castigate"[1] almond-trees and pomegranates) and adds a certain power and strength to the tree.[2] This is done by making cuts in the fig-tree[3] when it begins to sprout, to keep it from getting shady and thickening its juice, and by stripping the vine, pinching off the largest leaves throughout the summer, as some recommend, and by thinning the topmost roots[4] and by pulling the branches to one side and lowering the vine into the ground.[5] Spading moreover is here an emergency measure,[6] exposing the roots to wind and cold to let

the tree will bear. The same result is obtained if one prunes some of the roots . . . With figs in addition to pruning the roots they also smear ashes around the base and slit the trunk and say that the tree bears better." *Cf. CP* 1 17. 10; 2 14. 1, 4.

[4] *Cf HP* 2 7. 6: "The result is similar if one prunes away some of the roots. This is why this is done to the surface roots of the vine when it suffers from luxuriance."

[5] *Cf. HP* 4 13. 5: "But if in the vine when part of the roots are removed the trunk can survive, as some say, and the whole nature of the plant remains similar and bears similar fruit under these circumstances for any length of time, the vine would be the longest-lived of all. They say that one must proceed as follows when it appears to have entered into a decline: one pulls the branches to one side and gathers the fruit for that year; next one digs down on the other side and clears away all the roots there, filling the hole with brushwood and covering it with earth . . ."

[6] For the more ordinary purposes of spading *cf. CP* 3 10. 1.

ζας, ὅπως ἀπερῶσιν[1] πνεύματι καὶ ψύχει (καθάπερ ἔνιοι κελεύουσιν, πάγον εὐλαβουμένους, καὶ μετὰ ταῦτα δὲ τὴν κατεργασίαν ἀποδιδόναι)· πάντα γὰρ ταῦτα τῆς τροφῆς ἀφαιρεῖ καὶ συναύξει τὴν δύναμιν.

ἐναντίως δὲ τοῖς ἐξησθενηκόσι διὰ τὴν ἔνδειαν ἡ κατεργασία[2] καὶ ἡ ἄλλη θεραπεία τήν τε δύναμιν ἅμα καὶ τὴν εὐτροφίαν ἀποδίδωσιν.

9.12 τὴν δὲ ψώραν οἴονταί τινες γίνεσθαι καὶ ἄλλως, οἷον ὅταν ὕδωρ ἐπὶ Πλειάδι γένηται μὴ πολύ· τότε γὰρ ἀναζυμοῦται, καὶ ἀναθερμαινόμενα καὶ διίησιν[3] ἔξω, καθάπερ τὰ ἐξανθήματα· ἐὰν δὲ πολὺ γένηται, ἀποκλύζεται τὰ αἴτια, τάχα δὲ καὶ διαδίδωσιν εἰς τὰ ἐντὸς καὶ παύει· συμβαίνει δὲ τότε καὶ τὰ ἐρινὰ[4] καὶ τοὺς ὀλύνθους ἀπορρεῖν, δικνεῖται γὰρ εἰς ταῦτα ἡ ὑγρότης.

τῆς μὲν οὖν ψώρας ταῦτ' αἴτια τῇ συκῇ λέγουσιν.

[1] ego : //////ῶσιν U : ἐθισθῶσιν u.

[2] u : -αν U.

[3] καὶ διίησιν ego (διΐησιν Schneider) : καὶ δίεισιν U N : δίεισιν aP.

[4] N aP (ἔρινα U) : ἐρινεὰ u.

them drain off (as some recommend us to do as a precaution against frost, and after that to cultivate the tree as usual). For all this reduces the amount of food and helps to increase the power of the tree.

The opposite is done when the trees have become weak through want of food: cultivation and other procedures restore the power of the trees together with their good feeding.

Explanations by Others

9.12 Some suppose that scab arises in another way,[1] when there is light rain at the rising of the Pleiades.[2] For then the trees ferment, and as they get heated transmit the fermenting parts to the surface, like pustules. But if the rain is heavy the fermenting parts responsible for the scab are washed off, or perhaps the tree transmits them to its interior and there puts an end to the fermentation (at this time moreover dropping of the *eriná* and the *ólynthoi* occurs,[3] since the water penetrates to them).

These then are the causes given by some for scab in the fig.

[1] Than by excess of food (*CP* 5 9. 10).

[2] April 30 (Euctemon), May 9 (Eudoxus).

[3] *Cf. HP* 4 14. 5 (of the fig-tree): "Scab arises mainly when there is only light rain at the rising of the Pleiades. But if the rain is heavy the scab is washed off, and it happens then that both the *eriná* and the *ólynthoi* drop." (For these see notes 2 and 3 on *CP* 5 1. 5.)

9.13 τῇ δὲ ἀμπέλῳ τοῦ τραγᾶν, ὅταν ἢ ὑπὸ πνεύματος βλαστοκοπηθῇ, ἢ ὅταν τῇ ἐργασίᾳ συντμηθῇ,[1] ἢ τρίτον ὅταν ὑπτία[2] τμηθῇ· συμβαίνει <γὰρ>[3] πλείω τὸν ἀθροισμὸν γινόμενον μᾶλλον εἰς τὴν βλάστησιν σφοδροτέρως ὁρμᾶν, ὥστε μὴ δύνασθαι καρπογονεῖν.

τοῦ δὲ ῥυάδας[4] ἐπιγίνεσθαι δύο αἴτια· ἢ ὅταν ἐπινέφῃ[5] κατὰ τὴν ἀπάνθησιν, ἢ ὅταν κρειττωθῇ. τοῦτο δ' εἰ ἀληθές, ἔοικε κατὰ μὲν τὴν ἀπάνθησιν ὑγρότερος ὢν ὁ ἀὴρ ὥσπερ ἀπερυσιβοῦν, ἡ δὲ κρείττωσις οἷον ἀντισπᾶν καὶ μεθιστάναι τὴν τροφήν· ὥστε ἐξ ἀμφοτέρων εὔλογον ἀπορρεῖν τὰς ῥυάδας,[6] καὶ τὰς ἐπιμενούσας [ὡς] εἶναι μικράς [εἶναι].[7]

ἡ δ' ἄμβλωσις τῶν ὀφθαλμῶν ὅταν ῥιγώσῃ[8] ταὐτὸν καὶ παραπλήσιόν ἐστιν ὅπερ ἐπὶ τῶν ἀγρίων προβλαστάνειν ἀρχομένων ἐὰν ἐπιγένηται

§9.13: *Cf. HP* 4 14. 6.

[1] U : συμπάθῃ Schneider (from *HP* 4 14. 6).
[2] *HP* 4 14. 6: ὕπτια U.
[3] aP. [4] ego: ῥυάδα U.
[5] ego (ἐπινιφθῇ *HP* 4 14. 6) : ἐπινείφῃ U.
[6] ego : ῥαγάδας U : ῥᾶγας *HP* 4 14. 6.
[7] ego (*cf. HP* 4 14. 6) : ὡς μικρον εἶναι U (Heinsius reads μικρὰς, Schneider deletes ὡς). [8] U : -ῶσι u.

9.13 The causes given for the "he-goat" in the vine are these: the shoots are broken off by the wind, or the vine is cut too short in pruning, or third is pruned with the cut facing up.[1] The result is that more food is accumulated[2] and this moves with a stronger impetus to the production of vegetative growth, so that the vine is unable to produce fruit.

Two causes are given for the ensuing of grape-drop: the sky is cloudy at the time when the flower is shed, or the vine gets lusty.[3] If this account is true, it appears that (1) at the time of dropping the flower the air is too moist and infects the vine with rust (as it were); and that (2) the lustiness exercises a counter-attraction (as it were) and diverts the food. So as a result of both conditions it is reasonable that the vines so affected should shed, and that the remaining grapes should be small.

The miscarriage of the buds when the vine is chilled[3] is the same and similar to what occurs in wild trees when cold weather ensues as they are beginning to get on with sprouting.[4] This also hap-

[1] The loss of shoots and parts results in a greater accumulation of food, none being used to feed the missing parts. When the cut faces upward there is less drainage away of fluid, and so more food is available.

[2] *Cf. HP* 4 14. 6, reproduced here practically verbatim.

[3] *Cf. HP* 4 14. 6.

[4] *Cf. CP* 5 8. 3.

χειμών· πάσχει δὲ τοῦτο καὶ ἡ πρὸ ὥρας ἄνθησις· ἐπ' ἀσθενεῖ[1] γὰρ πνεύσας ἄνεμος ψυχρὸς ἀπέκαυσεν.[2]

ταῦτα μὲν οὖν τῶν δένδρων αὐτῶν ἐστι νοσήματα καὶ πάθη.

10.1 τὰ δὲ τῶν καρπῶν, οἷον τῶν μὲν βοτρύων ὁ καλούμενος καμβρός.[3] τοῦτο δ' ὅμοιον τῇ ἐρυσίβῃ, γίνεται γὰρ ὅταν, ἐπούσης ὑγρότητος μετὰ τὰς ψεκάδας, ἐπικαύσῃ σφοδροτέρως ὁ ἥλιος (ὅπερ συμβαίνει καὶ ἐπὶ τῶν οἰνάρων).

μηλέας δὲ καὶ ἐλαίας καὶ συκῆς καὶ σκωληκοῦνται οἱ καρποί. καὶ τὰς μὲν μηλέας καὶ συκᾶς ἀεὶ διαφθείρουσιν οἱ σκώληκες· τὰς δὲ ἐλάας, ἐὰν μὲν ὑπὸ τὸ δέρμα γένωνται, διαφθείρουσιν, ἐξεσθίουσι γάρ· ἐὰν δὲ τὸν ὑπὸ[4] τὸν πυρῆνα διαφάγωσιν,[5] ὠφελοῦσιν, ἐντὸς γὰρ ὄντες οὐχ ἅπτονται τῆς σαρκός. κωλύονται δ' ὑπὸ τῶν πνευμάτων[6]

§10.1: *Cf. HP* 4 14. 10.

[1] Gaza, Schneider : ἀσθενεία U.
[2] Schneider : ἀπέκλυσεν U.
[3] U : κραμβὸς N P : κράμβος a.

pens when the vine flowers before its season: a cold wind, blowing on the flower when it is weak, blasts it.

These then are the diseases and disorders of the trees themselves.

Diseases of the Fruit

10.1 Those of the fruit are for example the so-called *kambros*[1] in grape clusters. This is similar to rust, since it occurs when water is on the fruit after the light drizzles and the sun has burnt the surface with more than usual intensity[2] (this also happens with the leaves of the vine).

In apple, olive and fig the fruit as well[3] gets infested with grubs. In the apple-tree and fig-tree the grubs always ruin the fruit; in the olive they ruin the fruit if they get out into the flesh, since the flesh is consumed, but if they eat through the fruit under the stone they improve the olive, since they are then inside the stone and do not touch the flesh. They are kept from this[4] by the winds when it rains

[1] The word is not elsewhere attested.

[2] For rust similarly produced *cf. CP* 4 14. 3.

[3] As well as the whole tree (or the roots): *cf. CP* 5 9. 4; *HP* 4 14. 2–3.

[4] Coming up into the flesh.

---

[4] *τον ὑπὸ* U (Schneider deletes with *HP* 4 14. 10) : *τῶν ὑπὸ* u : *τὰ ὑπὸ* N aP.

[5] Itali : -*φύγωσιν* U.

[6] *τῶν πν.* U : *τῷ δέρματι* Schneider (from *HP* 4 14. 10).

[εἶναι][1] ὕδατος ἐπ' Ἀρκτούρῳ γενομένου,[2] καταψυχόμενοι γὰρ φεύγουσιν.

νοσεῖ δὲ πολλάκις καὶ τὰ σῦκα καὶ οἱ ἄλλοι καρποί.

10.2 τὸ δ' αἴτιον, ὡς ἁπλῶς εἰπεῖν, ἐν δυοῖν· ἢ γὰρ ἀπ' αὐτοῦ τοῦ δένδρου καὶ τῆς τούτου διαθέσεως (ὥσπερ ὅταν ἐν ταῖς ἀμυγδαλαῖς ἡ ὑγρότης κομμιδώδης ὑπογένηται), ἢ ἀπὸ τῶν κατὰ τὸν ἀέρα ξυμβαινόντων· τῇ δ' οὖν ἐλάᾳ καὶ τὸ ἀράχνιον[3] ἐμφύεται δι' ὑγρασίαν τινὰ τοῦ ἀέρος τοῦ περὶ αὐτάς, ὃ καὶ διαφθείρει τὸν καρπόν· ἡ δὲ ἐκ τοῦ ἕνου βλάστησις ὑδάτων ἐπιγινομένων γίνεται, δι' ἣν ἀποβάλλει τὸν καρπόν, ἐνταῦθα τῆς τροφῆς ῥεούσης,[4] ἅτε μὴ[5] ἰσχύοντος τοῦ καρποῦ, τῆς δ' ὑγρότητος ἠθροισμένης[6] τῆς βλαστήσεως· χείριστον δέ, ἐὰν ἀνθούσαις ἐφύσῃ[7] καὶ ἐλαίαις καὶ ἀμπέλοις καὶ τοῖς ἄλλοις· ἀπορρεῖ γὰρ τὰ ἄνθη καὶ οἱ καρποὶ δι' ἀσθένειαν.

[1] ego.
[2] u : -ων U N aP.
[3] u : ἀραχυνιον U.
[4] Gaza, Schneider : γε οὔσης U.
[5] Scaliger : καὶ U : καὶ οὐκ Itali : μήπω Schneider (*nondum* Gaza).

at the rising of Arcturus, since they are chilled and avoid the surface.

Figs too and other fruit are often victims of disease.

The cause, to put it simply, is twofold: either the tree itself and its disposition is responsible (as when a gummy exudation comes out under the almond shell), or what happens in the air. At all events the "cobweb" grows on olive-trees because of a certain wetness in the air round them, and it ruins the fruit.[1] Sprouting from last year's wood occurs when rains follow the formation of the fruit, and this makes the tree drop its fruit, as the food flows to the new sprouts, since the fruit is not strong, and the fluid for the sprouting has been collected. Worst of all is when rain falls on the olive, vine and the rest when they are in flower, since both flowers and fruit are so weak that they drop off.[2] 10.2

[1] *Cf. HP* 4 14. 10: "There also occurs another disease affecting olives which is called 'cobweb,' for this grows on the tree and ruins the fruit."

[2] *Cf. HP* 4 14. 8: "Worst of all is when rain falls on some when they are in bloom, as olive and vine; for the fruit is so weak that it drops too."

---

[6] ἠθροισμένης <. . .> Heinsius.

[7] ἐφύσῃ P[c] (*impluerit* Gaza): ἐμφυσῆι U: ἐμφυσήσῃ u: ἐμφύσῃ N aP[ac] (?).

10.3 ἐνιαχοῦ δὲ ἴδια πάθη συμβαίνει, καθάπερ ἐν Μιλήτῳ καὶ Τάραντι περὶ τὰς ἐλάας. ἐν Μιλήτῳ μὲν γάρ, ὅταν περὶ τὸ ἀνθεῖν ὦσιν, νοτίου ἀέρος ὄντος καὶ εὐδιεινοῦ,[1] κάμπαι γενόμεναι κατεσθίουσιν αἱ μὲν τὰ[2] φύλλα, αἱ δὲ τὰ ἄνθη, ἕτεραι [προς][3] τῷ γένει. βοηθεῖ δὲ πρὸς ταῦτα, ἐὰν ἐπιγένηται καῦμα, διαρρήγνυνται[4] γάρ. ἐν Τάραντι δὲ περὶ τὴν ἄνθησιν ὁτὲ μὲν ἀπερυσιβοῦνται διὰ τὴν ἄπνοιαν, ὁτὲ δὲ πνεῦμά τι πνέον ἐκ τῆς θαλάττης ὁμιχλῶδες καὶ παχύ, προσίζον ἐν τοῖς ἄνθεσιν, ἀπόλλυσιν τὰ ἄνθη τῇ ὀσμῇ·[5] διὰ τοῦτο, καλλίστων ὄντων καὶ μεγίστων τῶν δένδρων, ἐλάχιστος παρ᾽ αὐτοῖς ὁ καρπός.

10.4 ὅλως δὲ ἕκαστοι τῶν τόπων ἰδίας ἔχουσι κῆρας, οἱ μὲν ἐκ τοῦ ἐδάφους, οἱ δὲ ἐκ τοῦ ἀέρος, οἱ δὲ ἐξ ἀμφοῖν. ἐκείνη δὲ αἰτία κοινὴ πᾶσιν, ἡ ἀπὸ τῶν πνευμάτων κατὰ τὰς χώρας ἑκάστας, ὅσα θερμὰ

§10.3: *HP* 4 14. 9.

[1] u : -δειν- U.
[2] aP : *τὰρ* U[cc] (*τὰ* intended) : *γὰρ* U[ac] N.
[3] Heinsius (*οὖσαι* Wimmer).
[4] Schneider : *-νυται* U.
[5] U : *ἅλμῃ* Schneider (from *CP* 2 7. 5).

10.3 In some places affections peculiar to the region occur, as with the olives at Miletus and Tarentum. Thus at Miletus, when the trees are at the time of flowering and there are southerly winds and clear skies, bend-worms are produced and some devour the leaves, others of a different kind the flowers. A remedy for this is an ensuing hot spell, since the grubs then burst.[1] At Tarentum the olive trees at blossom time sometimes get rust because of the absence of wind, and sometimes a certain sea wind, foggy and thick, blows steadily on the flowers and destroys them with its odour. Hence although the trees are very fine and tall the harvest at Tarentum is of the smallest.[2]

10.4 In general each type of country has its own peculiar plague, some having it from the ground, some from the air, some from both. But a causation common to all is the following: local winds[3] that blow

[1] *Cf. HP* 4 14. 9: "At Miletus when the olive trees are at the time of flowering they are eaten by bend-worms, one kind eating the leaves and another the flowers, and they strip the trees. The grubs are produced if the winds are southerly and the weather clear; if heat spells ensue the grubs burst."

[2] *Cf. CP* 2 7. 5 and *HP* 4 14. 9: "At Tarentum the olive trees always promise an abundant crop, but at the time of flowering it is most often lost."

[3] *Cf. HP* 4 14. 11: "Different winds in different regions are of a nature to destroy and 'burn.'"

τοῦ θέρους πνεῖ,[1] καὶ τοῦ ἦρος ψυχρὰ σφόδρα, τῶν δένδρων ἄρτι βλαστανόντων, καὶ ὑπὸ τὴν ἄνθησιν· τὸ γὰρ ὅλον ἀληθὲς καὶ ἐπὶ τούτων, ὅτι δι' ὑπερβολὴν καὶ ἔνδειαν τροφῆς καὶ καύματος καὶ ψύχους νοσοῦσιν.

10.5 ἔτι δ' ἂν μὴ κατὰ καιρὸν τὰ πνεύματα καὶ τὰ ὕδατα τὰ οὐράνια γένηται. συμβαίνει γὰρ ὁτὲ μὲν ἀποβάλλειν γενομένων <ἢ>[2] μὴ γενομένων ὑδάτων (ὥσπερ τὰς συκᾶς), ὁτὲ δὲ χείρους γίνεσθαι σηπομένους καὶ καταπνιγομένους ἢ πάλιν ἀποξηραινομένους παρὰ τὸ δέον (ἐπεὶ καὶ καύματα[3] ἔνια καὶ βότρυν καὶ ἐλάαν ἀποκάει καὶ ἄλλους καρπούς).

τῶν <δὲ>[4] σκωλήκων ἐν πολλοῖς[5] διαφέρουσι μὲν καὶ μορφαί,[6] οὐ μὴν ἀλλ' ἐκείνη μείζων ἡ διαφορά, τὸ μὴ δύνασθαι τὰ[7] ἐξ ἑτέρου δένδρου καὶ καρποῦ μετατεθέντα[8] ἐν ἑτέρῳ γένει σῴζεσθαι (τοῦτο δ' εὔλογον, ἑκάστῳ γὰρ ἐκ τῆς οἰκείας ὕλης ἡ τροφή). πλὴν ἴδιον τὸ περὶ τὸν[9] κεράστην καλούμενον ξυμβαῖνον· τοῦτον γάρ

§10.5: *Cf. HP* 4 14. 8 and 4 14. 5.

[1] u : *πίνει* U.
[2] Schneider from *HP* 4 14. 8 : *λεγομένων* U.

very hot in summer and very cold in spring, when the trees are just sprouting or when they are in bloom. For what is true of the tree in general is also true of the fruit: disease arises because of excess and deficiency of food and of heat and cold.[1]

10.5 The fruit also gets diseases if wind and rain do not occur in season. Thus when rain falls or does not fall it sometimes happens that the trees (as the fig) drop their fruit, and sometimes that the fruit deteriorates by decomposing or becoming choked, or again by becoming too dry. Indeed certain hot spells cause both grape-clusters and olives and other fruits to wither and drop.

As for the grubs, the shapes too differ in many trees and fruits, but the greatest difference is this: the grubs from one tree or fruit cannot be transferred to another kind and survive. This is reasonable, since each gets its food from the matter that produced it[2] (except that the case of the so-called "horned worm" is peculiar, for it is reported to

[1] *Cf. CP* 5 8. 2; 5 9. 7.
[2] *Cf. CP* 3 22. 4.

---

[3] *καύ||ματα* u : ||*ματα* U. [4] M aP.
[5] *ἐν πολλοῖς* U : *πολλοῖς* N : *πολλοὶ* aP.
[6] U : *μορφαῖς* u.
[7] u : *τὸ* U.
[8] u : *μετα|θέντα* U^c (*ἐξ . . . μετα|* in an illegible erasure).
[9] u : *περιττὸν* U.

φασι καὶ ἐν τῇ ἐλαίᾳ γίνεσθαι καὶ εἰς τὴν συκῆν ἐντίκτειν·[1] ἔχει δὲ ἡ συκῆ καὶ ἐξ ἑαυτῆς σκώληκας καὶ τοὺς[2] ἐντικτομένους τρέφει, πάντες δ' εἰς κεράστην ἀποκαθίστανται· φθέγγονται <δὲ>[3] οἷον τριγμόν.

περὶ μὲν οὖν νοσημάτων ἱκανῶς εἰρήσθω.

11.1 περὶ δὲ[4] φθορᾶς ἁπλῶς τῶν δένδρων ἐκεῖνο πρῶτον δεῖ διελεῖν,[5] ὅτι εἰσὶν αἱ μὲν κατὰ φύσιν, <αἱ δὲ παρὰ φύσιν>·[6] κατὰ φύσιν μὲν αἱ γήρᾳ[7] καὶ δι' ἀσθένειαν αὐάνσεις, ὥσπερ ἀποπνεόντων καὶ ἀπομαραινομένων αὐτομάτων· παρὰ φύσιν δὲ αἱ ἀπὸ τῶν ἔξωθεν. τούτων δὲ αἱ μὲν βιαιότεραι[8] φαίνονται, καθάπερ ἐὰν κοπὲν ἢ πληγέν, αἱ δ' ἧττον, αἱ[9] διὰ χειμῶνας ἢ πνεύματα, πασῶν δὲ ἥκισθ' αἱ[10] διὰ νόσον, ὥσπερ ἐπὶ τῶν ζῴων·

[1] ἐλαίᾳ . . . ἐκτίκτειν omitted by U (expunged?).
[2] U cc from ἔτι.
[3] *HP* 4 14. 5.
[4] δε U css : U t omits.
[5] Schneider : διέλθεῖν U.
[6] Schneider.
[7] Schneider : γηραιαὶ U.
[8] u aP (-α N) : βιοτεραι U.
[9] ἧττον αἱ Wimmer : ἥττονσε U : ἧττον u.
[10] ego (ἥκιστα αἱ Wimmer) : ηκιστα U.

be also produced in the olive tree and breed its young[1] in the fig-tree. The fig-tree has grubs of its own making and also supports the grubs that are hatched in it, but all end up as horned worms. They make a sound like squeaking).

For the discussion of diseases let this suffice.

Death of the Tree Itself: Natural, Unnatural and Intermediate

11.1 Concerning the death of the trees themselves[2] we must first make the distinction that some forms of it are natural, some unnatural. Withering from old age and weakness is natural, when of its own accord (as it were) the tree dissipates its fluid and lets its heat die down, whereas death coming from the outside is unnatural.[3] Of the forms of death that come from the outside some appear more violent than the rest, as when a tree is chopped or struck, some less, as those due to cold weather or winds, and least violent of all appear those due to disease

[1] *Cf. HP* 5 4. 5, where the horned worm is said to breed its young in timbers, just as it does in trees.

[2] And not just the loss of the fruit.

[3] So at *CP* 5 18. 1 the death of the seed from evaporation of its heat and fluid is natural; all other forms of death, by grubs, liquefaction or other departures from nature, are unnatural.

ἐγχρονίζουσι γὰρ αὗται μάλιστα, διὸ καὶ οὐδὲ φαίνονται παρὰ φύσιν, ὁμοίως ἔν τε ζῴοις καὶ φυτοῖς, εἴπερ ἐκείνη τῆς φύσεως ἡ ἔκλειψις.

11.2 ἔστι δέ τις καὶ ἑτέρα παρὰ ταύτας, ἡ διὰ τὴν εὐκαρπίαν καὶ πολυκαρπίαν. τοῦτο μὲν οὖν (ὥσπερ ἐλέχθη) φύσει βραχύβιον, ἐξαναλισκομένης ἐνταῦθα τῆς οὐσίας, ὅσα πολυκαρπήσαντα ἀφαυαίνεται (καθάπερ ἐλέχθη), συμβαίνει γὰρ τοῦτο πλείοσιν· ἐπεὶ καὶ ἐλαῖαι αἱ νέαι δοκοῦσι φθείρεσθαι δι' εὐκαρπίαν, ὅτι, οὔπω τετελεωμέναι,[1] τὴν τῆς αὐξήσεως τροφὴν ἐξαναλίσκουσιν εἰς[2] τοὺς καρπούς. ταύτην δ' οὐ τῶν κατὰ φύσιν ἄν τις θείη τὴν φθοράν, οὐδ' ὡς ἀπὸ τῶν ἔξωθεν καὶ βιαίως, ἀλλ' ὡς ἀπ' αὐτῶν διὰ τὴν ἐπὶ πλέον ὁρμὴν τοῦ συμμέτρου πρὸς τὸν καρπόν, ᾗ[3] συναίτιός πως καὶ ὁ ἀὴρ καὶ ἡ τοῦ ὅλου κατάστασις (εὐκαρπία γὰρ γίνεται τοιαύτη διὰ τὴν τοῦ ἀέρος εὐκρασίαν).

ταύτην μὲν οὖν εἴτε κατὰ φύσιν, εἴτε παρὰ φύσιν, εἴτε καὶ μέσην τινὰ χρὴ λέγειν, οὐδὲν διαφέρει.

[1] ego : -εσμέναι U.
[2] u : ἧς U : οὗ N aP.
[3] ᾗ u : ἡ U.

(as in animals), for these take the most time, and for this reason do not even appear unnatural, whether in animals or in plants, inasmuch as the wasting away mentioned above[1] belongs to their nature.

11.2 There is a further form of death in addition to these, the one due to excellence and abundance of fruit. The tree that dies of this (as we said)[2] is by nature short-lived, since the substance is expended on fruit in all trees that wither away after producing a heavy crop (as was said),[3] for this occurs in a number of different trees. So young olives are held to die on account of a good crop, because at a time when their growth is not yet finished they expend the food for it on the fruit. One would not count this form of death as a natural one[4] nor yet as coming from the outside and occurring by violence, but rather as coming from the trees themselves because of the excessive impetus toward production of fruit, a cooperating cause (in a way) being the air and the climate, since fruitfulness of this kind occurs because of the well-tempered character of the air.

Whether, then, we are to call this death natural or unnatural or else intermediate between the two makes no difference.

[1] In the second sentence of this paragraph.

[2] *CP* 2 11. 1.

[3] *CP* 2 11. 2–3.

[4] That is, as a withering away from weakness or old age (*CP* 5 11. 1).

11.3 τάχα δὲ καὶ γένει τινὶ δένδρων ἔνιαι κατὰ φύσιν, αὐτομάτως τε γινόμεναι, καὶ οὐ κακουμένοις ἀλλ' εὐθενοῦσιν,[1] οἷον τῆς πεύκης ὅταν αἱ ῥίζαι δᾳδωθῶσιν· πάσχουσι μὲν γὰρ τοῦτο δι' εὐτροφίαν δι'[2] ὑπερβολήν, ἅμα δὲ τῇ δᾳδώσει, τὴν τροφὴν οὐ διιεῖσαι,[3] φθείρονται. καὶ ἔοικε παραπλήσιον τὸ συμβαῖνον εἶναι καὶ ἐπὶ τῶν ζῴων ὅταν ὑπερπιανθῶσιν·[4] οὐ δυνάμενα γὰρ ἕλκειν τὴν πνοὴν οὐδ' ὅλως τῷ πνεύματι χρῆσθαι διὰ τὴν σύμφραξιν καὶ τὴν πύκνωσιν, ἐκεῖνά τε ἀποπνίγονται καὶ αἱ πεῦκαι (διὸ καὶ οὐ κακῶς οἱ ὀρεοτύποι[5] τοὔνομα τέθεινται· φασὶ γὰρ ἀποπνίγεσθαι τὴν πεύκην διὰ τὴν πιότητα)· πάντα γὰρ (ὡς ἔοικε) δεῖται πνεύματός τινος, ἢ μανώσεως, ἢ πόρων.

τὰς μὲν οὖν τοιαύτας μᾶλλον ἄν τις θείη φυσικάς· ἐκείνας δὲ ὁποτέρως δεῖ προσαγορεύειν μηδὲν διαφερέτω (καθάπερ ἐλέχθη).

[1] U : εὐσθενοῦσιν u N : εὐσθενούσης aP.

[2] <καὶ> δι' aP.

[3] ego : διεῖσαι U.

[4] ego : -πανθῶσιν U : -παχυνθῶσιν u.

[5] ego (*cf. HP* 3 3. 7; 3 12. 4; 4 13. 1 : ὀρείτυποι Heinsius) : ὀρθότυποι U.

---

[1] *Cf. HP* 3 9. 5: "The people of Ida say that a disease

A Natural Form of Death Confined to a Certain Kind of Tree

11.3 Perhaps however there are also forms of death that are natural for a certain kind of tree, since they arise of their own accord, and in trees that are under no hardship but thriving, as in the pine when the roots turn to torch-wood. This happens to the trees because they are well-fed to excess, and as soon as it occurs they give no passage to the food and perish.[1] And what occurs in animals as well when they get over-fat appears to be close to this: being unable to draw breath or make any use of respiration at all because of the blockage and thickening, they too are choked, and so is the pine. This is why the wood-cutters have given to the phenomenon a name that is by no means unapt, and say that the pine is "choked" by its fatness, for all living things require, it seems, some sort of breathing or openness of texture or passages.

Death of this sort one would rather count as natural; but whether we should call the form of death mentioned earlier[2] natural or unnatural, we shall (as we said)[2] consider to be indifferent.

occurs in the pine of such a sort that as a result, when not only the heart-wood but the outer wood also of the trunk gets to be torch-wood, the trees 'choke' as it were. This, one might conjecture, arises spontaneously, since all of it turns to torch-wood." [2] *CP* 5 11. 2.

12.1 περὶ δὲ τῶν παρὰ[1] φύσιν λεκτέον (ἐπείπερ αἱ κατὰ φύσιν ἁπλαῖ τινές εἰσιν καὶ πανεραί), τούτοιν[2] δὲ σχεδὸν ἐν δυοῖν αἰτίαιν[3] τῶν κατὰ τὸν ἀέρα γινομένων, ὑπερβολῇ[4] ψύχους τε καὶ καύματος· αἱ γὰρ δι' ἔνδειαν τροφῆς (οἷον ἢ λειψυδρίαν ἢ χώρας κακίαν) ἕτερον εἶδος ἔχουσι, καὶ φανεραί.

τῶν <δ'>[5] εἰρημένων αἱ μὲν ὑπὸ καύματος ἐλάττους, ἂν μή τις αὐχμὸς ὑπερβάλλων γένηται, καὶ οὗτος δὲ μᾶλλον τἆλλ' ἀπόλλυσιν ἢ τὰ δένδρα, ταῦτα δ' ἂν συνεχὴς γίνηται καὶ πλείω χρόνον· ἀλλοῖα δὴ[6] τὰ ἐπέτεια καύματα γίνεται κατὰ τὸν οἰκεῖον καιρόν, οἷον ὑπὸ Κύνα· μόνα γὰρ ἀπόλλυται τὰ φυτά, ἀστροβολούμενα, καὶ ἐὰν ἄρα τι τύχῃ πεπηρωμένον ἢ ἀσθενές (οὐδὲ γὰρ ταῦθ' ὑπομένει).

[1] ego : κατα U.
[2] U : τούτων Scaliger.
[3] U : *causae* (nom.) Gaza : αἰτίαι Scaliger : αἱ αἰτίαι Schneider.
[4] N aP : -ῃ U.
[5] Dalecampius (*sed . . . profecto* Gaza).
[6] ego (ἀλλ' οὐχ οἷα Schneider) : ἀλλ οἷα δὴ U.

Unnatural Death from the Weather

12.1 But we must speak of unnatural forms of death (the natural ones[1] being of a simple sort and evident) that come under these two causes (one might say) belonging to occurrences in the air: excess of heat and excess of cold. As for forms of death due to deficiency of food (such as to want of water and to poor land), these are of a different type and evident.[2]

From Hot Weather

Of the forms of death just mentioned those due to hot weather are the less frequent, unless there has been an excessive drought, and this is more fatal to other plants than to trees, but it is fatal to trees if it is unbroken and prolonged. So annual hot weather at its proper season, as in the dog days, is of a different character, for then it is only immature trees that are killed by getting sunscorch[3] and an occasional tree that is maimed or weak (these too being unable to resist).

[1] That is, death by withering away from weakness and old age (*CP* 5 11. 1).

[2] *Cf. CP* 5 8. 2 (of diseases due to want of food, which may be regarded as an internal or external cause).

[3] *Cf. CP* 5 9. 1.

12.2 ὑπὸ δὲ τοῦ χειμῶνος πλείω καὶ κατὰ πλείους[1] τρόπους· ὁτὲ μὲν γὰρ αἰθρίαι[2] καὶ πάγων σφοδρότητες ἐκπηγνύουσιν, ὁτὲ δ' ἄνεμοι ψυχροὶ πνεύσαντες[3] ἀπέκαυσαν,[4] ἐνίοτε δὲ καὶ ἀπνοίας οὔσης, ἢ μετρίας πνοῆς, οὐ καθαροῦ δὲ τοῦ ἀέρος, ἀλλὰ θολεροῦ καὶ ἐπινεφοῦς· καὶ σχεδὸν οὕτως ἥ γε τῶν δένδρων γίνεται πῆξις, ἐάν τε αἴθριος, ἐάν τε μὴ αἴθριος ὁ οὐρανὸς ᾖ. γίνεται δὲ περὶ μὲν Ἀρκαδίαν καὶ Εὔβοιαν αἰθρίας καὶ πνεύματος μᾶλλον, περὶ δὲ Θετταλίαν καὶ τούτους τοὺς τόπους ἀπνοίας καὶ ἐπινεφοῦς, οὐ μὴν ἀλλὰ καὶ πνευμάτων ὄντων συμβαίνει τὸ πάθος (ὅταν γὰρ τὰ ὄρη λάβῃ χιόνα τὰ περικείμενα καθ' ἑκάστους τόπους· ἀφ' ἑκάστου γὰρ αἱ ἀπόπνοιαι καὶ τὸ ψῦχος τὸ ἀποκᾶον)· ὁτὲ δὲ καὶ νηνεμίας οὔσης εἰσδυόμενος ὁ ἀήρ.

12.3 ὃ δὲ λέγουσιν, ὡς ἧττον γίνεται περὶ Εὔβοιαν καὶ τὴν Βοιωτίαν ὅταν ἡ Ὀρχομενία λίμνη πλη-

[1] Schneider : πλειω U.
[2] ego (αἱ αἰθρίαι Scaliger) : αἱ ὀρεῖαι U.
[3] Wimmer : πνεύματος U.
[4] Wimmer : ἐπεκαυσαν U.

---

[1] *Cf. HP* 4 14. 13 (At Panticapaeum): "The frosts occur in clear weather . . ."

From Cold Weather

Cold weather on the other hand kills more trees and kills them in more ways, for sometimes clear weather with an intense frost freezes them,[1] and sometimes cold winds arise and sear them[2]; occasionally they freeze also when there is no wind, or only a moderate breeze, and the air is not clear but turbid and cloudy. Under such conditions (one may say) the freezing, at least of trees, occurs, whether the sky is clear or not. In Arcadia and Euboea freezing tends rather to occur with a clear sky and a wind, whereas in Thessaly and neighbouring parts with no wind and a cloudy sky. Still it nevertheless also occurs with winds (since it occurs when snow has fallen on the mountains surrounding the various districts, for winds come from each with the cold that sears), but at other times the air penetrates the trees in a calm. 12.2

As for the report that there is less such freezing in Euboea and Boeotia when the lake of 12.3

[2] *Cf. HP* 4 14. 11: "Local winds in different countries are of a nature to destroy and sear: so at Chalcis in Euboea when the Olympias [the wind from Mt. Olympus] blows cold shortly before or after the winter solstice, for it sears the trees and makes them more withered and dry than they could become from long exposure to the sun, which is why they are said to suffer from *kauthmós* ('burning')."

ρωθῇ, τάχα δ'[1] ἄν τις ἀμφοτέρως λάβοι· καὶ ὡς κατὰ συμβεβηκός, ὅτι ἔπομβρα συμβαίνει τότε μᾶλλον εἶναι τὰ ἔτη καὶ χιόνα μὴ πίπτειν· καὶ ὡς ὑδαρεστέραν[2] τὴν ἀναθυμίασιν γινομένην διὰ τὸ ἀπὸ πλέονος, ἡ τοιαύτη δ' ἧττον ψυχρὰ καὶ κακοποιός· καὶ φανερὸν ἔν τισι τόποις γέγονεν, ἀλεεινότεροι γὰρ γεγένηνται λιμνωθέντων τῶν πλησίον.

12.4 πνεῖ δὲ τὰ πνεύματα <τὰ>[3] ἀποκαίοντα περί γε τοὺς κατὰ τὴν Ἑλλάδα τόπους ἀπὸ δυσμῶν (ὥσπερ ὁ Ὀλυμπίας, ὁ ποιῶν ἄλλοθί τε καὶ ἐν Χαλκίδι τὸν καλούμενον καυθμόν[4]), ἐν Θετταλίᾳ γὰρ ἀμφοτέρωθεν πνέοντες ἐκπηγνύουσιν· ἡ δ' ὥρα τῆς πνοῆς μάλιστά πως περὶ τροπὰς ὑπὸ τὰς τετταράκοντα· τότε γὰρ καὶ ὁ ἀὴρ ὅλως[5] ψυχρότατος.

[1] U N : τάχ' aP.
[2] Schneider : ὑδρεστέραν U.
[3] Schneider.
[4] Heinsius : κλαυθμὸν U.
[5] U ar N aP : ὅλος U r.

---

[1] That is, lake Copais. Before it was finally drained (from 1883 on) it rose and fell irregularly, depending on whether the underground outlets became obstructed.

Orchomenus[1] is full, it may perhaps be taken in both of the following ways: (1) the fullness of the lake is merely incidental, since the lake fills when the weather during the year produces rain rather than snow; (2) or the fullness of the lake leads to a more watery exhalation coming from the larger surface, and such an exhalation is less cold and harmful. This has become evident in some districts, for they have become warmer after lakes were formed in the neighbourhood.

Cold: Winds

12.4 Blasting winds, at any rate in Greece, blow from the west,[2] like the Olympias which at Chalcis and elsewhere causes the so-called *kauthmós*[3] (for in Thessaly winds blowing from the east as well kill the trees). The season for these winds is around the winter solstice during the forty days,[4] the air in general being coldest then.

[2] From the region where the sun sets in summer (Aristotle, *Met.*, ii. 6 [363 b 23–25]), that is, northwest.

[3] "Scorch"; *cf. HP* 4 14. 11, cited in note 2 on *CP* 5 12. 2. The Olympias is named from Mt. Olympus, approximately northwest of Chalcis.

[4] *Cf. HP* 4 14. 13. These are the forty days when the sun appears to linger at the tropic, and the days do not get noticeably shorter or longer: *cf.* Geminus, *Elements of Astronomy*, chap. xvii. 29 (p. 152. 6 Manitius [Leipzig, 1898]).

τὰ μὲν οὖν γινόμενα ταῦτ᾽ ἐστίν· ἡ δ᾽ αἰτία σχεδὸν (ὡς καθόλου γε[1] εἰπεῖν) φανερά· τὸ γὰρ θερμόν, ἐξελαυνόμενον ὑπὸ τοῦ ψύχους, συνεξάγει καὶ τὸ ὑγρόν, ὥστε διαπνεῖσθαι.

12.5 συμβαίνει δὲ τοῦθ᾽ ὡς ἐπὶ τὸ πλέον ἐν τοῖς ὑπὲρ γῆν,[2] τὰ δὲ κατώτατα, περὶ τὰς ῥίζας, ἀπαθῆ, πολλάκις δὲ καὶ αὐτοῦ τι τοῦ στελέχους· οὐ μὴν ἀλλ᾽ ἐνίοτε διικνεῖσθαι[3] καὶ πρὸς τὰς ῥίζας, ὥσθ᾽ ὅλον ἐξαυαίνειν τὸ δένδρον. ἁπλῶς γὰρ ἄνωθεν ἡ ἀρχὴ καὶ ἡ παρείσδυσις τοῦ ψύχους, εἶτ᾽ ἀπὸ τούτων, ὥσπερ ὀχετῶν τινων, καταβαίνει πρὸς τὰ κάτω· διὸ καὶ παρ᾽ οἷς συμβαίνει τὸ πάθος κατακρύπτουσι τὰς ἀμπέλους, καὶ τὰς συκᾶς θαμνώδεις ποιοῦσιν· οὐδὲν δὲ δεῖ πολλῆς γῆς,

12.6 ἀλλὰ μετρία τις ἐποῦσα[4] δύναται διατηρεῖν. ἔνιοι δὲ μόνον παρὰ <τὰ> ἄκρα τῶν[5] κλημάτων καὶ τὰς κράδας αὐτάς, καθάπερ ἔν τε τῷ Πόντῳ καὶ περὶ Μηδίαν,[6] οἷον ἐμφράττοντες τὰς ἀρχάς· ἐὰν γὰρ ταῦτα συγκλεισθῇ, καθάπερ ἄλλ᾽ ἄττα[7] γυμνά, οὐδὲν πάσχουσιν, διὰ τὸ μὴ ἔχειν εἴσοδον,

[1] Scaliger : τε U.
[2] U : γῆς Eucken.
[3] U : διϊκνεῖται Schneider. [4] Schneider : ἐπιοῦσα U.
[5] παρὰ . . . τῶν ego (τὰ ἄκρα τῶν Itali : *extrema* Gaza) : παρα ἐκράτων U.

These then are the facts. The cause (one may say), at least to put it generally, is evident: the heat, driven out by the cold, takes the fluid along with it, so that the trees are dried out by the evaporation.

12.5 This happens mostly in the part of the tree above ground, whereas the lowest parts round the roots are unharmed, and often even some of the trunk as well.[1] Nevertheless the cold sometimes also reaches the roots and so withers the whole tree. For broadly speaking the cold begins by making its entrance above, and then descends through these upper parts, as through channels, to the parts below. This is why the people of places where the freezing occurs cover the vines with earth and give the fig-trees a shrub-like habit. No great amount of earth is required; a thin cover can save them.

12.6 Some put earth along the tips only of the vine-twigs and along the twigs only of the fig, as in Pontus and Media, blocking off (as it were) the entrances, since if these are closed (just as with other cases of covering an exposed part) the trees are unharmed, the cold

[1] *Cf. HP* 4 14. 12: "The freezing begins directly from the trunk, and in general (so to say) these parts are attacked more and earlier than the parts below."

---

[6] ego : *μήδειαν* U (U spells with *-ει-* at *CP* 5 18. 3; *HP* 4 4. 3; 8 11. 6; 9 1. 3; 9 7. 2; with *-ι-* at *HP* 4 4. 1; 4 4. 2).

[7] *καθ' άπερ ἄλλ' ἄττα* U : *καὶ περὶ τὰ ἄλλα τὰ* Coray : *καίπερ τἆλλα ὄντα* Wimmer.

δικνεῖται γὰρ ἀπὸ τῶν ἄνω πρὸς τὰ κάτω καὶ τὰς ῥίζας.

ἡ δ' ἰδιότης ἰσχυρά, καθάπερ ὑφ' ἡλίου καὶ χρόνου πολλοῦ, διὰ τὴν σφοδρότητα καὶ τὴν ἰσχὺν τοῦ ψύχους· ἐξάγει γὰρ μᾶλλον καὶ ἀθροώτερον τὸ ὑγρὸν ἅμα τῷ θερμῷ.

12.7 τὸ δὲ πονεῖν μάλιστα τῶν τόπων ὅσοι κοῖλοι καὶ αὐλῶνές εἰσιν, καὶ ὅσοι περὶ τοὺς ποταμούς, καὶ ὅλως τοὺς ἀπνευστοτάτους, οὐκ ἄλογον· ἵσταται γὰρ μάλιστα ἐνταῦθα φερόμενον πνεῦμα,[1] καὶ πλεῖστον διατρέχει χρόνον, ὥστε καὶ μάλιστ' ἀπεργάζεται· καὶ γὰρ ὅπου μὴ διὰ πνεύματα τὸ πάθος, ἀλλὰ διὰ παρουσίαν καὶ στάσιν τινὰ τοῦ ἀέρος γίνεται (καθάπερ ἐν Μακεδονίᾳ τε καὶ Θετταλίᾳ ἐνιαχοῦ καὶ περὶ Φιλίππους), ἐνταῦθ' οἱ κοῖλοι τόποι μάλιστα πονοῦσιν, ἐφεστηκὼς γὰρ ὁ ἀὴρ πήγνυται καὶ πήγνυσι, καὶ τὸ ὅλον πλείω χρόνον ἐργάζεται. κατὰ δὲ τοὺς ὑπτίους οὐδὲ γίνεται πῆξις ὅλως, ὥσπερ ἡ[2] τοῦ ὕδατος, ἡ[3] πνεύματος· κωλύει γὰρ ἡ κίνησις.

§12.7: *HP* 4 14. 12.

[1] τὸ πνεῦμα Schneider.
[2] N aP : ᾗ U : ᾖ u.

having no entry, since it reaches the lower parts and roots by passing through the parts above.

A peculiar character of this freezing is its strength; the effect is like that of prolonged exposure to the sun, so great is the intensity and strength of the cold, since its extraction of the fluid (and the heat along with it) is more thorough and less gradual.[1]

12.7 That hollows, valleys and riversides, that is, in general the least ventilated places, should suffer most[2] is not unreasonable. For it is here most of all that wind lingers in its course and these it takes longest to traverse, and in consequence operates with the greatest effect. Indeed even in countries where the freezing is not due to winds but to a stationary presence (as it were) of the air, as in Macedonia, certain parts of Thessaly and at Philippi, it is the hollows that suffer most severely, since the air rests on them and gets frozen and freezes the trees in turn and in general operates on them longer. But on slopes there is no freezing of the wind any more than of the water, since the movement prevents it.

[1] *Cf. HP* 4 14. 11, cited in note 2 on *CP* 5 12. 2.

[2] *Cf. HP* 4 14. 12: "Among localities it is hollows, valleys, those lying along rivers, and in general the ones least ventilated, that suffer most."

---

[3] ego : ἢ U.

12.8 ἀλλ᾿ ἐκεῖνο θαυμαστὸν καὶ λόγου δεόμενον, ὅτι οὐ τὰ ἀσθενέστατα μάλιστα πονεῖν εἴωθεν, ἀλλ᾿ ἐνίοτε τὰ ἰσχυρὰ μᾶλλον· ἐλάαν γὰρ καὶ συκῆν οὐδὲν ἄλογον ἀποκαυθῆναι, τῆς μὲν γὰρ μετέωροι, τῆς δὲ καὶ μαναὶ αἱ ῥίζαι, ὥστε καὶ ἄνωθεν τὸ ψῦχος δικνεῖσθαι· καὶ ἡ ἄμπελος δὲ ἔχει τινὰ αἰτίαν ἐκ τῆς φύσεως, καὶ ἐκ τῆς ἑλκώσεως τῆς περὶ τὴν τομήν. ἀλλὰ τὸ τὸν κότινον μᾶλλον πονῆσαι τῆς ἐλάας ἄτοπον, καὶ τὸ τὴν ῥόαν μηδὲν[1] παθεῖν, ἀσθενῆ πρὸς τοὺς χειμῶνας οὖσαν, 12.9 ἐκπήγνυσθαι[2] γὰρ τάχιστα δοκεῖ. καὶ γὰρ εἰ διαφέρουσιν αἱ ἐκπήξεις αἱ ὑπὸ τῶν πάγων καὶ τῶν πνευμάτων ὅλως, οὐκ ἄλογον ἀμφοτέρων τὰ ἀσθενέστατα μάλιστα ὑπακούειν, εἰ μὴ ἄρα τὸ αὐτὸ γίνεται καὶ ἐνταῦθα τῷ πρότερον εἰρημένῳ· μᾶλλον γὰρ ἐφίσταται καὶ ἐμμένει τοῖς πυκνοῖς ἢ τοῖς

[1] U^c (δεν ss.) : μην U^t.
[2] Schneider : ἐμπήγνυσθαι U.

---

[1] *Cf. HP* 4 14. 12: ". . . of the trees the fig suffers most, the olive next. The wild olive, which is stronger, suffered more than the olive, which excited astonishment, whereas the almond trees were not affected at all. The apples and pears and pomegranate trees were also unaffected, which was another cause for astonishment." Theophrastus is

But another matter is astonishing[1] and requires explanation: that it is not the weakest trees that habitually suffer, but sometimes the strong trees suffer more. For there is nothing unreasonable in the circumstance that the olive and the fig were blasted, since olive roots are shallow and fig roots are open in texture besides, and so the cold can pass to them from above; so too the blasting of the vine can be explained from its nature and from the wounding incidental to pruning. But that the wild olive should have suffered more than the olive is strange, and that the pomegranate should not have suffered at all, considering its weak resistance to cold weather (for it is held to be very quick to freeze). For even if the freezing brought on by frost[2] and that brought on by winds are quite different, it is not unreasonable that the weakest trees should succumb most readily to both, unless we have here the occurrence mentioned earlier[3]: the cold lingers and abides longer in trees of close texture than in 12.8 12.9

referring to the great freeze caused by the Olympias at Chalcis in the archonship of Archippus (*HP* 4 14. 11; an Archippus was archon in 321/0 B.C., another in 318/7 B.C.).

[2] Theophrastus speaks of "frost" (*págos*) of the air (*CP* 5 12. 7), of water (*CP* 5 12. 11) and of the earth (*CP* 5 13. 1–2). It is an increase in the direction of rigidity: the earth becomes stiff, the water is congealed and the air is still.

[3] *CP* 5 12. 7.

μανοῖς, τὰ δὲ μανὰ διίησιν[1] (ὃ καὶ τοῦ μὴ ἐκπήγνυσθαι τὰς μηλέας αἴτιον, ἢ[2] ἧττόν γ'[3] ἑτέρων, ἀσθενεῖς οὔσας, ὥσπερ καὶ περὶ Θετταλίαν, τῇ γὰρ μανότητι διίησιν καὶ οὐκ ἀποστέγει), τὸ δὲ μὴ χρονιζόμενον, μηδὲ ἀθρόον, οὐδ' ἐργατικόν, ὥσπερ οὐδ' ἐν τοῖς κεραυνοῖς. δεῖ δὲ δυοῖν[4] θάτερον· μὴ[5] δέχεσθαι, καθάπερ τὰ πυκνὰ καὶ ἰσχυρά (τοιοῦτον γὰρ ἡ ἄπιος καὶ τὰ ἄγρια δὲ καὶ τὰ ἄκαρπα ἔτι μᾶλλον, οἷον πτελέα ὀστρύη), ἢ δεξάμενον διιέναι καὶ μὴ κατέχειν.

ὑπὲρ μὲν οὖν τούτων ἐνταῦθά που τὸ αἴτιον.

12.10 ἡ δὲ τοῦ πνεύματος φορά, καθ' ὃν ἂν γένηται τόπον, ἐπέκλυσεν οἷον ποταμός τις ῥυείς, ὥσπερ ἐν τοῖς λοιμοῖς, διὸ καὶ τὰ ἁπτόμενα καὶ τὰ σύνεγγυς[6] ἀπαθῆ πολλάκις, ἐνίοτε δ' οὕτως εἰς ἀκριβὲς διεῖλεν[7] ὥστε τῶν ἐπὶ τοῦ αὐτοῦ κλήματος ὀφθαλμῶν οἱ μὲν ὑγιεῖς, οἱ δὲ πεπηγότες εἰσίν· πολλάκις δὲ κεκρυμμένης ὑπὸ χιόνος ὅλης τῆς ἀμπέλου (γίνεται γὰρ τοῦτο ἐν τοῖς ψυχροῖς

[1] U^c (η ss.) : διΐσιν U^ac : δίεισιν u.
[2] Wimmer : ἧ U N : ἧ ἢ P : ἦν a.
[3] Wimmer : δ' U N aP^c : P^ac omits.
[4] N : δυοῖ U : δυεῖν aP.
[5] ἢ μὴ aP.
[6] U^c : σύγγυς U^ac.

those of open texture, the trees of open texture letting it through. (This moreover is the reason why apple trees do not freeze, or at least freeze less than others, despite their weakness, as in Thessaly for instance: their open texture lets the cold through and does not keep it out.) But what does not remain long or come all at once is not effective, any more than it is in thunderbolts.[1] The tree to escape must do the one thing or the other: it must not admit the cold, as trees that are close textured and strong do not admit it (for such are the pear and wild trees too, and still more the ones that bear no fruit, like the elm and hop-hornbeam[2]), or else admit it but let it through and not detain it.

In these trees, then, we must look for the cause here.

12.10 The onrush of the wind inundates the place it comes to, like a river in spate, just as in pestilences. This is why of two trees standing next to one another or close to one another the one is blasted, the other unharmed; and sometimes the discrimination is so nice that on the same vine twig some buds are sound, others frozen. And often, when the vine is covered with snow (for this happens in cold

[1] *Cf.* Aristotle, *Meterologica*, iii. 1 (371 a 17–29).

[2] At *HP* 3 10. 3 its fruit is called "small, oblong like a barleycorn and yellow."

---

[7] ego : διῆλεν U[c] (-ή- U[ac]) : διῆλθεν u.

τόποις), ἐὰν λάβῃ[1] γυμνὰ τὰ ἄκρα κλήματα, ταῦτα ἀπέκαυσεν, καὶ ἀπὸ τούτων διαδίδωσι πρὸς τὰς ῥίζας, ὥστ' ἐνίοτε <τὴν μὲν πεπηγέναι, τὴν δὲ>[2] ζῆν.

12.11 περὶ[3] δὲ τὸν Πόντον ὑπὸ τῶν πνευμάτων ἔκπηξις γίνεται ὅταν <αἱ λεπίδες>,[4] αἰθρίας οὔσης, καταφέρωνται· ταῦτα δέ ἐστι πλατέα ἄττα, φερόμενα μὲν φανερά, πεσόντα δ' οὐδὲν διαμένει· δῆλον δὲ ὅτι πῆξίς τις τῆς ὑγρότητος ἐν τῷ ἀέρι, καθάπερ τῆς πάχνης. ὅταν οὖν τοιοῦτος ὁ ἀὴρ προσπίπτων ὑπὸ τοῦ πνεύματος ᾖ, καὶ ταῦτα συγκαταφέρηται, κατὰ λόγον ἡ ἔκπηξις γίνεται.

τὰ δὲ πνεύματα τοῖς μὲν ψυχροῖς φύσει τῶν τόπων ἐγχώρι' ἄττ'[5] ἂν εἴη (καθάπερ ἐν τῷ Πόντῳ καὶ τῇ Θρᾴκῃ), τοῖς δ' ἀλεεινοτέροις ἐκ τῶν ἔξωθεν ἡ ἐπιφορά (καθάπερ ἐν Εὐβοίᾳ)· δηλοῖ δὲ καὶ ἡ φύσις, ὡς <οὐκ>[6] αὐτόθεν καὶ [οὐκ][6] ἐκ

§12.11: *Cf. HP* 4 14. 13.

[1] λάβῃ U[r] N aP : λάβῃ τὰ U[ar].
[2] ego : Schneider supposes a lacuna after ῥίζας.
[3] Heinsius (*in* Gaza) : ἐπὶ U.
[4] ego (*quaedam in modo squamarum* Gaza : λεπίδες after οὔσης Schneider).
[5] ἐγχώριάττ' u : ἐνχω- U.
[6] Schneider.

regions), if the wind finds the tips of the twigs exposed it blasts them and transmits the cold from these to the roots, so that sometimes one vine is frozen when the other lives.

12.11 In Pontus freezing by the winds occurs when in clear weather there is a fall of "scales." These are certain flat objects that are seen as they fall but disappear once they have fallen.[1] Evidently we have here a type of freezing of the water in the air, just as we get the freezing that produces hoarfrost. So when the air that is borne against the tree by the wind is of this character,[2] and these "scales" come down with it, freezing of the tree is in order.

The freezing winds would be indigenous ones in the regions that are naturally cold, as Pontus and Thrace, whereas in warmer regions they bring the cold from abroad, as in Euboea. The nature of the wind moreover shows that it does not come from the region itself and proceed from some steady and

[1] *Cf. HP* 4 14. 13: "In Pontus at Panticapaeum trees are frozen under two circumstances: sometimes by cold if the year is wintry [*i.e.* by wind-borne cold], sometimes by frosts if these last a long time. Both occur mainly near the winter solstice in the forty days. The frosts occur in clear weather; whereas the cold that freezes the trees occurs mainly when during clear weather the 'scales' fall. These are like motes, only flatter, and are visible when they are falling but disappear when they have fallen; in Thrace they freeze solid."

[2] Cold enough to keep the "scales" frozen.

στασίμου τινὸς καὶ ὅλου πνεύματος, ἀλλ' ᾗ ἂν τοῦτο τύχῃ, ῥυέν· οὐδὲ γὰρ ἂν ἦν ἀκέραια τὰ πλησίον.

τῶν μὲν οὖν πνευμάτων καὶ αὕτη[1] τις[2] ἰδιότης.[3]

13.1 ὁ δὲ τῆς γῆς πάγος χαλεπώτατος ὅταν περιβεβοθρωμένα καὶ γυμνὰ λάβῃ τὰ δένδρα, μάλιστα δὲ ἐὰν καὶ ὕδωρ ἐνεστηκός· ἐὰν γὰρ διαμένῃ πλείω χρόνον, ἐξέπηξεν, εἰς ἀσθενεῖς τε καὶ γυμνὰς τὰς ῥίζας εἰσδυόμενος· ἐπεὶ καὶ ὅλως ἂν διειργασμένην λάβῃ τὴν γῆν, χαλεπώτερος, μανῆς γὰρ οὔσης, συνικνεῖται[4] μᾶλλον (ἀπαθεστέρα δέ,[5] ἐὰν κεκοπρισμένη τύχῃ, θερμὴ γὰρ ἡ κόπρος οὖσα βοηθεῖ· εὐλόγως δὲ καὶ χιόνος πεσούσης)· καὶ ἐὰν ἀνεζυμωμένης τῆς γῆς ἐπιγένηται, ψύχει[6] γὰρ καὶ πάχνη, καὶ πάγος[7] ἐκπηγνὺς[8] καὶ διαδύεται[9]

[1] Scaliger (*eam* Gaza) : αυτη U.
[2] u : της U.
[3] U cc (ς from τα).
[4] ego (ἐσικνεῖται Schneider) : ἐνϊκνεῖται U.
[5] U cc (-έρα δ from -ερονὰι) : ἀπαθέστεραι δὲ N aP.
[6] ego : ψυχη U.
[7] U c aP : πάχος U ac N.
[8] ego (ἐκπήγνυσι Schneider) : ἐκπηγνύουσι U r (from -ση).
[9] Schneider : διαλυεται U.

unfragmented kind of wind, but spills out in a mass of random channels,[1] since otherwise trees close to the blasted ones would not have escaped unharmed.

This, then, is a further peculiarity of freezing winds.[2]

Freezing: (1) *Of the Ground*

13.1 Freezing of the ground is worst when it catches the trees with the holes dug round them and the roots exposed (and especially when it also finds standing water in the holes); for if the frost lasts for some time it freezes the tree, entering the roots when they are weak and exposed. Indeed if the frost finds the ground tilled at all it is worse, since it reaches the roots better when the earth is loose (the earth on the other hand is less affected if it has been manured, since manure by its heat counteracts the cold; so too, as we might expect, when snow has fallen). Frost is also worse if it comes when the ground is in ferment, since even hoarfrost chills it then, and crippling frost also penetrates it, owing to

[1] *Cf.* Theophrastus, *On Winds*, chap. iii: "... what moves in a narrow current and with greater vehemence is colder; what moves onward to a great distance is more spread out and relaxed ..."; chap. vi: "... each wind for regions far from its origin is irregular and scattered."

[2] The other is the strength and rapid working of the cold: *CP* 5 12. 6.

διὰ τὴν μανότητα[1] καὶ αὐτὴν τὴν γῆν πήγνυσιν.

13.2 ὁ δὲ τῆς γῆς πάγος ὀλεθριώτερος τοῖς δένδροις, ἅπτεται γὰρ μᾶλλον τῶν ῥιζῶν· ὁ δὲ τοῦ ὕδατος καὶ ἧττον. χαλεπώτερος δὲ καὶ οὗτος κἀκεῖνος, ὅταν ἀνιῇ,[2] καὶ πάλιν ἐπιπηγνύηται, καὶ τοῦτο ποιῇ[3] πολλάκις, ἐξαιρεῖται γὰρ τὴν δύναμιν· ὁ δὲ συνεχής, ἐγκατακλείσας τὸ θερμόν, οὐχ ὁμοίως, πλὴν ἐὰν ὑπερβάλῃ[4] τῷ χρόνῳ. καὶ χαλεπώτερος δ' ὅλως ὁ κάτωθεν πάγος ταῖς μὴ εὐγείοις, μηδὲ πυκναῖς,[5] μηδ' ἐνίκμοις· οὐδὲ γὰρ ὁμοίως θερμαίνει, καὶ διαδύεται[6] δὲ πορρωτέρω καὶ ἅπτεται τῆς ῥίζης.

13.3 συμβαίνει δὲ τὰς μὲν τῶν βλαστῶν ἐκπήξεις, καὶ ἁπλῶς τῶν ἄνω, πολλάκις γίνεσθαι, τὰς δὲ τῶν ῥιζῶν καὶ ὅλως[7] τῶν δένδρων, ὀλιγάκις καὶ παρ' ὀλίγοις. αἴτιον δὲ τὸ ἔχειν προβολὴν καὶ οἷον ἀποστέγασμα τοῦ ψύχους τὴν γῆν, εἰς ἣν καὶ συνελαύνεται τὸ θερμόν.

[1] u : ματαιοτητα U.
[2] Schneider (ὕῃ Gaza [*pluvia*]) : ἂν εἴη U N : ἀνείη aP.
[3] ποιῇ aP : ποιεῖ U N.
[4] ὑπερβάλῃ U : ὑπερβάλλῃ Wimmer.
[5] u : πυκνοῖς U.

its loose condition, and freezes the very earth.

Freezing: (2) *Of the Water*

Freezing of the ground is more fatal to trees than freezing of the water, since it attacks the roots more effectively. Both are worse when a thaw is followed by a second freezing and this is repeated several times, since this drains the tree of its power. Unbroken frost, on the other hand, shuts the heat in and is not so fatal, except when it exceeds in duration. Again frost from below[1] is worse in general for land that is not deep or close in texture or moist, since it does not bring the same warmth[2] and moreover penetrates further and attacks the root. 13.2

Freezing: (3) *From the Air*

It happens that the shoots and upper parts in general are frequently killed by frost, whereas the roots and indeed the entire tree is seldom so killed and only in a few countries. The reason is that the roots have the earth as a shield and as it were a cold-proof cover, and further that the heat is rounded up and driven into the earth. 13.3

[1] That is, from ground and water.
[2] By counter-displacement.

---

[6] Gaza (*irrepit*), Scaliger : διαλύεται U.
[7] U : ὅλων Schneider.

ἐνίοτε δ' οὐκ ἀπόλλυται οὐδὲ τὰ ἄνω, ἀλλὰ μᾶλλον ἐπικάεται. καὶ τοῦτ'[1] οὐκ εὐθὺς ἀφαιρεῖν δεῖ· πολλάκις γὰρ ἅμα τῇ ὥρᾳ διεβλάστησεν, καὶ αὐτὰ τὰ φύλλα, δοκοῦντα αὖα εἶναι, πάλιν ὑγράνθη καὶ ἐγένετο χλωρά. διὸ καὶ οὐκ ἄτοπον, εἴ τι

13.4 που συνέβη τοιοῦτον ὥστε, ἐλάας αὐανθείσης, καὶ αὐτῆς καὶ τῶν φύλλων, πάλιν ἀναβλαστῆσαι· οὐ γὰρ ἦν αὔανσις, ἀλλὰ φαινομένη τις, ξηρότητι καὶ τῇ ἄλλῃ χρόᾳ, πρὸς δὲ τὴν ἀρχὴν οὔτε τὴν τῶν φύλλων διῆκεν, ἔτι δὲ ἧττον τῶν βλαστῶν καὶ τῶν ἀκρεμόνων. ὁμοίως δὲ καὶ ἐπ' ἄλλων τινῶν τοῦτο ξυμβαίνει, μάλιστα δ' εἰκὸς ὧν τὸ φύλλον σαρκῶδες, καὶ αὐτὰ τῇ φύσει θερμά· τὰ γὰρ ἀσθενῆ καὶ λεπτὰ τοῦτ' οὐ πάσχει, καθάπερ τὰ τῶν μυρρίνων, ἀλλὰ καὶ τάχιστα ἐπικάεται· λεπτὰ γὰρ καὶ αὐτὰ τὰ κλωνία καὶ ἁπαλὰ τῇ φύσει, καὶ ὅλον τὸ δένδρον οὐ θερμόν, διὸ καὶ ἐκπήγνυσθαι[2] τάχιστα. ἡ γὰρ αὖ δάφνη, καίπερ οὖσα μανή, καὶ διαμένει διὰ τὸ θερμόν, ἡ δὲ ῥόα καὶ ἡ συκῆ μαναί τε καὶ ὑγραί, καὶ οὐχ ὁμοίως θερμαί.

[1] U : ταῦτ' Schneider. [2] U : ἐκπήγνυται Schneider.

---

[1] *Cf. HP* 4 14. 12: "All the olives that shed their leave[s] revive; those that do not are completely destroyed. But i[n] some regions when the olive has been partly nipped an[d]

Occasionally not even the upper parts are killed, but rather only nipped. One should not proceed to strip away the nipped portion at once, since it has often come out at the return of spring and the very leaves that were taken to be withered have become moist and turned green. Hence it is not strange if somewhere such a thing as the following has occurred: an olive tree had been withered, both the tree and the leaves, and came out again.[1] For this was no true withering, but only seemed so from the dryness and colour, and the cold did not reach the starting-points either of the leaves or still less of the shoots and branches. The like happens with other trees as well, most understandably those in which the leaf is fleshy and the tree itself hot by its nature, since it does not happen in weak trees with thin leaves, such as those of the myrtle. Myrtle leaves in fact are the quickest to be nipped, the twigs themselves being also naturally thin and tender, and the whole tree not being hot (which is why it is reported to be very quick to freeze). The bay, on the other hand, in spite of its open texture, actually survives because it is hot; whereas the pomegranate and fig are open in texture and fluid and not hot to the same degree.[2]

13.4

ts leaves have withered the tree has sprouted again without shedding them and the leaves have revived."

[2] Olive, myrtle and bay are evergreen, fig and pomegranate deciduous.

13.5 ταχεῖα δὲ ἡ ἐκ τῶν ἐκπηγνυμένων ἀναβλάστησις,[1] ὅτι συμβαίνει τὴν ῥίζαν ἰσχυρὰν γίνεσθαι καὶ πλήρη, συνηθροισμένης τῆς ἔνης τροφῆς, ἣν οὐ διέδωκεν, ἔτι δὲ τὴν πρότερον οὐ πολλὴν[2] καὶ ἰσχυρὰν διανεμομένην, τότ' εἰς[3] εὔλογον ἀναλίσκεσθαι· ὥστ' εὐλόγως καὶ ἡ αὔξησις ταχεῖα καὶ ἡ καρποτοκία.

τοῦτο μὲν οὖν ὁμολογουμένως γίνεται παρὰ πᾶσιν.[4]

13.6 ὃ δέ τινες θαυμάζουσιν, ὅτι ἡ μὲν χιὼν οὐκ ἐκπήγνυσιν, ἡ δὲ πάχνη, μετριωτέρα[5] τῆς χιόνος οὖσα,[6] οὐδὲν ἄτοπον· πρῶτον μὲν ὅτι ἐπιμένει, ἡ δ' οὐκ,[7] ἀλλ' ἀποτήκεται ἀπὸ τῶν κλημάτων καὶ τῶν βλαστῶν, ἡ δὲ πάχνη ταῦτ' ἀποκάει· ἔπειτα καὶ ἡ διάθεσις αὐτὴ τοῦ κλήματος· ἡ μὲν γὰρ

[1] Wimmer : ἅμα βλάστησις U.
[2] N aP : πολὺν U.
[3] εἰς τὴν βλάστησιν Itali after Gaza.
[4] u : πᾶσαν U.
[5] u : μεριωτέρας U.
[6] u : οὔσης U N aP.
[7] ἐπιμένει, ἡ δ' οὐκ ego : ἐπι μὲν ἐπί δοῦ U (with signs of corruption over μ and δ) : ἐπὶ μέν, ἐπὶ δ' οὐ N : ἡ μὲν οὐκ ἐπιμένει aP.

The frozen trees come out again very rapidly because it turns out that the root gets strong and full, since last year's food, which the tree has not distributed, is still collected in it, and that the food, doled out before the tree froze in no great quantity and strength, is now expended to good purpose, so that with good reason growth as well and bearing of fruit are rapid.[1] 13.5

This recovery, then, is a matter of general agreement.

An Apparent Oddity: Hoarfrost is More Injurious than Snow

Some authorities however find it surprising that whereas snow does not cause freezing, hoarfrost does, although more moderate in its coldness than snow, a matter in which there is nothing odd. In the first place hoarfrost remains on the vine and snow does not, but melts away from the twigs and shoots, whereas hoarfrost blasts them. In the second place there is also the condition of the vine-twig itself: 13.6

[1] *Cf. HP* 4 14. 13: "The frozen trees, when not totally destroyed, come out again very rapidly, so that the vine bears fruit (as in Thessaly) with no retardation."

ἀβλαστοῦς,[1] ἡ δὲ διεβλαστηκότος[2] ἄρτι πίπτει, ὅτε ἀσθενέστατον, ἐνίοτε δὲ ἀνοιδοῦντος πρὸς τὴν βλάστησιν, ὅτε[3] οὐχ ἧττον (ὡς εἰπεῖν) ἐπίκηρον (ἐπιπάττει[4] γὰρ διυγραινόμενον ἤδη καὶ μανούμενον)· ἔτι καὶ λεπτοτέρα τῆς χιόνος, ὥστε δι' ἄμφω σφοδροτέραν τὴν πῆξιν εἶναι.

13.7 ἡ δὲ χιὼν ὅλως οὐδ' ἂν ἐπιμείνειεν[5] ἐπὶ τοῖς κλήμασιν, μὴ κατακρυπτομένης ὅλης, ὅταν δὲ τοῦτο πάθῃ, σκεπάζει τῷ ἐπιμένειν, ἐγκατακλείουσα τὴν θερμότητα καὶ ἀποστέγουσα τὴν ἔξω, καθάπερ καὶ τὴν γῆν. τὸ δ' ὅλον καὶ τμητικωτέρα[6] δοκεῖ ἡ πάχνη τῆς χιόνος εἶναι, διὸ καὶ τὰς νεοὺς[7] οἴονταί τινες βελτίους ταύτην ποιεῖν, διαχεῖν γὰρ τὰς βώλους διαδυομένην καὶ δάκνουσαν τῷ συνεστάναι μᾶλλον. λεπτοτέρα δ' ἐστὶ τῆς χιόνος ὅτι ἡ μὲν ἐκ νέφους, καὶ οἷον ἀφρός τις, ἐμπεριειληφυῖα πνεῦμα, ἡ δ' αὐτὴ[8] καθ' αὑτὴν συνεστηκυῖα, καὶ ἐκ λεπτοτέρου τινὸς ἀέρος καὶ

[1] a : -ους U : -οὺς u N P.

[2] a (διαβεβ- Schneider : βεβ- Wimmer) : βιεβλαστηκότος U : διεβλαστηκότως N P.

[3] Gaza (*quo tempore*), Schneider : ὅτι U.

[4] ego : ἐπὶ πάσι U.

[5] u : ἐπιμένειεν U N aP.

[6] U^{c} : τμηκωτέρα U^{ac}.

snow falls when the twig has not yet sprouted, hoarfrost when the twig has just come out and is at its weakest, and occasionally when the twig is swelling in preparation for sprouting, at a moment when the twig (one might say) is no less liable to injury, for the hoarfrost sprinkles it as it is already becoming more fluid and open in texture. Again, hoarfrost does this having even finer particles than snow, with the result that for both reasons the freezing it causes is more intense.

13.7 Snow on the other hand will not remain on the twigs at all unless the whole vine is covered with it; but when this happens the snow shelters the vine by remaining on it, shutting in and sealing off the warmth above ground just as it does with the ground.[1] Again hoarfrost is held to possess in general a sharper cutting edge than snow (which is why some believe that hoarfrost improves ploughed land not yet sown by breaking up the clods, penetrating and cutting them by reason of its firmer consistency).[2] It has finer particles than snow because snow comes from a cloud and is (so to speak) a kind of foam, since it contains *pneuma*, whereas hoarfrost has no such admixture and comes from a type

[1] *Cf. CP* 3 23. 4.
[2] *Cf. CP* 3 20. 7.

---

[7] Schneider : *νέους* U.
[8] *ἥδ' αὐτὴ* aP : *τη* (*τῆ* u) δ' *αὐτῆ* U : *τῆ* δ' *αὖ τῆ* N.

ὑγροῦ.

τούτων μὲν οὖν ταύτας ὑποληπτέον τὰς αἰτίας.

14.1 αἱ δ' ἐκπήξεις ὅλως πότερον διὰ τὴν παχύτητα τοῦ ἀέρος ἢ διὰ τὴν λεπτότητα γίνονται, καὶ εἰ δι'[1] ἄμφω, διὰ πότερον μᾶλλον, ἀπορήσειεν ἄν τις.

ᾗ[2] μὲν γὰρ οὐδὲ τοῦ ἀέρος γίνεται πῆξις ὅταν μὴ αἴθριος ὁ ἀήρ, οὐδ'[3] ἂν τούτου [ἀέρος][4] δόξειεν, ὁ γὰρ αἴθριος λεπτότερος.

ἅμα δὲ καὶ διαδυτικώτερος[5] μᾶλλον ὁ λεπτός· ἡ δ' ἔκπηξις εἰσδυομένου καὶ τέμνοντος.

[1] u aP : ἢ δι' U : εἴη δ' N. [2] ego : εἰ U.
[3] ego : μηδ' U.
[4] τούτου ego (τοῦ παχέος [Schneider adds ἀέρος] Scaliger) τοῦ ἀέρος U.
[5] διαδυκώτερος (sic) Scaliger : διαλυτικώτερος U.

---

[1] *Cf.* Aristotle, *On the Generation of Animals*, ii. 2 (735 b 19–21): "The cause (*sc.* of the thickening and whitening of various substances) is the admixture of *pneuma* (*i.e.* gas) which makes the volume greater and brings out the white colour, as in foam and snow, snow too being foam." *Cf.* also Aristotle, *Meteorologica*, i. 10 (357 a 13–19): "From the vapour rising in the daytime that fails to be lifted to a great height because of the small quantity, in comparison to the water raised, of the fire that raises it, we have

of air and of fluid that have finer particles.[1]

We must suppose, then, that the matters treated here have the causes mentioned.

A Problem: Does Thick Air or Thin Cause Frost?

14.1 Here one might pose a problem: is freezing due in general to thickness of the air or to thinness, and if to both, to which of them more?

The Case for Thin Air

On the view that there is no freezing by the air unless the air is clear,[2] it would appear that there is then no freezing by thick air either, since clear air is thinner.

Then too, thin air is the more penetrating, and freezing comes about when the air penetrates and cuts its way in.

when the vapour sinks downward at night on being cooled, what is called dew and hoarfrost. It is called hoarfrost when the vapour is frozen before it can collect into water again . . ., and dew when the vapour has collected into water . . ."; *ibid.* i. 11 (357 b 12–24): "For from there (*sc.* the region of the clouds) come three solids condensed by the chilling: water, snow and hail . . . When the cloud is frozen, it is snow; when the vapour, it is hoarfrost."

[2] This is at odds with the statement (*CP* 5 12. 2) that freezing occurs whether the sky is clear or not.

ἔτι δ' εὐψυχότερος (καὶ εὐπαθέστερος ὅλως) ὁ λεπτός (διὰ τοῦτο γὰρ καὶ τὰ ὕδατα προθερμανθέντα[1] ψύχεται καὶ πήγνυται θᾶττον, ὅτι λεπτύνεται τῇ θερμότητι).

14.2 συνεπιμαρτυρεῖν δὲ καὶ οἱ τόποι δοκοῦσιν οἱ ἐναίθριοι λεγόμενοι· πλείω γὰρ ἐκπήγνυται καὶ πλεονάκις ἐν τούτοις, ἐνιαχοῦ μὲν καὶ μικρὰ[2] [δὲ][3] πάνυ διεχόντων (ὥσπερ ἐν Κορίνθῳ τὸ Κράνειον[4] καὶ τὸ Ὀλύμπιον· σκληραὶ γὰρ αἱ αἰθρίαι σφόδρα περὶ τὸ Κράνειον, ὥστε καὶ τοῖς φυτοῖς καὶ τῇ αἰσθήσει δῆλον εἶναι), ἐνιαχοῦ δὲ καὶ ἐπὶ πλείονι διαστήματι.

καὶ ὅλως οἱ πρότερον οὐκ ἐκπηγνύντες τόποι παχέος ὄντος τοῦ ἀέρος, νῦν ἐκπηγνύουσιν, καθάπερ οἱ περὶ Λάρισαν τὴν ἐν Θετταλίᾳ·[5] τότε μὲν γάρ, ἐνεστηκότος ὕδατος πολλοῦ καὶ λελιμνωμένου τοῦ πεδίου, παχὺς ὁ ἀὴρ ἦν, καὶ ἡ χώρα θερμο-

§14.2–3: Pliny, *N. H.* 17. 30.

[1] N aP : -θερμαθέντα U.
[2] U : μικρὸν Wimmer.
[3] Schneider (or δὴ).
[4] κρανιον U.
[5] u : θατταλιᾳ U.

Again, thin air is more easily chilled (and in general is more easily affected); for this is why water
too, when it has first been warmed, is more quickly
chilled and frozen[1]: it is thinned by the heat.

14.2 The so-called "clear-weather"[2] localities are considered to lend their support to these views, since in
hese places freezing affects more trees and is more
frequent than elsewhere. In some countries the dis-
ances between such places are very small indeed
as at Corinth between Craneion[3] and Olympion,[4]
he fine weather being extremely harsh at Craneion,
so harsh that the effects are not only seen in the
plants but evident to sense), whereas in other coun-
ries the distances are greater.

One can go further: districts where formerly,
when the air was thick, there was no freezing, are
now subject to frosts, as the country around Larisa
n Thessaly, where formerly, when there was much
tanding water and the plain was a lake, the air was
hick and the country warmer; but now that the

[1] *Cf.* Aristotle, *Meterologica*, i. 2 (348 b 30–349 a 3).

[2] The word occurs only here. Presumably the localities
ere comparatively free from rain and fog. *Cf.* W. M.
eake, *Travels in the Morea*, vol. iii (London, 1830), p. 261:
"t is difficult to account for the extreme unhealthiness of
orinth in the summer and autumn, as the situation
eems such as to expose it to the most complete ventila-
on. The dews are said to be particularly heavy."

[3] An eastern suburb and favourite residential district.

[4] Unknown.

τέρα· τούτου δ' ἐξαχθέντος καὶ ἐνίστασθαι κωλυθέντος, ἥ τε χώρα ψυχροτέρα γέγονε, καὶ <αἱ>[1]

14.3 ἐκπήξεις πλείους. σημεῖον δὲ λέγουσιν, ὅτι τότε μὲν ἦσαν ἐλᾶαι καὶ ἄλλοθι καὶ ἐν αὐτῷ[2] τῷ ἄστε μεγάλαι καὶ καλαί, νῦν δὲ οὐδαμοῦ· καὶ αἱ ἄμπελοι τότε μὲν οὐκ ἐξεπήγνυντο, νῦν δὲ πολλάκις ὅτι δὲ τὸ ὕδωρ οὐκ ἀεὶ ψυχροτέρας ποιεῖ, ἀλλ ὅπερ καὶ τὸ πρότερον ἐλέχθη, σημεῖον τὸ περ <Αἶνον> γενόμενον·[3] αὕτη γὰρ ἀλεεινοτέρα δοκεῖ νῦν [δε] γεγονέναι,[4] πλησιαίτερον ὄντος τοῦ Ἕβρου.

ταύτῃ μὲν οὖν δόξειεν ἂν ὁ λεπτὸς ἀὴρ πηκτικώτερος εἶναι.

τῇδε[5] δὲ πάλιν ὁ παχύτερος· ἀκινητότερο γάρ, ὁ δ' ἀκίνητος εὐπηκτότερος.

[1] Schneider.
[2] αὐτῷ aP : ἑαυτῶι U (-ῷ N).
[3] περὶ Αἶνον γενόμενον Palmerius (*circa* [a blank 15–15–11 letters G ur G ce G bu; no blank G ed G ve G mo G ch *evenerat* Gaza) : περιγενομενον U.
[4] Heinsius (*esse* Gaza) : δε γεγονε U.
[5] ego : τῇ U.

---

[1] *Cf.* Strabo, *Geography*, ix. 5. 19 (440 C): ". . . the men Larisa . . . held the parts of the plain that were mo

water has been drained away and prevented from collecting,[1] the country has become colder and freezing is more common. In proof the fact is cited that formerly there were fine tall olive trees in the city itself and elsewhere in the country, whereas now they are found nowhere, and that the vines were never frozen before but often freeze now. That water does not always make a country colder, but rather the change that we mentioned first,[2] is proved by what has happened at Aenos, where the city is now considered warmer, when the Hebros has moved closer.[3]

14.3

These considerations, then, would make it appear that thin air has the greater freezing power.

The Case for Thick Air

But the following considerations would make this appear truer of thick air: it moves less, and still air is the more easily frozen.

prosperous, except for a depression by lake Nessonis, into which lake the river [the Peneus] used to flow and so deprive the Larisaeans of some of their arable land; but the Larisaeans later remedied this by raising embankments"; *cf.* also *ibid.* ix. 5. 2 (430 C).

[2] *CP* 5 14. 2 (drainage of the country around Larisa).

[3] Strabo, *Geography*, vii. 51, 51*a* (Loeb edition, ed. H. L. Jones, vol. iii, pp. 372, 374) speaks of the Hebrus as having two mouths. Perhaps the present passage indicates a westward shift of the nearer one.

14.4 ἔτι δὲ αἱ στύγες ἐν τοῖς τοιούτοις γίνονται[1] τόποις, αἵπερ μάλιστα εἰσδύονται εἰς τὰ σώματα (φυλάξασθαι γὰρ οὐκ ἔστιν οὐδ' ἐν τοῖς στρώμασιν κατακείμενον).

ἐπιβεβαιοῖ δὲ καὶ τὸ ἐν τοῖς ὄρεσιν ἧττον ἢ ἐν τοῖς πεδίοις ἔκπηξιν γίνεσθαι· λεπτότερος γὰρ ὁ ἀὴρ καὶ εὐκινητότερος.

οἱ δὲ καθ' ἕκαστα τόποι καὶ ἐπὶ τούτων πίστιν ἔχουσιν· τῆς τε γὰρ Θετταλίας περὶ Κίερον[2] μάλισθ' (ὡς εἰπεῖν) ἔκπηξίς[3] ἐστιν (ὁ δὲ τόπος 14.5 κοῖλος καὶ ἔφυδρος[4]). ἔν τε Φιλίπποις πρότερον μὲν μᾶλλον ἐξεπήγνυντο, νῦν δ', ἐπεὶ καταποθὲν[5] ἐξήρανται τὸ πλεῖστον ἥ τε χώρα πᾶσα κάτεργος γέγονεν, ἧττον πολύ, καίτοι λεπτότερος[6] ὁ ἀὴρ δι' ἄμφω, καὶ διὰ τὸ ἀνεξηράνθαι τὸ ὕδωρ, καὶ διὰ τὸ κατειργάσθαι τὴν χώραν· ἡ γὰρ ἀργὸς ψυχροτέρα καὶ παχύτερον ἔχει τὸν ἀέρα διὰ τὸ ὑλώδης[7] εἶναι καὶ μήτε τὸν ἥλιον ὁμοίως

§14.5: Pliny, *N. H.* 17. 30.

[1] u : γινον U.
[2] Benseler : κίθρον U.
[3] u N : -πλ- U aP.
[4] aP (ἔ- u) : ὕφυδρος U N.
[5] ego (καταποθεὶς <. . .> Schneider) : καταποθεὶς U.

Further, the spells of "bone-chilling" cold occur in localities with thick air, and this is the cold that penetrates the body most, for one cannot guard against it even by keeping to one's bedclothes. 14.4

Further confirmation is the fact that crippling frost occurs less in the mountains than in the plains: the mountain air is thinner and more mobile.

The testimony of the particular localities supports this view as well[1]: so in Thessaly frost occurs most of all (one might say) at Cierus,[2] and the place is in a hollow with much ground water. And at Philippi there was formerly more freezing of trees; but at present, now that the water has for the most part disappeared underground and been dried out, and the whole country has come under cultivation, there is much less. Yet the air is now thinner on both accounts: the water has dried up and the country become cultivated. For uncultivated land is colder and its air thicker because such land is wooded[3] and 14.5

[1] Just as such testimony supported the other view (*CP* 5 14. 2–3).

[2] The same as the Pierius of Theophrastus, *On Winds*, chap. vii. 45: *cf.* Friedrich Stählin, *Das hellenische Thessalien* (Stuttgart, 1924), p. 130, note 8, and pp. 130–132.

[3] *Cf.* Arius Didymus (*Doxographi Graeci*, ed. Diels, p. 854 a 27–28): "Hence some waters become warm when trees are cut down."

---

[6] aP : -ον U N.

[7] Schneider (ὑλήεις Scaliger) : ὕλης U.

διικνεῖσθαι μήτε τὰ πνεύματα διαπνεῖν, ἅμα δὲ καὶ αὐτὴν ἔχειν ὑδάτων συρροὰς καὶ συστάσεις

14.6 πλείους. ὃ καὶ περὶ τὰς Κρηνῖδας[1] ἦν,[2] τῶν Θρᾳκῶν[3] κατοικούντων· ἅπαν γὰρ τὸ πεδίον δένδρων πλῆρες ἦν καὶ ὑδάτων.

<ὥσθ'> ὁπότε[4] νῦν μᾶλλον <ἢ>[5] πρότερον ἐκπήγνυσιν, ἐξηραμμένων τῶν ὑδάτων, οὐ[6] τὴν λεπτότητα τοῦ ἀέρος αἰτιατέον (ὥς τινές φασιν).

αἱ μὲν οὖν αἰτίαι ὑπὲρ ἑκατέρου, καὶ τοιαῦται τινες.

ἴσως δέ, ἀμφοτέρων γινομένων ἐκπήξεων (τοῦτο γὰρ φανερὸν ἐκ τῶν εἰρημένων), διαφέρει καὶ τόπος τόπου καὶ[7] τῷ μᾶλλον παχύνεσθαι καὶ λεπτύνεσθαι· ὁ γὰρ ὑδατώδης καὶ θολερὸς οὐχ

[1] Schneider : κρηνίδας U.
[2] Itali (*fuisse proditum* Gaza) : ή U : ἦ u.
[3] Gaza (*thraces*), Scaliger (θράκων) : θράκην U.
[4] ὥσθ' ὁπότε ego (*quum igitur . . . locis his* Gaza : ὁπότ οὖν Heinsius) : ὁποτε U.
[5] Basle ed. of 1541 (*quam* Gaza).
[6] U : Heinsius omits : οὗ Schneider.
[7] καὶ ἀὴρ ἀέρος Schneider.

---

[1] *Cf.* Diodorus, *The Library of History*, xvi. 8. 6 (event of the year 358 B.C.: Philip takes Potidaea and enslave

the sunlight cannot reach it as well or the winds carry the moisture away, and because at the same time the land itself has a number of places where water collects and stands. And such was the case at Crenides[1] when the Thracians inhabited the country: the whole plain was covered with trees and lakes. 14.6

So that whenever the air causes more freezing now than before, after the water has dried up, we are not (as some persons do)[2] to assign the cause of the change to the thinness of the air.

The causes, then, support each of the two contentions, and are of the sort that we have seen.[3]

Solution

Perhaps, since both kinds of destructive frosts occur (this being evident from the discussion), there is also a difference between the districts in the greater degree of the thickening (and thinning) of the air: thus air that is full of water and overcast[4]

the inhabitants): "After that he passed to the city of Crenides and increased it by a great number of new settlers and changed its name to Philippi, naming it after himself; and he also so improved the gold mines around the city, which had been very simple and ordinary, that they were able to bring him a revenue of more than a thousand talents."

[2] The persons who so explained the case of Larisa (*CP* 14. 2). [3] They cited specific localities.

[4] *Cf. CP* 5 14. 1 with note 2.

ὁμοίως ἐργατικός, οὐδ' αὖ πάλιν ὁ λεπτός, καὶ γὰρ εὐκίνητος καὶ οὐκ ἔμμονος,[1] ἡ πῆξις δὲ εἰς χρόνον.

14.7 ἔτι δὲ καὶ τὰ ἐπιγινόμενα δεῖ ποῖ' ἄττα εἶναι, καὶ τὴν ὅλην[2] κατάστασιν, καὶ εἴ τι ἄλλο τῶν ἔξωθεν· εἰ γὰρ αἱ μεταβολαὶ ταχεῖαι πρὸς τὰς ἀνέσεις, κωλύοιντ' ἂν αἱ πήξεις.

ὡς δ' ἁπλῶς εἰπεῖν, εὐπαθέστερος μὲν εἰς τὸ πάσχειν ὁ λεπτός· ὅταν μέντοι καταψυχθῇ, ὁ παχύτερος ψυχρότερος, ὥσπερ καὶ θερμότερος· ἔμμονος γὰρ μᾶλλον ἡ θερμότης καὶ ἡ ψυχρότης, ὥσπερ καὶ ἐν τοῖς ἄλλοις τοῖς σωματωδεστέροις. διὸ καὶ εἴ που μὴ ὑδατώδης, ἀλλ' ἅμα τῷ πάχει ξηρός, οὐκ ἄλογον εἰ μᾶλλον ἐκπηκτικός.

ἀλλὰ γὰρ περὶ μὲν τούτων ἱκανῶς εἰρήσθω.

14.8 ἡ δὲ τοῦ καύματος ὑπερβολὴ τὰ μὲν φυτὰ καὶ τὰ παντελῶς νέα <ἂν>[3] φθείροι[4] διὰ τὴν ἀσθένειαν, ὥσπερ καὶ τὰ ἐπέτεια, τὰ δ' ἐρριζωμένα

[1] οὐκ ἔμμονος Heinsius (*minus ... stabilis* Gaza): ο[ὐ] κεκομμένος U.
[2] ὅλην U : τοῦ ὅλου Schneider.
[3] ego. [4] U : -ει u.

---

[1] *Sc.* "as thick air that is dry" (*cf. CP* 5 14. 7). Anything between the extremes can be called "thick" or "thin

does not operate so effectively to cause freezing,[1] nor on the other hand does thin air work so well, since it is mobile and does not remain at its task, and freezing takes time.

14.7 Furthermore the circumstances following the frost must have a certain character, and so too the climate in general and other matters external to the trees; thus if there is a rapid shift to milder weather he freezing of the trees would be prevented.

To speak broadly, thin air is the more readily affected so as to undergo the change; on the other hand, thicker air, once chilled, is the colder, just as t is warmer once it has been heated. For its warmth and cold is more lasting, just as in other substances that have more body. This is why in places where the air is not watery, but dry as well as hick, it is not unreasonable that it should be more apt to freeze and kill.

Let this treatment of the question suffice.

Cold More Deadly than Heat

14.8 Excessive hot weather on the other hand, while it may destroy cuttings and very young trees by reason of their weakness (just as it destroys annuals), oes not destroy trees that are well rooted and of

s one chooses; so too with "fluid" and "dry." So before raining the thick air at Philippi was dry enough to cause eezing; at Larisa it was too wet to do so.

καὶ ἔχοντα μέγεθος οὐ φθείρει, διὰ τὸ μὴ δύνασθαι
διαφύεσθαι,[1] μηδὲ ὁμοίως θιγγάνειν τῆς ἀρχῆς
ἀλλ' εἴπερ, τοὺς βλαστοὺς καὶ τοὺς καρποὺς ἐπι-
καίειν.[2]

καὶ τὸ ὅλον ἴσως ἀλλοτριώτερον τῇ φύσει τὸ
ψυχρόν·[3] ἐπεὶ ὥς γ' ἁπλῶς εἰπεῖν ἐπὶ πάντων
γίνεται [ἡ][4] φθορὰ ταῖς ὑπερβολαῖς ὅσα συνεργεῖ
14.9 πρὸς τὸ ζῆν. καὶ γὰρ χώρα τις ἄφορος ἡ μὲν διὰ
λυπρότητα,[5] ἡ δὲ διὰ πιότητα·[6] καὶ ἀὴρ κωλύει
περισκελὴς ὢν ἐφ' ἑκάτερα· καὶ ὕδατος ἔνδεια
καὶ πλῆθος, φθείρει γὰρ καὶ τὸ στάσιμον ἐὰν
ὑπεραίρῃ τὸ μέγεθος τοῦ δένδρου, καθάπερ ἐν ταῖς
ἐπομβρίαις καὶ τοῖς[7] λιμνουμένοις τόποις, ὃ καὶ
περὶ Φενεὸν[8] ξυνέβαινεν· ὅπου δὲ ἀπορροὴ γίνεται, κἂν ἔχῃ βάθος, διαφέρουσι μᾶλλον, βοηθεῖ
γὰρ ἡ κίνησις καὶ τῷ ἱκανὸν ἀεὶ <πρὸς> τὸ προσπῖπτον[9] εἶναι.

[1] U aP (διαδύεσθαι Itali) : διαφύεσθαι N.
[2] U : ἐπικαίει Heinsius (*adurit* Gaza).
[3] Itali (*frigiditas* Gaza) : των ψυχρων U. [4] ego.
[5] U N : λιπαρότητα aP. [6] ego : λεπτοτητα U.
[7] u : τους U. [8] u : φενέον U.
[9] ego (*contra praesentem alluviem* Gaza : πρὸς τὸ πίπτε
v) : τὸ προσπίπτον U : πρὸς τὸ πίπτειν N aP.

---

[1] That is, the roots: *cf. CP* 5 12. 5.

any size because it cannot pervade them or reach their starting-point[1] to the same extent as cold; instead if it harms grown trees at all, it can only scorch their shoots and fruit.

Indeed cold perhaps is in general more alien to the nature of a plant than heat; in fact (to put it broadly) everything that furthers life[2] will kill by its excess. So one soil fails to bear because it is too lean, another because it is too fat; air impedes growth when it is harsh to either extreme; and lack and abundance of water does so too, since even standing water kills if it rises higher than the tree, as occurs during the rains in places where a lake is formed, as used to happen at Pheneüs[3] (but where there is an outflow, and the lake, though deep, recedes, the trees are better able to survive, since the recession counteracts the inundation because the tree has the resources to withstand each flood as it comes).

14.9

[2] Heat is necessary to life: *cf.* Aristotle, *On the Parts of Animals*, ii. 3 (650 a 2–7).

[3] *Cf. HP* 3 1. 2; 5 4. 6. *Cf.* Strabo, *Geography*, viii. 8. 4 (389 C): "Eratosthenes says that at Pheneüs the river called Anias turns the land in front of the city into a lake, and flows down into certain 'sieves.' When these are occasionally blocked, the water overflows into the plains, and when they are opened up again it rushes in a mass from the plains and discharges into the Ladon and the Alpheüs, with the result that even the land of the district of Olympia around the temple was once flooded, and the lake made smaller . . ."

τούτων μὲν οὖν ταύτας ὑποληπτέον τὰς αἰτίας.

15.1 λοιπὸν δὲ δὴ[1] εἰπεῖν περί τε τῶν βιαίων παθῶν καὶ εἴ τις ἄλλη μὴ ὑπὸ τοῦ ἀέρος καὶ τῶν τῆς φύσεως ἀλλ' ὑφ' ἡμῶν γίνεται, καθάπερ ἥ τε διὰ τὸν περιφλοϊσμὸν[2] καὶ διὰ τὴν ἕλκωσιν τῶν φυτῶν καὶ ὅσα παραβαλλόμενα παρὰ τὰς ῥίζας αὐαίνει, καθάπερ τὰ τῶν κυάμων κελύφη καὶ εἴ τι ἄλλο τοιοῦτον ἕτερον.

καὶ πρῶτον ὑπὲρ τῶν κυάμων λέγωμεν· φθείρει[3] γὰρ τὰ τῶν κυάμων κελύφη περιβαλλόμενα ταῖς ῥίζαις καὶ τοῖς βλαστοῖς οὐ πάντα, ἀλλὰ τὰ ἀρτίως[4] ἀναφυόμενα, ταῦτα γὰρ ἀσθενέστερα· φθείρει δ' ὅτι τῇ σκληρότητι καὶ <τῇ>[5] ξηρότητι ἀφαιρεῖται τὴν τροφήν, τὴν μὲν ἕλκοντα αὐτά, τὴν δ'[6] ἀποστέγοντα· μὴ τρεφόμενα γὰρ φθείρεται.

15.2 καὶ ταῦτα μὲν καὶ εἴ τι τοιοῦτον ἕτερον, ὥσπερ

§15.1: Clement *Strom.* iii. 24. 3 (p. 207. 2–4 Stählin-Früchtel); Apollonius, *Mir.* xlvi.

[1] ego (δ' ἐστὶν Wimmer) : δε δεῖ U.
[2] u : τῶν περιφλοϊσμῶν U. [3] u : φθεῖρεται U.

Here, then, we must suppose that the causes are the ones given.

Destruction Due to Man

It remains to discuss violent types of destruction[1] and any other types that are not brought by the air and what belongs to the nature of the tree but by ourselves, as destruction by removal of the bark[2] and by wounding the young tree[3] and by the application of materials to the roots that cause withering,[4] such as bean pods and the like. 15.1

Killing by the Application of Bean Pods

Let us first deal with the bean pods: applied to the roots and shoots they destroy not all trees, but only the ones just growing up, since these are weaker. The pods destroy them by taking the food away by reason of their hardness and dryness, absorbing some of it themselves, and shutting out the rest, for when the trees get no food they perish.

The bean pods and the like destroy the tree by 15.2

[1] *CP* 5 16. 1.
[2] *CP* 5 17. 1.
[3] *CP* 5 17. 5.
[4] *CP* 5 15. 6.

---

[4] ego : ἄρτι U[r] a : ἀρτίοις (?) U[ar] : ἄρτια N P.
[5] Wimmer. [6] τήνδ' aP : τῇ δ' U N.

ἐναντία πρὸς τὴν βλάστησιν ὄντα, φθείρει.

ἔνια δὲ καὶ τῶν οἰκείων καὶ συνεργούντων, ἐὰν πλείω συνενεχθῇ[1] τῶν συμμέτρων ἢ ἰσχυρότερα ἢ μὴ κατὰ καιρόν, οἷον ἡ κόπρος ἢ συνεχῶς ἢ πλείων ἢ ἰσχυροτέρα παραβαλλομένη, καθάπερ ἡ σκυτοδεψική·[2] πάντα γὰρ (ὡς εἰπεῖν) αὕτη[3] δοκεῖ φθείρειν ἄκρατος οὖσα καὶ μὴ καλῶς κραθεῖσα·[4] καὶ ὅσαι δὲ θερμαὶ καὶ ξηραὶ καὶ ἰσχυραὶ

15.3 τὸ ὅλον, καὶ μὴ οἰκεῖαι πρὸς ἕκαστον. εἰσὶν γάρ, ὥσπερ εἴπομεν, αἱ πρὸς τὰς φυτείας ἁρμόττουσαι, καὶ οὐχ ὥσπερ τὸ ὕδωρ πᾶσι κοινόν. ἀλλὰ τοῦτο ἐνίοτε τῷ πλήθει διαφθείρει, σῆπον τὰς ῥίζας, καὶ λίαν ἐκμεθύσκον· ἐὰν δὲ δὴ νέα τύχῃ, καὶ μὴ ἄγαν ὄντα φίλυδρα, καθάπερ ἡ κυπάριττος[5] καὶ τἆλλα τὰ ξηρά, καὶ μᾶλλον, καὶ ἐὰν δή τις μὴ κατὰ καιρὸν ἢ τούτοις ἢ τῇ διακαθάρσει ἢ τῇ σκαπάνῃ χρήσηται· πάντα γὰρ συναίτια γίνεται φθορᾶς· ὁ δὲ καιρὸς καὶ πρὸς αὐτὰ καὶ πρὸς τὰ

[1] u : συνεχθῆι U. [2] u : σηντοδεψικῆ U.
[3] u : αὐτὴ U. [4] Scaliger : κρατηθεῖσα U.
[5] Gaza, Itali : περίκυττος U.

---

[1] *Cf.* *CP* 3 9. 2.
[2] *Cf.* *HP* 2 7. 4: "Manure does not suit all trees equally

being hostile (as it were) to its sprouting.

Help that Kills When Excessive or Inopportune

But even things that favour the tree and lend it aid will destroy it if accumulated in too great quantity or strength or at the wrong time, as manure applied either uninterruptedly or in too great quantity or possessing too great power, as tanner's manure. For this manure when applied undiluted and when improperly diluted is held to destroy just about all kinds of tree[1]; and so do all manures that are too hot and dry and in a word too strong, and that are not suited to the particular kind of tree.[2] For (as we said)[3] there are certain manures that are suited to planting different trees, and manure is not like water good for all. But water sometimes by its great quantity destroys a tree, decomposing the roots and fuddling them with fluid. It is even more destructive if the tree happens to be young and not very fond of water, like the cypress and other dry trees; and again it is harmful if one resorts to manuring or watering at the wrong time, or to pruning or spading, for all of these turn out to contribute to its destruction. And the right moment is relative not only to the trees themselves but also to the char- 15.3

nor does the same manure suit all; for some require that it should be pungent, some that it should be less so, and some that it should be very light."

[3] *CP* 3 9. 5.

ἐπιγινόμενα παρὰ τοῦ θεοῦ κατὰ τὰς ὥρας.

αὗται μὲν ἐν τοῖς οἰκείοις αἱ φθοραί, δι' ὑπερβολὴν ἢ δι'[1] ἔλλειψιν τροφῆς ἢ ἀκαιρίας ἔργων.

15.4 αἱ δ' ἀπὸ τῶν φυτευομένων,[2] ἢ παραβλαστανόντων αὐτομάτων, τῷ ἀφαιρεῖσθαι τὰς τροφάς, θᾶττον δ' ἐὰν ἰσχυρότερα καὶ πλείω, καθάπερ τὰ ἄγρια, καὶ ὅσα δὴ πολύρριζα καὶ πολύτροφα, καὶ ἐπισχίζοντα καὶ περιπλεκόμενα καὶ καταπνίγοντα, καὶ ἐμφυόμενα, καθάπερ ὁ κιττός. ἐπεὶ καὶ ἡ ἰξία[3] δοκεῖ, καὶ ὅλως τὰ ἐμβλαστάνοντα, φθείρειν· ὁ δὲ κύτισος[4] καὶ τὸ ἅλιμον τῇ τε πολυτροφίᾳ καὶ τῇ ἁλμυρίδι τῇ περὶ αὐτά· ἰσχυρότερον δὲ[5] τὸ ἅλιμον διὰ τὸ πλείω [ἔχειν].[6]

15.5 φθοραὶ δὲ καὶ ἄλλοις ὑπ' ἄλλων εἰσὶν ἴδιαι, καθάπερ ἐν τοῖς ἐλάττοσιν· καὶ γὰρ ἡ ὀροβάγχη[7]

§15.4: *HP* 4 16. 5.

[1] u aP : δ' U N.
[2] U : παραφυτευομένων Schneider.
[3] u (ἰεζία N aP) : ἰζία U.
[4] Schneider : κιττὸς U.
[5] Gaza (*sed*), Schneider : τε U.
[6] ego.
[7] Gaza (*orobancha*) : ὀροβάκχη U.

acter of the seasonal weather that follows.

These forms of death, then, are due to what favours the tree, and arise from excess or deficiency of food or carrying out agricultural procedures at the wrong moment.

Death by Neighbouring Plants

15.4 Destruction coming from neighbours that are planted or grow up spontaneously near by is due to their removing the tree's food; and the destruction is more rapid if the neighbours are stronger and more numerous, as is the case when they are wild, or when they have many roots and take much food, or branch out and entwine about the tree, choking it, or grow into it, like ivy. Indeed mistletoe too, and in general all plants that sprout in the tree, are held to kill it. Tree-medick and tree-purslane kill by their great consumption of food and by their salinity; tree-purslane is the stronger because it has more.[1]

15.5 There are also cases where a special victim is singled out, as among the smaller plants: so the

[1] *Cf. HP* 4 16. 5: "Trees are also killed by one another because the one takes the food of the other and because it impedes it in other matters. Ivy too is bad when it grows next the tree, and so is tree-medick, for it destroys practically all kinds of trees; but tree-purslane is still stronger, for it kills tree-medick."

καλουμένη φθείρει τὸν[1] ὄροβον τῷ περιπλέκεσθαι καὶ καταλαμβάνειν, καὶ τὸ αἱμόδωρον[2] τὸ βούκερας, εὐθὺς τῇ ῥίζῃ παραφυόμενον, καὶ ἄλλαι δ'[3] ἄλλων. καὶ ὅσα δὴ συγγεννᾶται καθ' ἕκαστον σπέρμα, οἷον αἶρα[4] καὶ αἰγίλωψ καὶ πυροῖς καὶ κριθαῖς, καὶ ἡ ἀπαρίνη φακοῖς, καὶ ἕτερα δ' ἑτέροις· ἅπαντα δὲ τῷ τὰς τροφὰς ἀφαιρεῖσθαι, τάς τε ἐκ τῆς γῆς καὶ τὰς ἀπὸ τοῦ ἡλίου καὶ τοῦ ἀέρος.

καὶ τούτων μὲν σχεδὸν φανεραί τινες αἱ αἰτίαι.

15.6 ἡ δ' ὑπὸ τοῦ ἐλαίου[5] καὶ τῆς πίττης καὶ τοῦ

§15.6: *HP* 4 16. 5.

[1] u : το U (τὸ N aP).
[2] Hindenland (from *HP* 8 8. 5) : λειμοδωρον U.
[3] ego (ἄλλα δι' Schneider) : ἄλλα δ' U.
[4] N aP (αἶρα u) : αι and a blank of two letters U.
[5] G ed (*oleum*) : ἡλίου U (*sol* G ur G ve G ch G ce G bu G mo).

---

[1] Dodder. [2] *Cf. HP* 8 8. 5.

[3] *Cf. HP* 8 8. 3–5: (if darnel does not actually come from the decomposition of wheat and barley) "it is in any case fondest of growing in wheat, just as Pontic 'black-wheat' and the seed of purse-tassels, and other plants growing among other seed-crops. So haver-grass is held to prefer barley . . . ; and practically for each plant there is another that grows up with it and is mingled with it, whether

so-called "vetch-choker"[1] kills vetch by entwining round it and holding it fast, strangleweed kills fenugreek as soon as it grows by its root,[2] and there are other killers that have other special victims. Then there are the cases of the plants generated with this or that seed-crop, as darnel and haver-grass with both wheat and barley, bedstraw with lentils, and others with others. All kill by taking away the food, both the food that comes from the ground and that which comes from the sun and the air.[3]

The causes here are (one may say) of the evident sort.

Killing by Oil, Pitch and Fat

Killing by oil, pitch and fat (for these too destroy 15.6

owing to the country, which is not unreasonable, or for some other reason. Some plants are quite evidently the common destruction of a number of others, but because they are most at their ease among one kind of plant rather than another, they seem to be peculiar to the former, as 'vetch-choker' to vetch and bedstraw to lentils. But 'vetch-choker' overpowers vetch more than it does the rest because vetch is weak; and bedstraw gets most food among lentils. In a way it is close to 'vetch-choker' because it covers the whole victim and holds it fast as with tentacles, for this is how it 'chokes' and got its name. But the plant that kills as soon as it comes up from the root of cummin and fenugreek, the plant called *haimódōron* [strangleweed], is more restricted in its victims . . . No other plant is made to wither by it except fenugreek."

στέατος (καὶ γὰρ καὶ ταῦτα φθείρει, καὶ μάλιστα τὰ φυτὰ τὰ νέα· καὶ[1] οὐκ ἐῶσιν ἅπτεσθαι καὶ περιπλέκουσιν), ἐν ἐκείναις οὖσα[2] ταῖς αἰτίαις· ὅτι θερμὰ καὶ λεπτὰ τὴν φύσιν ὄντα, διαδύεταί τε πόρρω καὶ πυκνοῖ καὶ ἐπικάει τὸν φλοιόν (σημεῖον δ', ὅτι καὶ σκληρύνεται καὶ ἀφίσταται)· πονήσαντος δὲ καὶ ἀποσκληρυνθέντος καὶ τούτου καὶ τοῦ ἐντός, οὐ δύναται διιέναι ἡ τροφή. τοῦ δ' ἐπικάειν καὶ διαδύεσθαι πόρρω κἀκεῖνο σημεῖον· οἱ γὰρ ἡμεροῦντες, τοῦ θέρους ἐπὶ τὰ ὑπολείμματα τῶν ῥιζῶν ἔλαιον ἢ πίτταν ἐπιχέουσιν, ἢ τῷ στέατι ἀλείφουσιν, ἅπερ ὅλως[3] ξηραίνουσιν, καὶ μάλισθ' ἡ πίττα, διὰ τὸ ἰσχυροτάτη[4] [εἶναι].[5]

16.1 λοιπαὶ δὲ τῶν φθορῶν ὥσπερ[6] βίαιοι λεγόμεναι· αὗται δὲ γίνονται πληγῇ ἢ περιαιρέσει

[1] διὸ καὶ Schneider.
[2] ego (ἐστὶ Wimmer : εἰσὶ Heinsius) : οὔσαις U.
[3] u : ὅλω U. [4] u : ἰσχυροτάτην U.
[5] ego.
[6] U : *quae* Gaza : αἵπερ Scaliger : αἱ Schneider.

---

[1] *Cf. HP* 4 16. 5: "Oil is most effective with young trees and trees just starting to grow, for these are weaker, and this is why we are told not to let them touch it."

[2] *Cf. HP* 4 16. 5: "Killing of a tree by oil is due rather to addition than to removal, for oil too is bad for all trees,

trees, especially the young cuttings; and we are warned not to let the cuttings touch these substances[1] and to wrap the cuttings up) works by coming under the following causes: the things are by their nature hot and thin and so penetrate deep and thicken and scorch the bark (this is shown by the bark getting hard and getting detached), and when both the bark and the interior have suffered and hardened the food is unable to pass through. Another proof that these substances scorch and penetrate is this: persons engaged in reclaiming land pour oil or pitch on the remnants of the roots[2] in summer or smear them with fat, and these dry them up completely, especially pitch, by reason of its superior strength.[3]

Violent Death: Its Forms

16.1 What remain are the forms of death that are called (one may say) violent.[4] These arise from (1) a blow[5] or (2) the stripping off of certain parts[6] or

and to kill what is left of the roots oil is poured on them."

[3] Shown by its adhesiveness.

[4] The examples of "violent death" given in *CP* 5 8. 1 and 5 11. 1 were death by a blow or by chopping; here death by removal or subtraction (suggested by the death by addition of *CP* 5 15. 6; *cf. HP* 4 16. 5, cited in note 2 on *CP* 5 15. 6) has been added; hence "one may say."

[5] *CP* 5 16. 1–4.

[6] *CP* 5 17. 1.

τινῶν ἢ κολούσει ἢ τὸ ὅλον ἀφαιρέσει.

ἔνια μὲν γάρ, ἑλκούμενα βαθύτερον, ἀπόλλυται, διὰ τὴν ξηρότητα καὶ ἀσθένειαν· ὁ δὲ φοῖνιξ καὶ τιτρωσκόμενος εἰς τὸν ἐγκέφαλον, ἐν τούτῳ γὰρ ἡ ζωὴ καὶ ἡ βλάστησις· ὅταν οὖν ἀναξηρανθῇ, ἢ τὸ ὅλον ἀλλοιωθῇ, παρεισπεσόντος ἀέρος τε καὶ ἀλλοτρίου θερμοῦ, διαφθείρεται, καὶ διαφθειρόμενος διίησιν εἰς τὰ κάτω.

τούτου μὲν οὖν καὶ ἰδία τις[1] ἡ φύσις, ὥστε καὶ ἐμφανὲς εἶναι τὸ κύριον τοῦ ζῆν.

16.2 ἔνια δ' οὐ πρὸς πληγὴν ἀπαθῆ μόνον, ἀλλὰ καὶ ξύλων ἐξαιρουμένων ἐκ τοῦ στελέχους, οἷον ὅσα φύσει καὶ εὐβλαστῆ καὶ ὑγρά, καθάπερ πτελέα, πλάτανος, τὰ πολλὰ τῶν παρύδρων· ἡ δὲ πεύκη καὶ δᾳδοκοπουμένη σῴζεται, μέγα δὲ ταύτῃ καὶ ἡ

[1] u : της U.

---

[1] *CP* 5 17. 5.

[2] *CP* 5 16. 2; 5 17. 1–7. *Cf. HP* 4 15. 1–4 16. 5: "It remains to discuss the cases where trees perish from stripping off certain parts. Death arising from stripping the bark off all round is common to all . . . What is called topping of trees is fatal only to fir, silver fir and Aleppo pine . . . Most trees perish also if the trunk is split . . .;

cutting back[1] or in general from removal.[2]

(1) *Blow with a Wound*

Thus some trees perish on receiving a fairly deep wound, by reason of their dryness and weakness[3]; and the date-palm also perishes when wounded in the "head," since its life and its ability to sprout lie here.[4] So when the head is dried out (or altered in general) by the invasion of air and foreign heat, the head perishes, and in perishing allows these to pass to the parts below.

In this tree, then, the nature is of a special sort, so that the part that controls life is evident.

(1) *Blows with Removal of Wood*

16.2 But some trees are not only unharmed by blows but even by the removal of wood from the trunk, as all that by nature are both good sprouters and full of fluid, such as the elm, plane and most trees that grow by water. The pine can even survive the removal of torchwood[5]; here its oiliness is of great help.

some even are killed if they receive a fairly large and deep wound . . . All trees are killed if the roots are cut off . . ."

[3] They have no adhesive quality in their sap or wood: *cf. CP* 5 16. 4.

[4] *Cf. CP* 1 2. 3.

[5] *Cf. HP* 4 16. 1: "But some trees are not affected, as the pine when the torchwood is removed and the trees from which resins are gathered, as silver fir and terebinth, for here too the cut and wounding goes deep."

λιπαρότης. ἅπαντα δὲ ταῦτα καὶ τἆλλα, μέγεθος ἔχοντα, ὑπομένει, διὸ καὶ ἐκσηπόμενα σῴζεται καὶ ζῇ. ἐπεὶ καὶ τὴν πεύκην γ' οὐ νέαν[1] δᾳδουργοῦσιν,[2] ἀλλ' ὅταν ἐν ἀκμῇ, καὶ πορρωτέρω, γένηται· νέα γὰρ οὐκ ἔχει διὰ [δὲ][3] τὸ μὴ πέττειν μηδὲ συναθροίζειν τὸ ὑγρόν, ἀλλ' εἰς τὴν βλάστησιν[4] καὶ τὸ μῆκος καταναλίσκειν· ἅμα γάρ πῃ <τῇ>[5] εἰς βάθος αὔξῃ καὶ ἡ τοιαύτη διάθεσις καὶ δύναμις ἔοικεν ἀκολουθεῖν.

16.3 ὅσα δὲ καὶ τετραινόμενα[6] καὶ κολαζόμενα βελτίω γίνεται καὶ καρπιμώτερα, δῆλον καὶ ταῦθ' ὑπομένειν.[7]

πληγὴν μὲν οὖν καὶ διαίρεσιν καὶ τὰ τοιαῦτα

[1] Wimmer : γοῦν ἐὰν U.
[2] Heinsius : δαδουργῶσιν U. [3] Gaza, Heinsius.
[4] U^{cc} (from βλάστην). [5] πῃ τῇ ego : πηι U : τῆι u.
[6] ego : τιτρωμένα U : τετρωμένα u.
[7] Heinsius (*posse sufferre* Gaza) : -ει U^{c} : -η U^{ac}.

---

[1] That is, the oily trees from which resin is obtained; see the preceding note.

[2] That is, the trees of the preceding sentence.

[3] *Cf. HP* 9 2. 7: (after torchwood has been cut from the pine three times) "the tree, which has decomposed, is thrown down by the winds because of the undercutting."

[4] *Cf. HP* 9 2. 8: "For the pines bear fruit from their earliest years, but produce torchwood much later, when they

All these[1] and the rest[2] endure such treatment when they have grown to some size (this is why they survive and live even when they have lost part of the trunk by decomposition).[3] Indeed the pine is not cut for torchwood when it is young, but only at its prime and later, since when young it has no torchwood,[4] owing to its not concocting the fluid to resin or forming an accumulation of it, but expending it on new growth and height, for it seems that the disposition and power that produces torchwood comes with the increase of the tree in lateral growth.

16.3 The trees also that improve and become more fruitful when a hole is drilled in them and they are "castigated"[5] evidently survive too.

(1) *Blow with Splitting of the Trunk*

Now such trees also accept a blow and the break-

are getting older."

[5] *Cf. HP* 2 7. 6–7: "With the almond they further drive in an iron peg and when they have made a hole replace the peg with one of oak and cover the spot with earth; and some call the procedure 'castigation,' on the ground that the tree had got out of hand. Some do the same with pear also and with others. In Arcadia they also speak of 'correcting' the sorb . . . And they say that when this is done the ones that do not bear will bear and the ones that fail to concoct their fruit will concoct it properly. They say that the almond even changes from bitter to sweet if you dig round the trunk and make a hole in it . . ."

δέχεται· σχίσιν δὲ τοῦ στελέχους πρὸς τούτοις ἄμπελος καὶ συκῆ καὶ ῥόα καὶ μηλέα, τὰ δὲ ἄλλα ἀπόλλυται (ὅσα γὰρ αὖ πληγέντα καὶ σχισθέντα συμμύει πάλιν καὶ συμφύεται, ταῦθ' ὥσπερ ὑγιασθέντα[1] ζῇ, καὶ οὐχ ὥσπερ ἐκεῖνα, διεσχισμένα). τῆς δ' ὑπομονῆς αἴτιον ἡ ὑγρότης καὶ ἡ φύσει μανότης· τροφήν τε γὰρ ἱκανὴν λαμβάνουσιν, καὶ οὐκ ἀναξηραίνεται·[2] διὰ τὴν σχίσιν δ' ὑπὸ[3] τοῦ ψύχους οὐδὲν πάσχουσιν.

16.4 εἰ δ' οὕτω δεῖ λαβεῖν τὴν σχίσιν ὥστε ἀνέχεσθαι μὲν ἃ μόνα δοκεῖ τῶν ἡμέρων τοῦτο δύνασθαι (συκῆ, ἄμπελος, ἐλαία, ἀμυγδαλῆ) διὰ τὸ συμμύειν τάχιστα[4] (διὸ καὶ δύσσχιστα[5]) τῷ κολλώδη τὴν ὑπόστασιν ἔχειν. καὶ γὰρ ὁ ὀπὸς καὶ τὸ τῆς ἐλάας τοιοῦτον, καὶ πᾶν τὸ λιπαρόν, τὸ δὲ τῆς ἀμπέλου ξύλον αὐτὸ τοιοῦτον·[6] σημεῖον δέ, ὅτι καὶ σχιζό-

[1] aP : ὡς ὑπερυγιασθέντα U N. [2] U : -νονται Heinsius.
[3] δ' ὑπο U : οὐδὲ ὑπὸ Heinsius after Gaza : ὑπὸ δὲ Wimmer.
[4] τάχιστα <. . .> Wimmer.
[5] N : δύσχιστα U aP. [6] U ac : τοιοῦτο U c.

---

[1] *Cf. HP* 4 16. 1: "Most trees are also killed if the trunk is split, for none are held to endure this except the vine, fig pomegranate and apple . . ."

[2] *Cf. HP* 5 3. 4: "Willow wood and vine wood are also viscous. This is why shields are made of them, for the

ing of their surface. But vine, fig, pomegranate and apple allow in addition to this a splitting of the trunk, whereas the rest are killed by it.[1] (As for the trees that after being struck and split close up the wound and grow together again,[2] these live because they have been healed, as it were, and not like the others in a split condition.) The survival is due to the abundant fluid and natural open texture of the trees, for they get sufficient food and the fluid is not dried out; and the split does not expose them to harm by the cold.[3]

16.4 But if we are to take the "splitting" in a sense that lets only those cultivated trees endure it that are considered to have this ability—fig, vine, olive and almond[4]—the cause of the endurance is their being the swiftest to close the wound (which makes them hard to split), owing to their having a glutinous deposit. Thus the saps of fig and olive (and all oily saps) are of this character,[5] and in the vine the wood itself is glutinous (proof of this is the quick

wood closes up after a blow . . ." No doubt the same was true of the living trees.

[3] *Cf. CP* 5 12. 9 for the open texture of the apple as protecting it from freezing.

[4] Presumably some authority had given this list of the trees that survive splitting (which differs from the list in *CP* 5 16. 3). Theophrastus then interprets "splitting" in a way that will account for just the trees listed here.

[5] The almond exudes gum (*HP* 9 1. 2).

μενα[1] τὰ κλήματα, καὶ τῆς ἐντεριώνης ἐξαιρουμένης, ταχὺ συμφύεται· καὶ[2] τούτου γε μᾶλλον ὁ κάλαμος, καὶ γὰρ συνέρχεται αὐτόματος· φασὶν δὲ καὶ τὴν ἄπιον σχίζεσθαι. περὶ μὲν οὖν τούτων σκεπτέον.

17.1 ἡ δὲ τοῦ φλοιοῦ περιαίρεσις κοινὴ πάντων (ἢ τῶν πλείστων) ἐστὶ φθορά (περὶ ἧς εἴρηται πρότερον)· εἴτε γὰρ ἐν τοῖς κυρίοις ἐστὶ τοῦ ζῆν, εὐλόγως ἂν γίνοιτο (καθάπερ τινές φασιν), εἴτε καὶ ἀπὸ τούτων πυκνουμένων δικνεῖται πρὸς τὸ ὅλον, εὔλογον καὶ οὕτως.

[1] u : σχιζόμε U.
[2] Gaza (*atque*), Scaliger : ἢ U.

---

[1] In the operation described at *CP* 5 5. 1.
[2] This statement is not found elsewhere.
[3] Perhaps in connexion with the operations described in *HP* 2 7. 6 ("If a tree bears no fruit but runs to new growth, they split the part of the trunk that is underground and insert a stone to make the split break and say that the tree then will bear") and *HP* 2 7. 7 ("Some do the same [*sc.* drive a peg into the trunk] with the pear as well and others").
[4] *HP* 4 15. 1–4.
[5] *Cf. HP* 4 15. 1: "For a form of killing common to all trees is the removal of the bark all around; for every tree

coalescence of split vine-twigs when the pith has been removed[1]); and reed[2] does this better than the wood of the vine, since it comes together of its own accord. It is said that the pear too can be split.[3] We must investigate these matters.

(2) *Removal: Stripping the Bark*

17.1 Stripping the bark all around is a form of killing (discussed earlier)[4] effective with all or most trees. For whether the stripping is of parts that control life[5] (as some assert), death would be reasonable; or whether the parts exposed get thick and the thickening spreads to the whole tree,[6] death is reasonable with this process too.

practically speaking is held to be killed by it except *Arbutus andrachne*, and this too is killed if one presses very hard on the flesh . . . , except if one can do this with cork-oak, for they say that the tree actually is stronger when the bark is removed, but this is evidently the bark outside that comes down as far as the flesh, as with *Arbutus andrachne.* Indeed the bark is also stripped from cherry, vine and linden . . . , but this is not the vital or innermost bark, but that on the surface . . ."

[6] *Cf. HP* 4 15. 2: "For some trees endure for some time, as fig, linden and oak. Some say that these live and are not killed, and that the elm and date-palm also live . . . , but that in the rest a kind of callus forms which has a nature of its own." For the effect of the thickening *cf. CP* 5 15. 6 (on the effect of oil): "when both the bark and the interior have suffered and hardened the food is unable to pass through."

ἡ δὲ μήτρα, μέχρι μέν τινος ἐξαιρουμένη, οὐ φθείρει,[1] δι' ὅλου δέ, φθείρει (καθάπερ οἱ περὶ Ἀρκαδίαν φασὶν καὶ πεύκην καὶ ἐλάτην καὶ ἄλλο πᾶν[2]). οὐκ ἄλογος δ' οὐδὲ ταύτης ἡ αἰτία, καὶ ταῦτ' <ἂν>[3] εἴη τῶν εἰρημένων (ὑγρότατον γὰρ

17.2 δοκεῖ καὶ ὥσπερ μάλιστα εἶναι ζωτικόν). σημεῖον δ', ὅτι καὶ τὰ ξύλα τὰ ἔμμητρα[4] διαστρέφεται, κατειργασμένα ἤδη, μέχρι οὗ ἂν τελέως ἀναξηρανθῇ, διὸ καὶ ἔνσχιστα καὶ οὐκ ἔμμητρα ποιοῦσι τὰ τῆς ἐλάτης καὶ τῆς πεύκης.

ταύτης δ' οὖν[5] μέχρι μέν τινος ἐξαιρουμένης, οὐκ ἄτοπον διαμένειν τὸ δένδρον, ὥσπερ καὶ τοῦ φλοιοῦ μέχρι τινὸς ἀφαιρουμένου, τελέως δ' ἐξαιρεθείσης, αὐαίνεσθαι, καθάπερ ἀρχῆς τινος ἢ καὶ

[1] U^r N aP : φθείρεται U^ar.
[2] *HP* 4 16. 4 : ἄλλὅ|τι ἂν U.
[3] ego. [4] u N : ἔμμετρα U aP.
[5] Schneider (*itaque* Gaza) : δ' οὐ U N : δὲ aP.

---

[1] *Cf. HP* 4 16. 4: "When the core is removed hardly any tree, practically speaking, is killed; this is shown by the existence of many hollow trees among those of some size. The Arcadians say that when this is done up to a point the tree lives, but that when the core is removed from the entire tree both the pine and silver fir and all others die."

[2] In the preceding paragraph.

[3] *Cf. HP* 5 5. 2 (of warping in timber): "Warping is the occupying by the wood of vacated positions as the core

(2) *Removal: Of the Core*

Removal of the core up to a point does not kill the tree, but it does when complete, as the Arcadians say it kills pine and silver fir and all others.[1] Here too the causation is not unreasonable, and these cases belong with those just mentioned[2]: the core is held to be the most fluid and (as it were) vital of all the parts. Proof of this vitality is that (until thoroughly dried) wood with the core left in will warp even after it has already been worked by the carpenter, which is why timber of silver fir and pine is prepared by splitting the trunk and not leaving the core inside.[3] 17.2

So it is not strange that whereas the tree should survive when the core is removed up to a point, just as it does when the bark is stripped up to a point,[4] it should wither when the core is taken out completely, as if deprived of some starting-point or fluid

moves. For it lives (it appears) for a long time. This is why they remove it at the same time from all the structures that go to make a door, and especially from the door-frame, to keep them from warping; and for this reason they split the wood."

[4] *Cf. HP* 4 15. 4: "In all trees the stripping of the bark must be of some breadth, especially in the strongest; indeed if one removes only a very narrow strip, there is nothing strange in the tree's not being killed . . ."

οἰκείας ὑγρότητος συμφύτου στερούμενον.[1]

17.3 περὶ δὲ τῆς ἐπικοπῆς[2] καὶ τῆς κολούσεως ἐν ὀλίγοις ἡ σκέψις, ὀλίγα γὰρ τὰ φθειρόμενα.

κατὰ μὲν τὴν ἐπικοπὴν[2] ἐλάτη, πεύκη, πίτυς, φοῖνιξ, ἔνιοι δὲ καὶ κέδρον καὶ κυπάριττόν φασιν. καλοῦσι δ' ἐπικοπήν,[2] ὅταν ἀφαιρεθείσης τῆς κόμης ἐπικόψῃ τις τὸ ἄκρον. οὐκ ἀλόγως δ' ἂν δόξειε φθείρεσθαι, ξηρά τε τῇ φύσει καὶ εὐθυπορώτατα μέν, μονόρριζα δ' ὄντα. καὶ γὰρ αἱ ἑλκώσεις πόνον παρέχουσι, καὶ εὐπαθέστερα ποιοῦσιν εἰς τὸ διικνεῖσθαι καὶ τὸ καῦμα καὶ τὸ ψῦχος, ἄλλως τε καὶ πανταχόθεν οὖσαι· καὶ διὰ τὴν ὀρθότητα καὶ διὰ τὴν εὐθυπορίαν ταχὺ διικνεῖται καὶ πρὸς τὰς ῥίζας, ὥστε πολλαχόθεν ἡ φθορά. 17.4 φύσει δὲ

[1] Heinsius (*destituta* [of *arbos*] Gaza) : στερουμένοις U : στερουμένης u.
[2] U^r N aP : -σκ- U^ar.

---

[1] *Cf. HP* 4 16. 1: "The form of killing called 'topping' a tree is found only in pine, silver fir, Aleppo pine and date palm; some add prickly cedar and cypress. For all these trees are killed and do not sprout if the foliage is stripped and the top above lopped off, just as all or some of them do when burnt"; *cf.* also *HP* 3 7. 1–2: ". . . pine and silver fir perish totally root and all in the same year even if only

of its own belonging to its nature.

(2) *Removal:* (a) *Topping and* (b) *Cutting Back*

The consideration of topping and cutting back is concerned with only a few different kinds of trees, for only a few are killed. 17.3

(a) *Topping*

Silver fir, pine, Aleppo pine and date-palm are killed by topping; some add prickly cedar and cypress. The term "topping" is used of the removal of the foliage followed by lopping off the top.[1] It would appear not unreasonable that these trees should be killed, since they are dry by their nature and have the straightest of passages and but a single root. The wounds too make them suffer and predispose them to allow passage to heat and cold, especially since the wounds are on all sides of the tree; and the heat and cold, because the tree is erect and its passages are straight, quickly pass all the way to the roots. So death comes from many sources. All are by nature without side-shoots,[2] not 17.4

he top is lopped off.... For when you remove all the branches and cut off the top, the tree [the silver fir] soon dies..."

[2] "Side-shoots" are shoots sent up from the root that grow alongside (that is, parallel to) the trunk.

ἀπαράβλαστα πάντα, καὶ διὰ τὴν ξηρότητα τῶν ῥιζῶν, καὶ διὰ τὴν εἰς τὸ ἄνω φοράν, καὶ ἔτι[1] *τῷ μονόρριζα,*[2] *καὶ ἐπὶ τὸ*[3] *βάθος*[4] *ἐνίας*[5] *ἔχειν· οὐδαμοῦ γὰρ οὐδὲν περιττόν, οὐδὲ παροχετευόμενον ἐκπίπτει τῆς τροφῆς, ὥστε μὴ εἶναι βλαστήσεως*[6] *ἀρχήν. ὅταν οὖν πάντα ταῦθ' ὑπάρχῃ, πᾶσιν σχεδὸν ἀναγκαῖον τὴν φθορὰν*[7] *εἶναι τελείαν.*

17.5 *αἱ δὲ κολούσεις φθείρουσιν ὅλως ὀλίγα· μόνον γὰρ* <*ἢ*>[8] *μάλιστα τὸ τῆς ἀμπέλου φυτὸν ἀπόλλυται, καὶ εἴ τι ἕτερον ἁπαλὸν καὶ ἀσθενὲς καὶ εὐθύπορον, ταῦτα γὰρ αἴτια τοῖς τοιούτοις τῆς φθορᾶς· χείρω δὲ ποιοῦσι πλείω, καὶ γὰρ ἡ ἀμυγδαλῆ πικρὰ γίνεται, καὶ ἡ ῥόα σκληροτέρα, καὶ ἕτερ' ἄττα μεταβάλλει. τὸ δ' αἴτιον εἴρηται πρότερον,*

[1] Gaza (*etiam*), Schneider : ἐπὶ U.
[2] τῷ μονοριζα U^{c} : τωνοριζα U^{ac}.
[3] U N aP : τοῦ u (-υ now crossed out).
[4] U^{r} N aP : βάθους U^{ar}.
[5] U : ἔνια Schneider.
[6] παραβλ- Heinsius.
[7] N : φορὰν U aP.
[8] Schneider.

---

[1] Cypress (*cf. HP* 1 6. 4) and prickly cedar (*HP* 3 6. 8 have shallow roots.

[2] Hence the sun cannot reach them and produce side

only because the roots are dry but also because the food moves rapidly upward, and further because the root is single and some[1] of the single roots go deep.[2] For nowhere does it have any extra food or food spilled from the channel, and so there is nowhere any place to sprout from. So with all these conditions present it is in all of them (one might say) a necessary consequence that the killing is total.

(b) *Cutting Back*

17.5 Cutting back kills but few trees outright, the young vine being the only or chief one to perish[3] and any other tree that is tender, weak and with straight passages, these being the characters responsible for the death of trees to which cutting back is fatal). A greater number of trees, however, are made worse by being cut back: so the almond gets bitter[3] and the pomegranate harder[4] and certain others undergo changes.[5] The cause has been

shoots: *cf. CP* 1 3. 4–5. [3] *Cf. CP* 2 15. 1.

[4] Perhaps referred to in *HP* 2 2. 9: "Trees also change because of the care given to their feeding and to the other procedures of husbandry, whereby a wild tree is turned into a cultivated one, and among cultivated trees themselves some are changed to wild, as pomegranate and almond."

[5] Not identified elsewhere; at *CP* 2 15. 2 cutting back is said to improve the Phocian pear and others; at *CP* 2 15. 3 changes for the worse in other trees than the almond are said to pass unnoticed by our sense.

ὅτι τῆς ἀρχῆς ἀλλοιουμένης, συναλλοιοῦται καὶ τὸ τέλος.

17.6 χαλεπαὶ δὲ καὶ[1] ἐπιβοσκήσεις, ὅτι συνεπικάουσιν ἅμα τῇ τομῇ καὶ ἀφαιρέσει, διὸ καὶ ὁ πόνος πλείων.

ἴδιον δὲ τὸ[2] περὶ τὸν φοίνικα καὶ τὴν πίτυν· ὅταν γάρ τι κολουσθῇ,[3] ταῦτ' οὐκ ἀπόλλυται μέν, ἄκαρπα δὲ γίνεται. τὴν δ' αἰτίαν παραπλησίαν ὑποληπτέον εἶναι τοῖς πρότερον, ὅτι τῆς ἀρχῆς ἀλλοιουμένης καὶ ἀσθενεστέρας γινομένης, ἀφαιρεῖται τῆς[4] δυνάμεως, καὶ μάλιστα τὸ ἔσχατον καὶ τελεώτατον, ὁ καρπός. ἅμα δὲ καὶ ἡ μὲν βλάστησις ὑπάρχει, καὶ ὥσπερ ἤδη γέγονεν, ἐκεῖνο[5] δὲ δεῖ γενέσθαι· τὸ δ' ἐν γενέσει καὶ μελλήσει[6] τοῦ ὄντος ἀσθενέστερον. ἐπὶ τούτων δὲ μάλιστα

[1] καὶ αἱ Schneider.
[2] Schneider : τον U (τὸν N) : τῶν u : τὰ aP.
[3] U^ar : -ουθ- U^r N aP.
[4] τι τῆς Wimmer. [5] U : -ον Scaliger.
[6] ego (δὲ γενέσθαι καὶ μελλῆσαν Itali : δ' ἐν γενέσει καὶ μέλλον εἶναι Wimmer) : δὲ γενέσθαι καὶ μελλήσειν U.

---

[1] *CP* 2 14. 3; 2 15. 2; 2 16. 3; 3 9. 4; 3 17. 7; 3 24. 4.
[2] *Cf. HP* 4 14. 8: "Some mutilations do not result in th

nentioned before[1]: when the beginning is altered, he final product is altered as well.

(2) *Removal: Cropping by Animals*

17.6 Cropping by animals is also serious, because it corches at the same time that it cuts and removes, vhich is why the trees suffer more.

A Peculiar Case

The case of the palm and Aleppo pine is peculiar: vhen a certain part is cut off they do not die but ecome unfruitful.[2] We must suppose that the cau- ation is close to that of the occurrences mentioned arlier[3]: when the starting-point is altered in char- cter and gets weaker, the power of the tree is lost to degree, and especially the ultimate and most fully leveloped product of that power, the fruit. Then too, vhereas the vegetative growth is present and has lready completed (as it were) its coming into xistence, that other product, the final one, is still to e produced; but the thing engaged in becoming and till to be is weaker than the thing already existing. 'he occurrence is limited chiefly or only to these

eath of certain trees but in their unfruitfulness. Thus if ou remove the top of the Aleppo pine or the palm, the rees are both held to become unfruitful, and not to be illed outright." [3] *CP* 5 17. 5 (with note 4).

<ἢ>[1] μόνων, ὅτι ὀλιγόκαρπα καὶ βραδύκαρπα, κατά γε τὴν παρ' ἡμῖν φυτείαν· ἐν δὲ ταῖς οἰκείαις χώραις ἴσως οὐκ ἀκαρπίας,[2] οὐδὲ θαυμαστὸν εἰ μὴ ἀκαρπεῖ ταῦτ' ἐκεῖ κολουόμενα.

17.7 τὰ δὲ τῆς ἀρχῆς ὅτι μεγάλα, πολλαχόθεν μὲν φανερόν· εἰ δὲ μή, καὶ ἐκ τῶν βλαστήσεων ὅταν κακωθῶσιν, ἐν ἄλλοις τε καὶ οὐχ ἥκιστ' ἐν τῇ ἀμπέλῳ· κατεδεσθείσης[3] γὰρ ὑπὸ τῶν ἰπῶν,[4] οὐκέτι δύναται βλαστάνειν, ἀλλ' αὐτὴ ἡ τῆς βλαστήσεως ἀπορροὴ[5] παύεται καὶ ἀποσβέννυται, ᾧ φαίνεται ὥσπερ τυφλουμένη καὶ πεπηρωμένη[6] πως. ἰσχυρὸν δὲ κἀν[7] ἅπασιν ἡ ἀρχή, καὶ ἀσθενές, οὐ τὸν αὐτὸν δὲ τρόπον. ἐπεὶ ὅτι γε βλαστητικὸν ἄμπελος, καὶ <αἱ>[8] εἰς τὸν ὕστερον χρόνον ἐπιβλαστήσεις μηνύουσιν.

[1] Schneider. [2] U : ἀκαρπεῖ Itali after Gaza.
[3] ego (*exesis* Gaza : κατέδεται Heinsius : κατεδεσθεῖσα Wimmer) : κατ' ἔδεσθαι U.
[4] ego (κτηνῶν Heinsius) : δεινῶν (dot over ει u) U.
[5] U : ἐπιῤῥοὴ Schneider.
[6] ego (πηρουμένη Schneider) : πηρωμένη U.
[7] ego (ἐν Itali) : καὶ U. [8] u.

---

[1] *Ipôn* ("bud-worms") is a conjecture: *cf. CP* 3 22. 5–6 and Theophrastus, *On Stones,* chap. viii. 49: "In Cilicia there is a certain earth which on boiling becomes viscous. They smear their vines with this instead of bird-lime

trees because they have little fruit and bear it late (at least when planted in Greece; in their own countries the mutilation is perhaps no cause of unfruitfulness, and there is nothing surprising if when mutilated there they should not fail to bear).

17.7 (That the fate of the starting-point is of great importance is made evident by many other occurrences, and leaving these aside is even evident from injury to new growth, both in other trees and not least in the vine. Thus when the new growth has been devoured by the bud-worms[1] the vine loses the ability to sprout, and the very outflow that produces new growth stops and runs dry, with the result that the vine appears sterile[2] as it were and in a sense crippled.[3] In all other trees too the starting-point is something strong and also something weak, but not in the same way; in fact the sprouting power of the vine is shown by its additional sprouting later in the year.[4])

against the bud-worms."

[2] Literally "blinded"; a "blind" shoot at *CP* 3 2. 8 is one that produces no fruit; a blind knot at *CP* 3 5. 1 is one that has no bud (or fruit bud). The bud of a vine is in Greek an "eye" (*ophthalmós*).

[3] "Crippled" in Greek often means blind.

[4] *Cf. HP* 3 5. 4: "Sprouting in the dog days and at the rising of Arcturus, a sprouting which comes after the sprouting in spring, is common to all trees (one might say), but is most obvious in the cultivated, and among these above all in the fig, the vine and the pomegranate . . ."

18.1 φθοραὶ δὲ καὶ τῶν σπερμάτων εἰσίν τινες τῶν μὴ κατὰ φύσιν σπειρομένων, ὁμοίως τῶν τε δενδρικῶν καὶ τῶν σιτηρῶν καὶ τῶν ἄλλων, ὧν καὶ χρόνοι τινές εἰσιν ὡρισμένοι· πάντα γὰρ ζῇ μέχρι τινός, εἶτ' ἀπόλλυται (ζῆν δὲ λέγω δυνάμει). φθείρονται δὲ φυσικῶς, ἀποξηραινόμενα, καὶ ὥσπερ διαπνεούσης ἅμα τῆς θερμότητος καὶ ὑγρότητος.

αἱ δὲ ἄλλαι πᾶσαι παρὰ φύσιν, οἷον ὅσα θηριοῦται, καὶ ἀνυγραίνεται, καὶ ἄλλως πως ἐξίσταται.

18.2 διὸ καὶ τὰ μὲν πολύλοπα καὶ πολυχίτωνα καὶ λιπαρὰ καὶ δριμέα καὶ πικρὰ καὶ ὀστώδη καὶ ξηρὰ

---

[1] As soon as dropped by the parent plant.

[2] *Cf. HP* 8 11. 5: "For each kind of seed has a definite length of life for reproduction."

[3] *Cf.* Aristotle, *On the Generation of Animals*, ii. 3 (736 a 33–35): ". . . for the seeds and fetations of animals live no less than those of plants, and are fertile up to a point"; ii. 3 (736 b 8–12): "Now we must lay it down that seeds and fetations (evidently the ones not separated from the parent) possess the nutritive soul potentially, but not actually, until, like the fetations that are separated, they attract their food and do the work of such a soul"; ii. 3 (736

Death of the Seed in Storage
(1) *Natural*

18.1 There are also certain kinds of death of the seeds when they are not sown in the natural way,[1] of the seeds of trees equally with those of cereals and the rest. In each case the seed lasts a certain definite time, since all seeds live up to a point and then perish[2] (by "live" I mean potentially).[3] They perish in the natural way by drying out and by the loss through dissipation (as it were) of their heat along with their moisture.[4]

(2) *Unnatural*

All the other forms of death are unnatural, as when the seeds breed worms and liquefy and depart from their nature in some other way.

18.2 This is why[5] seeds with many peels and coats and seeds that are oily, pungent, bitter, bony and

b 14–15): ". . . they [the seeds and fetations] must have all the souls [*i.e.* the nutritive, the sensitive and the intellectual] potentially before they have them actually."

[4] *Cf.* the description of the natural death of a plant (*CP* 5 11. 1).

[5] The presence of unnatural forms of death means that seeds protected against them live out more of their natural span (the "definite time" of the preceding chapter), and that seeds without such protection perish prematurely.

πάντα πολυ<χρόνια, τὰ δ’ ὀλιγο>χρόνια,[1] ταχὺ γὰρ[2] ἐξίσταται, τὰ μὲν γὰρ ὑπ’ ἀλλήλων θερμαινόμενα, καθάπερ ὁ σῖτος καὶ τὰ χεδροπά, τὰ δ’ ὑπὸ τοῦ ἀέρος καὶ τῶν ἔξωθεν ὑγραινόμενα, καθάπερ τὰ τῶν λαχάνων καὶ τῶν στεφανωμάτων, οἷον γὰρ μαδᾷ καὶ εἰς διαβλάστησιν ὁρμᾷ.[3] ζῳοῦται δὲ θᾶττον[4] καὶ φθείρεται τῶν γε[5] χεδροπῶν τὰ τεράμονα· γλυκύτερα γάρ (ἐν τούτῳ δὲ ἡ ζῳοποιία), καὶ ἅμα θᾶττον ἐξίσταται δι’ ἀσθένειάν τε, καὶ διὰ[6] τὸ ὥσπερ ἐν πέρατι εἶναι· καὶ τοῦ σίτου δ’ ὡσαύτως ὁ γλυκύτερος.

18.3 μέγα δ’ οἱ τόποι διαφέρουσιν εἰς φυλακήν, ἐὰν ὦσιν ξηροὶ καὶ ψυχροί· διαμένουσιν γὰρ πλείω χρόνον, ὥσπερ ἔν τε Μηδίᾳ[7] καὶ Παφλαγονίᾳ, καὶ

[1] ego (*diu perdurant, reliqua* Gaza: πολυχρόνια Itali: πολυχρόνια· τὰ δ’ ἄλλα Schneider): πολύχρεια U.
[2] [γὰρ] Gaza, Schneider.
[3] ego (ὁρμᾶται Wimmer: ὁρμᾷ· τὰ δ’ οὐ μυδᾷ Schneider): οὐ μαδᾶ U.
[4] Gaza (*celerius*): θάτερον U.
[5] ego (Wimmer deletes): δε U. [6] u: δι U.
[7] ego: μηδεία U.

---

[1] *Cf. CP* 4 15. 3.
[2] *Cf. HP* 8 11. 1: “Some seeds sprout well but soon decompose, as bean, the readily cooked seed more than the other . . .”

dry are all long-lasting,[1] whereas the rest are of short duration, since they quickly depart from their nature, some being heated by one another (like cereals and legumes), some being liquefied by the air and things outside the pile of seeds (like the seeds of vegetables and coronary plants), since they become sodden (as it were) and have the impulse to sprout. Among legumes the readily cooked seeds breed worms sooner and perish,[2] for they are sweeter, and sweetness breeds animals; then too, they depart sooner from their nature both by reason of weakness and because they have (as it were) come to a limit.[3] The same holds of sweeter cereals.

18.3 The country is much better for keeping seeds if it is dry and cool. For seeds last longer then, as in Media and Paphlagonia,[4] and in these countries

[3] *Cf. CP* 4 12. 12.

[4] *Cf. HP* 8 11. 5–6 (of cereals): "For quick sprouting and for sowing in general the seeds one year old are considered best, those two and three years old inferior, and those still older practically infertile, although good enough for food; for each kind of seed has a definite life-span for reproduction. Yet seeds too vary in their powers because of the localities where they are stored. Thus at a certain spot in Cappadocia called Petra they are said to remain fertile and useful for sowing for sixty or seventy years . . . For the spot (they say) is not only elevated and well-ventilated and with constant breezes, but the inhabitants enjoy winds from east, west and south. In Media too and other elevated countries they say the seeds keep a long time in storage."

τούτων ἐν τοῖς ὑψηλοτάτοις χωρίοις, ὡσαύτως δὲ καὶ εἴ που ἄλλοθι τοιοῦτον· ἀμφότερα γὰρ ἐξείργονται τὰ φθείροντα, τό τε θερμὸν καὶ <τὸ>[1] ὑγρόν· ἐπεὶ καὶ ἡ διαπαττομένη γῆ τοῦτο ποιεῖ· ξηραίνει τε <γὰρ>[2] καὶ ψύχει.

18.4 τῶν δὲ δενδρικῶν ὅσα μὲν μαλακὰ καὶ σαρκώδη (καθάπερ ἡ ἀμυγδάλη[3] καὶ τὸ κάρυον καὶ ἡ βάλανος) τοῖς περιέχουσιν σῴζεται, τὰ δὲ ξηρὰ καὶ ξυλώδη (καθάπερ τὸ γίγαρτον καὶ τὰ τοιαῦτα) καὶ ἑαυτοῖς. ὡς δ' ἁπλῶς εἰπεῖν, πάντα πολυχρονιώτερα διὰ τὸ περιέχον· ἐπεὶ καὶ τὸ γίγαρτον καὶ ἡ κεγχραμὶς καὶ τὰ ἄλλα τὰ τοιαῦτα πολλῷ μᾶλλον ἀποξηραίνεται γυμνούμενα. μάλιστα δὲ διαμένει τῶν τοιούτων ὅσα πυρῆνι περιέχεται (καθάπερ τὸ τῆς ἐλάας) καὶ εἴ τι ξυλῶδες ἢ ὀστῶδες τυγχάνει (καθάπερ τὸ τοῦ φοίνικος καὶ ὁ κνῆκος καὶ τὰ ἄλλα τὰ κνηκώδη)· πυκνὰ γὰρ πάντα καὶ προ-

[1] Itali.
[2] aP.
[3] u N : -δαλῆ U aP.

---

[1] *Cf. HP* 8 11. 7: "There is again a sort of earth in some countries that is sprinkled over the wheat and preserves it, as the earth at Olynthus and at Cerinthus in Euboea . . ."

in the most elevated places, and similarly in other such places elsewhere, since here both of the destroying agents are kept out, heat and fluid. Indeed even the soil that is sprinkled on the seeds does this, for it dries and cools them.[1]

18.4 Among tree seeds[2] all that are soft and fleshy (such as almond, filbert and acorn) are preserved by the container,[3] whereas dry and woody seeds (as the grape-pit and the like) are also preserved by themselves; still, all seeds (broadly speaking) are made to last longer by the container,[4] since the grape-pit too and the fig-seed and the rest of this character are much more apt to dry out when stripped of the pulp. Of such[5] seeds those last best that are contained in a stone (like olive seed) and any that are like wood or bone[6] (like the date pit and safflower-seed and safflower-like seeds),[7] since all such structures are

[2] *Cf. CP* 4 1. 2 (with the note); 4 2. 1.

[3] The shell.

[4] The shell or pulp.

[5] That is, "bare" or "by themselves" (*HP* 1 11. 3): considered apart from any pulp or shell they may once have had.

[6] That is, where the seed inside is apparently not softer in texture than the woody or bony substance in which it is imbedded.

[7] *Cf. HP* 1 11. 3.

βολὴν ἔχοντα· τὸ δὲ σπέρμα τὸ ἐντὸς ὁτὲ μὲν κεχωρισμένον τι καὶ φανερόν, ὁτὲ δὲ ἀχώριστον καὶ ἀφανές (ὥσπερ τὸ τῶν φοινίκων). [περὶ δὲ χυλῶν καὶ ὀσμῶν ἐπειδη και ταῦτα τῶν φυτῶν οἰκεῖα] [1]

[1] aP (= opening of *C.P.* 6).
subscription: *θεοφράστου περι φυτῶν αἰτιων ε̄* U.

close in texture and constitute a shield. The seed inside is sometimes separate and plain to see, but sometimes unseparated and invisible, like that of the date-palm.[1]

[1] *Cf. HP* 1 11. 3.

Z

ΠΕΡΙ ΧΥΛΩΝ ΚΑΙ ΟΣΜΩΝ

1.1 περὶ δὲ χυλῶν καὶ ὀσμῶν, ἐπειδὴ καὶ ταῦτα τῶν φυτῶν οἰκεῖα, πειρατέον ὁμοίως ἀποδοῦναι τοῖς πρότερον τά τε συμβαίνοντα περὶ ἕκαστον εἶδος καὶ διὰ τίνας αἰτίας.

ἡ μὲν οὖν φύσις ποία τις ἑκατέρου[1] τοῦ γένους ἐν ἄλλοις ἀφώρισται, καὶ ὅτι μικτά πως ἄμφω

§1.1: Aristotle, *On Sense*, iv–vi.

[1] Gaza, Moreliana : ἑκατέρα U.

---

[1] U subscribes this book (and the whole extant *CP*) "Theophrastus on the Causes of Plants." *HP* 1 12. 1 and 1 12. 4 refer to the present discussion: "All these matters (*sc.* the kinds of flavour and odour and what odours are associated with what flavours) will be treated in greater detail when we discuss flavours, enumerating their species, the differences between flavours, and the nature and power of each" (*HP* 1 12. 1); "But we must endeavour to see the causes of these things (*sc.* differences of odour and

BOOK VI

FLAVOURS AND ODOURS[1]

Touching flavours[2] and odours, since these too belong to plants,[3] we must endeavour to set forth, just as in the preceding discussions, what happens with each type and for what reasons.[4] 1.1

Flavour and Odour: Their Nature

The nature of each of the two things has been distinguished elsewhere,[5] to this effect: both are mixed

flavour in different parts of the same plant) and the like later" (*HP* 1 12. 4).

[2] Editors (following Galen, *De Simpl.* i. 38 [vol. xi, pp. 449. 15–450. 3 Kühn]) carry through a distinction between *chymós* (the object of taste) and *chylós* (a plant juice). We follow the manuscript (U), translating *chylós* as "flavour" (or "flavour-juice"), *chymós* as "savour."

[3] *Cf.* Aristotle, *On Sense*, iv (442 b 23–26): "We have now discussed the gustible and savour; for the other affections of savours are properly investigated in the treatment of nature that is concerned with plants."

[4] *Cf.* Plato, *Phaedrus*, 270 C 10–D 7; 271 B 4–5, D 5–7.

[5] Aristotle, *On Sense*, v–vi. *Cf.* Theophrastus, *On Odours*, chap. i. 1: "Odours as a whole class come from mixture, like savours . . ."

κατὰ λόγον ἐστί· χυλὸς μὲν ἡ τοῦ ξηροῦ καὶ γεώδους[1] *τῷ ὑγρῷ ἐναπόμιξις,*[2] *ἢ ἡ <διὰ>*[3] *τοῦ ξηροῦ [δια]*[3] *τοῦ ὑγροῦ διήθησις ὑπὸ θερμοῦ (διαφέρει δ' ἴσως οὐδέν)· ὀσμὴ δὲ τοῦ ἐγχύλου*[4] *ξηροῦ τῷ διαφανεῖ (τοῦτο γὰρ κοινὸν ἀέρος καὶ ὕδατος)· καὶ σχεδὸν τὸ αὐτὸ πάθος ἐστὶ χυλοῦ τε καὶ ὀσμῆς, οὐκ ἐν τοῖς αὐτοῖς δ' ἑκάτερον.*[5]

ταῦτα μὲν οὖν οὕτω κείσθω κατὰ τὸν[6] *εἰρημένον ἀφορισμόν.*

1.2 *τὰ δ' εἴδη τῶν χυλῶν, ὡς μὲν εἰς ἀριθμὸν ἀπο-*

[1] *καὶ γεώδους* U[c] in an illegible erasure (*δια τοῦ ὑγροῦ* from the next line?).
[2] U : *ἐναπόμορξις* u.
[3] ego.
[4] ego : *ἐγχύμου* Beare : *ἐν χυλῶ* U.
[5] u : *ἑκάτερων* U.
[6] u : *το* U.

---

[1] This looks like Theophrastus' own addition; the ratio in savours in Aristotle, *On Sense*, is different: *cf.* note 1 on *CP* 6 6. 1. But *cf.* Aristotle, *On the Soul*, iii. 9 (426 a 27–b 7).

[2] Aristotle, *On Sense*, iv (441 b 15–19): "Then just as people who wash off colours and savours (*sc.* to obtain a dye or a juice) in the fluid cause the water to possess that colour or savour, so nature washes off the dry and earthy and, straining the fluid through the dry and earthy and

(with certain specifications) in a ratio,[1] flavour being the intermixture in what is fluid of the dry and earthy (or the straining through the dry of the fluid by heat[2]; it makes perhaps no difference), odour the intermixture in the transparent of the flavoured dry ("transparent" applying in common to air and water)[3]; and what has happened in flavour and odour is (one may say) the same, but has not happened in the two cases in the same things.[4]

Let these points, then, be laid down in conformity with the foregoing distinction.

The Different Species of Flavour and Two Ways of Defining Them

It is easy to give the species of flavours so far as 1.2

moving it by heat, gives the fluid a certain quality." *Cf.* Plato, *Timaeus*, 59 E 5–60 A 1 (when water is filtered through plants we speak of savours).

[3] Aristotle, *On Sense*, v (442 b 27–443 a 2): "One must think in the same way (*sc.* as was done about savour) about odours as well, for what the dry does in the fluid, this the savorous fluid does in another kind of thing, in air and water alike. We now apply the common term 'transparent' to these; but it is odorable not in its character of being transparent, but in that of washing and scouring off savorous dryness."

[4] Aristotle, *On Sense*, iv (440 b 28–30): ". . . we must speak of odour and savour. For the thing that has happened is (one may say) the same, but has not happened in the two cases in the same things."

δοῦναι, ῥᾴδιον, οἷον γλυκύς, λιπαρός, αὐστηρός, στρυφνός, δριμύς, ἁλμυρός, πικρός, ὀξύς· ὡς δὲ κατὰ τὴν οὐσίαν ἑκάστου, χαλεπώτερον· αὐτὸ γὰρ τοῦτο πρῶτον ἔχει τινὰ σκέψιν, πότερα[1] τοῖς πάθεσι τοῖς κατὰ τὰς αἰσθήσεις ἀποδοτέον, ἢ (ὥσπερ Δημόκριτος) τοῖς σχήμασιν ἐξ ὧν ἑκάστοις[2] (εἰ μὴ ἄρα καὶ ταῦτα συνάπτει πως εἰς τὰς δυνάμεις, κἀκείνων λέγεται χάριν), ἢ[3] καὶ εἴ τις ἄλλος τρόπος ἐστὶν παρὰ τούτους.

1.3 λέγω δὲ τοῖς πάθεσιν τοῖς κατὰ τὰς αἰσθήσεις, οἷον εἴ τις ἀποδοίη[4]·

[1] ego (πότερον Scaliger) : ποτε | γὰρ U : πότερον γὰρ u.
[2] U : ἕκαστοι Gaza (*singuli*), Itali.
[3] Gaza (*aut*), Schneider : εἰ U.
[4] u aP (ἀποδοῖν N) : ἀποδύει U.

---

[1] In Greek *austērós*, used of dry wine.

[2] Here and at *CP* 6 4. 1 Theophrastus avoids listing the bitter (Aristotle's "privative" flavour) last. For Aristotle's arrangement *cf. On the Soul*, ii. 10 (422 b 10–14): "As for the species of savours, just as with those of colours, the opposite ones are unmixed, sweet and bitter; the oily

the number goes: sweet, oily, dry-wine,[1] astringent, pungent, salty, bitter and acid[2]; it is harder to differentiate them by their essence. For at the very beginning there is a point that involves some investigation: the question whether we are to account for the differences by the different effects when the flavours are tasted, or (as Democritus does) by the several shapes out of which they are composed (unless the shapes are bound up in a certain way with the production of effects and are introduced to account for them),[3] or else by yet some other way there may be of accounting for the difference.

A Differentiation of Savours by their Effect on the Sensorium

1.3 By explaining the differences by the different effect when we taste them I mean such account as

comes next to the sweet, the salty to the bitter; and in between there are the pungent, dry-wine, astringent and acid. These, one may say, are regarded as the varieties of savours"; *On Sense*, iv (442 a 17–19): "The oily savour then is a savour belonging to the sweet, the salty and bitter are about the same, and the pungent, dry-wine [so in the passage just cited; but some MSS of *De Sensu* give "dry-wine, pungent"], astringent and acid are intermediate." *Cf. CP* 6 6. 1 note 1.

[3] *Cf. CP* 6 2. 1.

γλυκὺν μὲν τὸν διακριτικὸν τῆς ἐν τῇ γλώττῃ συμφύτου ὑγρότητος, ἢ χυμὸν λεαντικὸν ἢ λεπτὸν ἢ λεῖον·

στρυφνὸν δὲ τὸν ξηραντικὸν ἢ πηκτικὸν[1] *ἠρέμα ταύτης·*

δριμὺν δὲ τὸν [τηκτικὸν ἢ][2] *τμητικόν,*[3] *ἢ ἐκκριτικὸν τῆς ἐν τῇ συμφύτῳ ὑγρότητι θερμότητος εἰς τὸν ἄνω τόπον, ἢ ἁπλῶς χυμὸν καυτικὸν ἢ θερμαντικόν·*

ἁλμυρὸν δὲ τὸν δηκτικὸν καὶ ξηραντικόν·

πικρὸν δὲ τὸν φθαρτικὸν τῆς ὑγρότητας, ἢ τηκτικὸν[4] *ἢ δηκτικὸν*[5] *ἢ ἁπλῶς τραχὺν ἢ μάλιστα τραχύν·*

[1] u : πικτικὸν U. [2] ego : πηκτικὸν ἢ Heinsius.

[3] ego (δηκτικὸν Gaza [*mordendi*], Scaliger) : τακτικὸν U^c aP (τὸ κτικὸν N) : τηκτικὸν U^{ac}.

[4] Gaza (*liquescere facit*), Heinsius : πηκτικὸν U^c (η from ι).

[5] u : δι- U.

---

[1] Since Aristotle gives no details of the effect of savours on the sense-organ Theophrastus adapts from Plato (non-committally) the following series of effects. Two of the eight savours of *CP* 6 1. 2 are omitted, the oily and the acid, the latter perhaps by a fault of transmission. The oily is also omitted by Plato.

In Aristotle *Categories*, viii (9 a 28–b 9) the "passible qualities" sweetness, bitterness, astringency and the like (that is, the remaining savours), together with heat and cold, are said to be called "passible" not because the reci-

this[1]:

(1) Sweet: the savour with the capacity to expand the native fluid of the tongue,[2] or the savour with the capacity to make smooth or that has fine particles or is smooth[3];

(2) Astringent: the one with the capacity to desiccate or to solidify this fluid gently[4];

(3) Pungent: the one with the capacity to cut, or to separate out the heat in the native fluid into the region above, or simply the savour with the capacity to burn or heat[5];

(4) Salty: the one with the capacity to irritate and desiccate[5];

(5) Bitter: the one with the capacity to corrupt the fluid, or to melt or irritate, or simply the savour that is rough or roughest[5];

pients have undergone an affection, but because each in the process of being perceived produces an affection: so sweetness produces an effect in us when we taste it.

[2] *Cf.* Plato, *Timaeus*, 65 B 7–C 7 (the sweet has the contrary effect to all the rest and restores the vessels of the tongue to normal, by dilation if they have been contracted, by contraction if they have been dilated).

[3] *Cf.* Plato, *Timaeus*, 65 C 3 (it smooths the parts that have been roughened, coating them over). From this Theophrastus infers that it has fine particles.

[4] *Cf.* Plato, *Timaeus*, 65 C 6–D 4 (bodies that enter the vessels of the tongue and by becoming dissolved contract and desiccate them are when rougher astringent, when less rough in their effect of the dry-wine savour).

[5] *Cf.* Plato, *Timaeus*, 65 D 4–66 A 2, cited in note 3 on *CP* 6 1. 4–5.

αὐστηρὸν δὲ τὸν ῥυπτικὸν τῆς αἰσθήσεως ἢ τῆς ὑγρότητος τῆς ἐν αὐτοῖς[1] ἢ τῆς ἐπιπολῆς ὑγρότητος, <ἢ>[2] δηκτικὸν ἢ πηκτικὸν[3] ἢ ξηραντικόν, ἢ ἁπλῶς στρυφνότητά τινα ἠρεμαίαν[4] καὶ μαλακήν.

1.4 ἢ πάλιν, ὡς Πλάτων καθόλου τὰς διαφορὰς τῶν δυνάμεων ἀποδίδωσι τῷ συγκρίνειν καὶ διακρίνειν, κεχρημένας[5] τραχύτητι καὶ λειότητι, καθ' ἑκάστας δὲ διαιρεῖ τοῖς εἴδεσιν·

§1.4–5: Plato, *Timaeus* 65 C 1–66 C 7; Theophrastus, *De Sensibus*, lxxxiv (p. 525. 5–11 Diels).

[1] U : *αὐτῇ* u.
[2] ego.
[3] N aP (*πηκτικτόν* u) : *πικτικτον* U.
[4] u aP (*ἠ-* N) : *ἠρεμίαν* U. [5] u : *καὶ χρημένας* U.

---

[1] The vessels, which are the instruments of tasting of the tongue and extend to the heart (Plato, *Timaeus*, 65 C 7). Theophrastus has little interest in them and does not stop to explain.

[2] Plato, *Timaeus*, 65 C 1–6: "First then let us make clear so far as may be all that we left out in our earlier mention of savours, which are peculiar affections on the tongue. These too, like most effects, appear to arise through certain contractions and dilations, but in addition to these to make a somewhat wider use than the rest do of degrees of roughness and smoothness."

(6) Dry-wine: the one with the capacity to scour the sense-organ (or the fluid in them[1] or the fluid on the surface), or to irritate or solidify or desiccate, or simply a gentle and soft kind of astringency.

Plato's Account of the Differences

1.4 Or again, there is Plato's way of assigning the differences of the powers in general to their causing contraction and dilation, letting them make use of roughness and smoothness,[2] and for each of these distinctions of power distinguishing a different savour[3]:

[3] What follows is an abbreviated citation of Plato, *Timaeus*, 65 C 6–66 C 7: "For all things that, entering at the little vessels (which are as it were instruments of testing of the tongue, extending to the heart), falling upon the moist and tender parts of the flesh as earthy particles, by their own melting contract the little vessels and dry them out, when rougher appear astringent. But when they roughen less they appear with the dry-wine taste; and the bodies that both scour these vessels and wash off the whole region of the tongue, when they do this beyond the measure and go so far as to melt away some of the very nature, such as the powerful sodas, are all called under these circumstances bitter, whereas the bodies that fall short of the condition of soda, and make moderate use of their scouring, appear salty without rough bitterness and rather pleasing than otherwise to us. The particles that after having shared in the heat of the mouth are smoothed by the mouth, getting ignited by the heat and in turn themselves burning what heated them, and darting

ὅσα μὲν οὖν συνάγει[1] τὰ φλέβια καὶ ἀποξηραίνει, τραχύτερα μὲν ὄντα,[2] στρυφνά, <ἧττον δὲ τραχύνοντα>,[3] αὐστηρά· τὰ δὲ[4] τούτων τε ῥυπτικὰ[5] καὶ πᾶν τὸ περὶ τὴν γλῶτταν ἀποπλύνοντα. ὅσα δὲ ῥύπτει πέρα μὲν τοῦ μετρίου, ὥστε καὶ ἀποτήκειν αὐτῆς τι τῆς φύσεως, οἷον ἡ τῶν νίτρων δύναμις, πικρά· τὰ δὲ ὑποδεέστερα τούτων, καὶ ἐπὶ τὸ μέτριον τῇ ῥύψει χρώμενα, ἁλυκά, ἄνευ πι-

1.5 κρότητος τραχείας, καὶ φίλα μᾶλλον ἡμῖν. τὰ δὲ τῇ τοῦ στόματος[6] θερμότητι κοινωνήσαντα καὶ λεαινόμενα, καὶ συνεκπυρούμενα καὶ πάλιν ἀντικάοντα, φερόμενα δ' ὑπὸ κουφότητος

[1] Plato : κάει U.
[2] τραχύτερα μὲν ὄντα Plato : τραχύνοντα U.
[3] Plato. [4] τὰ δὲ ego : δὲ τὰ U.
[5] τε ῥυπτικὰ Plato : ῥυπτικώτερα U. [6] u : σπέρματος U.

(*continued*)
by their lightness upward to the senses of the head, and cutting whatever they encounter, because of these powers all such particles have been called pungent. The effect caused on the other hand by the particles reduced to fineness by decomposition, and entering the narrow vessels, and possessing an effectiveness adjusted to the earthy and airy particles present there, so as to set them in motion and be stirred up round one another, and in

Now the particles that contract the vessels and dry them out, when rougher, are (1) astringent, but, when their roughening is less, are (2) of the dry-wine taste. Others both scour the vessels and wash off the whole surface of the tongue: all that scour to excess, so that they melt away some of the very nature of the part, as do the powerful sodas, are (3) bitter; those that do not attain to such vigour and do not overdo their scouring, are (4) salty, and are pleasant to us rather than otherwise. Those that after sharing in the heat of the mouth become fine and are ignited by that heat and in turn burn the mouth, and by their lightness dart 1.5

this stirring to assume a circular stance (?), and entering other (parts?) to form other hollow structures that stretch as a film around the entering bodies, these . . . become hollow bodies of water . . . , and some . . . are called bubbles, others . . . frothing and fermentation, and the cause of all these effects is termed acid. An effect contrary to all those mentioned about these particles comes from a contrary reason: when the formation in liquids of the entering particles, being of a nature akin to the state of the tongue, smooths the roughened parts by coating them over, and takes the parts that have been contracted or expanded unnaturally and brings together the ones, but loosens the others, and establishes all so far as possible in their natural state, all such remedying of violent effects turns out pleasant and acceptable to everyone and is called sweet."

ἄνω πρὸς τὰς τῆς κεφαλῆς αἰσθήσεις καὶ τέμνοντα, δριμέα· τὰ δὲ προλελεπτυσμένα[1] ὑπὸ σηπεδόνος, εἰς δὲ τὰς στενὰς φλέβας εἰσδυόμενα, καὶ ἀνακυκῶντα καὶ ἀναζυμοῦντα καὶ ἀφρίζειν ποιοῦντα καὶ ἕλκειν, ὀξέα· τὸ δὲ σύμπασι τοῖς περὶ ταῦτα πάθος ἐναντίον οἰκεῖόν ἐστι τῆς γλυκύτητος· ὅταν λεαίνῃ ἐπαλεῖφον τὰ τραχυνθέντα, καθιστῇ[2] δὲ καὶ συνάγῃ[3] <τὰ>[4] παρὰ φύσιν κεχυμένα, τὰ δ' αὖ συνεστῶτα διαχαλᾷ, καὶ ἁπλῶς ὅτι μάλιστα ἱδρύῃ[5] κατὰ φύσιν, γλυκύ.

οὗτος μὲν οὖν ὁ τρόπος ταῖς δυνάμεσιν ἀφορίζει.

1.6 Δημόκριτος δέ, σχῆμα περιτιθεὶς ἑκάστῳ, γλυκὺν μὲν τὸν στρογγύλον καὶ[6] εὐμεγέθη ποιεῖ·

§1.6: Democritus, frag. A 129 (Diels-Kranz, *Die Fragmente der Vorsokratiker*, vol. ii[8], p. 113. 19–26; Theophrastus, *De Sensibus*, 65–67.

[1] Plato (Heinsius) : *προσλελεπτυμένα* U^{ar} : *προλ-* U^{r}.
[2] ego : *καθήστι* U : *καθίστησι* u N : *καθιστᾶ* aP.
[3] P : *-ει* U N a. [4] u. [5] Plato (Schneider) : δριμὺ ἢ U.
[6] u : *στρογγυλακαι καὶ* U.

upward to the senses of the head, cutting their way, are (5) pungent. And those that after being reduced to fine particles by decomposition enter the narrow vessels and stir them up and cause ferment and make them foam and warp are (6) acid. The effect contrary to all the effects here is what belongs to (7) sweetness: what smooths the roughened parts by coating them over and settles and brings together the parts unnaturally expanded, and again breaks up the parts that have come together, and in a word establishes them to the greatest degree in their natural state, is sweet.

This way, then, of accounting for the different savours distinguishes them by their powers.

Democritus' Explanation of the Differences by Figures

1.6 Democritus on the other hand assigns a figure to each, and makes

(1) the round and sizable savour sweet[1];

[1] *Cf.* Theophrastus, *On the Senses*, lxv: "The sweet savour is composed of shapes that are curved and not very small; hence in general they relax the body, and do not accomplish their effects with violence or all of them quickly."

στρυφνὸν δὲ τὸν μεγαλόσχημον, τραχὺν δὲ καὶ[1] πολυγώνιον καὶ ἀπεριφερῆ·

ὀξὺν δέ, κατὰ τοὔνομα, τὸν ὀξὺν τῷ ὄγκῳ καὶ γωνοειδῆ[2] καὶ καμπύλον καὶ λεπτὸν καὶ ἀπεριφερῆ·

δριμὺν δέ, τὸν περιφερῆ καὶ λεπτὸν καὶ γωνοειδῆ καὶ καμπύλον·

ἁλμυρὸν δέ, τὸν γωνοειδῆ καὶ εὐμεγέθη καὶ σκολιὸν καὶ οὐ σκαληνῆ·[3]

πικρὸν δέ, τὸν περιφερῆ καὶ λεῖον, ἔχοντα σκολιότητα, μέγεθος δὲ μικρόν·

[1] ego (τε καὶ Schneider) : δὲ τὸν U.

[2] Wimmer (γωνιοειδῆ Schneider) : κωνοειδῆ U.

[3] οὐ σκαληνῆ ego (*cf. CP* 6 10. 3 σκαληνές) : ἰσοσκαλῆ U : ἰσοσκελῆ u.

---

[1] *Cf. ibid.* lxvi: "The astringent savour is composed of shapes large and with many angles and is round least of all. For when these enter bodies they shut off the little vessels by blocking them up and so prevent fluids from collecting, and this is why they stop the bowels."

[2] *Oxýs* (literally "sharp") is the Greek for "acid."

[3] *Cf. ibid.* lxv: "Thus the acid savour is angular in shape with many wrinkles, small and thin. For owing to its pungency it penetrates everywhere quickly, and since it is rough and angular it effects a contraction and tight-

(2) the one that consists of large figures, but is rough, many-angled and non-curved, astringent[1];

(3) the savour that is "sharp" in its bodily conformation (as befits its name [oxýs])[2] angular, wrinkled, thin and non-curved, acid[3];

(4) the curved, thin, angular and wrinkled, pungent[4];

(5) the angular, sizable, crooked and with non-interlocking irregularities, salty[5];

(6) the curved and smooth, crooked but small in size, bitter[6];

ening, and hence heats the body by producing empty spaces in it; for what has the greatest amount of empty space is most heated."

[4] *Cf. ibid.* lxvii: "The pungent savour is small, round and angular and has no interlocking irregularities."

[5] *Cf. ibid.* lxvi: "Salty is the savour composed of corpuscles large and not round, but in some cases with interlocking irregularities, and for this reason not much wrinkled either (he means by 'interlocking irregularities' corpuscles that have irregular surfaces that fit one another and so become attached); large, because brine is on the surface (if they were small and kept being struck by what envelops them they would get mixed in with the whole body); not round because what is salty is rough, whereas what is round is smooth; and without interlocking irregularities because they do not form a plug, and this is why salt is friable." (*Cf.* 6 10. 3).

[6] *Cf. ibid.* lxvi: "The bitter savour is composed of small, smooth and round corpuscles, with a roundness that also contains wrinkles; this is why it is viscous and gluey."

λιπαρὸν δέ, τὸν λεπτὸν καὶ στρογγύλον[1] καὶ μικρόν.

οἱ μὲν οὖν τρόποι τοσοῦτον διαφέρουσιν.

2.1 τάχα δ' ἂν δόξειεν (ὥσπερ ἐλέχθη) καὶ οὗτος ἐκείνων εἶναι χάριν· αὐτῶν γὰρ τῶν δυνάμεων, οὕτως ἀποδιδούς, οἴεται τὰς αἰτίας ἀποδιδόναι δι' ἃς ὁ μὲν στύφει καὶ ξηραίνει καὶ πήγνυσιν, ὁ δὲ λεαίνει καὶ ὁμαλύνει καὶ καθίστησιν, ὁ δὲ ἐκκρίνει καὶ διαχεῖ καὶ ἄλλο τι τοιοῦτο δρᾷ.

πλὴν ἴσως ἐκεῖνα ἄν τις ἐπιζητήσειε παρὰ τούτων, ὥστε καὶ τὸ ὑποκείμενον ἀποδιδόναι ποῖόν τι· δεῖ γὰρ εἰδέναι μὴ μόνον τὸ ποιοῦν, ἀλλὰ καὶ τὸ πάσχον, ἄλλως τ' εἰ καὶ μὴ πᾶσιν ὁ αὐτὸς ὁμοίως φαίνεται (καθάπερ φησίν)· οὐθὲν γὰρ κω-

§ 2.1: Democritus, Frag. A 130 (Diels-Kranz, *Die Fragmente der Vorsokratiker*, vol. ii[8], p. 113. 27–35).

[1] u : στρογγυλοκαι (?) U.

---

[1] The oily savour is not mentioned *ibid.*; perhaps it has dropped out by accident.
[2] *CP* 6 1. 2.
[3] The astringent: *cf. CP* 6 1. 6 (number 2).
[4] The sweet: *cf. CP* 6 1. 5 (number 7).

(7) and the thin, round and small, oily.[1] To this extent, then, the ways of explanation differ.

The Theory Is Intended To Explain the Powers

2.1 But perhaps this latter way of explanation too would be considered (as we said)[2] to have the powers in view, since in accounting for the savours in this way Democritus believes that he is giving the reasons for the powers themselves, the reasons why one savour has the power of puckering, drying, and solidifying,[3] another that of making smooth and even and restoring to normal,[4] another that of separating out[5] and loosening[6] and so forth.

It does not Do So:
(a) *By Reason of an Omission*

Still one might perhaps require a fuller account of those actions from the proponents of this view, which would add the character of the subject operated upon,[7] since one must know not only the thing that acts, but the thing that is acted upon, especially if the same savour (as he says)[8] does not produce a similar impression on all (nothing

[5] As the pungent: *cf. CP* 6 1. 3 (number 3).
[6] As the sweet: *cf. CP* 6 1. 3 (number 1).
[7] *Cf.* Theophrastus, *On the Sense*, lxxii.
[8] *Cf. ibid.* lxiii.

λύει τὸν ἡμῖν γλυκὺν ἑτέροις τισὶ τῶν ζῴων εἶναι πικρόν, καὶ ἐπὶ τῶν ἄλλων δὲ ὁμοίως. δῆλον γὰρ ὡς ἐν τῷ αἰσθητηρίῳ τίς ἐστιν ἑτέρα διάθεσις, ἐπεὶ τό γε[1] τοῦ χυμοῦ σχῆμα ταὐτό,[2] καὶ ἐν ἄλλῳ ἔοικεν οὐ ταὐτὸ δύνασθαι[3] πανταχοῦ ποιεῖν. εἰ δὲ τοῦτο ἀληθές, αἰτίαν ὁμοιότητος εἶναι[4] τῶν ὑποκειμένων· διὸ δὴ καὶ λεκτέον ὑπὲρ αὐτῶν.

2.2 ἅμα δὲ κἀκεῖνο φανερόν, ὡς οὐ μία τις δύναμις τοῦ αὐτοῦ σχήματος, εἴ γε[5] τἀναντία πέφυκε ποιεῖν ἐν ἄλλῳ καὶ ἄλλῳ· τὸ μὲν γὰρ μὴ πάντα πρὸς τὴν αὑτοῦ[6] δύναμιν ἄγειν οὐχ ὁμοίως ἄτοπον (ὥσπερ οὐδὲ τὸ πῦρ κάειν)· εἰ δὲ καὶ ἐναντίως

[1] Schneider : τότε U. [2] ταυτὸ u : ταυτῶι U.

[3] Scaliger : δύναται U.

[4] αἰτίαν ὁμοιότητος εἶναι U : (*subiectorum*) *quoque similitudinem pro causa accepisse par est* Gaza : αἰτίαν τὴν ἀνομοιότητα (adding δεῖ ὑπολαβεῖν after ὑποκειμένων) Schneider : δεῖ τι ἀνομοιότητος εἶναι Wimmer.

[5] Heinsius (*quod* Gaza) : εἰ δε U.

[6] Moreliana : αὐτοῦ U.

---

[1] Such as solidifying–loosening, puckering–making even (*CP* 6 2. 1), or the presumably opposite actions that produce the opposite impressions of sweet and bitter (*CP* 6 2. *ad fin.*).

[2] A reference to the inability of the same figure to "carry out the same action . . . everywhere," mentioned three sentences earlier. *Cf.* also Aristotle, *On Sense*, iv (441 19–23): "And savour is this, the effect produced by

preventing the savour that is sweet to us from being bitter to some other animals, and similarly with the rest of the savours). For there is evidently here an unspecified difference of disposition in the second sense-organ, since the figure of the savour is the same, and appears not to have the power to carry out the same action in a new sense-organ everywhere. If this is true, it appears that a reason for sameness of action lies in the subjects, which is why they should be discussed.

2.2

At the same time this too is clear: the same shape has more than one power, since it is of a nature to carry out opposite actions[1] in two different sense-organs. That it should not bring everything to the actualizing of its own power[2] is not so great an oddity (any more than it is that fire should not burn things[3]); but if the same shape is of a nature[4] even

the aforementioned dry in that which is fluid, an effect with the capacity to alter potential tasting to actual tasting, for it brings what has the capacity of sensing to this (*sc.* to sensing), this being previously present potentially (since perceiving does not correspond to learning [*sc.* acquiring new knowledge] but to understanding [*sc.* applying knowledge already present; *cf.* Plato, *Euthydemus*, 277 E 5–278 A 7])."

[3] *Cf.* Theophrastus, *On Fire*, especially i. 4; ii. 10–11; iii. 20–29; iv. 34–36; v. 40–42; vi. 49.

[4] "To be of a nature" to do this or that implies positive action, not the mere failure to carry it out; *cf.* the use of "natural" and "nature" in the discussion of the "privative" flavours (*CP* 6 4. 2; 6 4. 3; 6 4. 4; 6 4. 6).

ἔνια διατιθέναι,[1] τοῦτο πλείονος λόγου δεῖται.

2.3 καίτοι κἀκείνου δεῖ τινα λέγειν αἰτίαν. ἐπὶ γοῦν τοῦ πυρός, ὅσα μὴ δύναται κάειν, ἢ τὸ ὕδωρ ὑγραίνειν, ἔστιν τις αἰτία καὶ λόγος· εἰ δὲ καὶ τοὐναντίον ἑκάτερον ἐποίει, μᾶλλον ἂν ἐδεῖτο καὶ πλειόνων.

τούτοις μὲν οὖν ὑπὲρ ἁπάντων τούτων λεκτέον· τοῖς δὲ κατὰ τὰς δυνάμεις ἐκεῖνο μόνον ἀναγκαῖον, ποῖόν τι τῶν αἰσθητηρίων ἕκαστον. δεῖ[2] γὰρ εἰδέναι καὶ τὴν τούτων φύσιν καὶ διάθεσιν, ἐπείπερ οὐχ ὁτιοῦν μόριον αἰσθητικόν· ἴσως δὲ καὶ ὅταν ἐκ πλειόνων ἕν τι γένηται, τὴν ἐκ θατέρου δύναμιν καὶ οὐσίαν οὐ δεῖ λανθάνειν.

ἄτοπον δὲ κἀκεῖνο τοῖς τὰ σχήματα λέγουσιν, ἡ

[1] U : διατίθεται Gaza (*efficiat*), Heinsius.
[2] Itali (*operae pretium est* Gaza) : ἀεὶ U.

---

[1] For the present discussion of "power" *cf.* Plato, *Phaedrus*, 270 C 10–D 7: "Must we not think as follows about any nature? First, whether the thing about which we are going to wish to be skilled ourselves and able to make others skilled is simple or of many kinds; next, if it is simple, we must consider its power, what power the nature naturally has with regard to what, in the direction of acting, or what power it has in the direction of being acted upon by what; and if it has several kinds, one must,

to bring about opposite states in certain things, the point needs fuller explanation than the mere failure of the shape to realise its power (and yet even for the last some cause should be given).[1] Thus with fire or water there is some cause and explanation for all cases where fire is unable to burn things or water unable to wet them. But if each performed the very opposite, there would be greater need for explanations and there would need to be more than one of them.[2]

2.3

The proponents of the theory of shapes, then, must discuss all these matters; the proponents of the theory of powers are here required only to discuss the character of the sense-organs in question, since we must also know the nature and disposition of these, inasmuch as not every part of the body is capable of sensation; again, perhaps, when a single result comes about through the interaction of a second participant we must not be unaware of the nature[3] and power of the new element.

(b) *By Reason of an Addition*

A further oddity for the proponents of the theory

after enumerating them, see in each what one saw in the unity, by what power it is of a nature to do what to what [τί <τί> D 5] itself, or by what to be done what to by what?"

[2] One must explain not only the failure to produce the one effect but also the production of the opposite.

[3] Literally "essence."

τῶν ὁμοίων διαφορὰ κατὰ μικρότητα καὶ μέγεθος

2.4 εἰς τὸ μὴ τὴν αὐτὴν ἔχειν δύναμιν. οὐ γὰρ ἔτι[1] τῆς μορφῆς, ἀλλὰ τῶν ὄγκων δυνάμεις·[2] ἃς[3] εἰς μὲν[4] τὸ διαβιάσασθαι (καὶ ἁπλῶς τὸ μᾶλλον καὶ τὸ ἧττον) τάχ' ἄν τις ἀπέδωκεν,[5] εἰς δὲ τὸ μὴ αὐτὸ[6] δύνασθαι μηδὲ ποιεῖν, οὐκ εὔλογον,[7] ἐπεὶ ἐν τοῖς σχήμασιν αἱ δυνάμεις. ᾗ[8] γὰρ ὅμοιος σχήματι,[9] ταὐτὸν [εἰ][10] τὸ ὑπάρχον, ὥσπερ καὶ ἐν τοῖς ἄλλοις· τὸ γὰρ τρίγωνον τὸ ποδιαῖον καὶ τὸ μυριόπουν ὁμοίως <ἴσον ἅπασαν>[11] [δύο ὀρθᾶν][12] καὶ [τὸ μυριάγωνον τέτταρσιν ὀρθᾶν· ἢ][13] τὸ

[1] ego : ἐπι U.
[2] ποιοῦσι τὰς (αἱ Wimmer) δυνάμεις Schneider.
[3] U : οὓς Schneider.
[4] Schneider : ἴσμεν U.
[5] ego (ἀποδῴη Schneider) : ἀποδέδωκεν U.
[6] τὸ αὐτὸ Schncider.
[7] u : εὔλον (?) U.
[8] ego : ἥ U : εἰ u.
[9] ego : ὁμοῖοσχήματε U : ὁμοιοσχήματα u N (-μονα aP).
[10] ego : εἴη u N : ἂν εἴη aP. (εἰ was a variant for ἥ five words before.)
[11] ego.
[12] ego : δ ο (from δι- U[cc]) ὀρθᾶν U : δύο ὀρθαῖς u : δύς ὀρθὰ N δυσὶν ὀρθαῖς aP.
[13] Wimmer.

---

[1] At *CP* 6 1. 6 the sweet is "sizable," the astringent has "large figures," the salty is again "sizable," and the bitter

of shapes is this: difference in size of similar shapes leads to their not possessing the same power.[1] For 2.4 here we have powers belonging no longer to the shape but to the bulk. Now one might perhaps have assigned them here to let them work to the point of making action violent and in a word different in degree; but to let them work to the point of stripping the figure itself of all power and action is not reasonable, since it is in the figures that the powers[2] reside. For insofar as the bulk is similar in figure, so far is its property the same, just as with the other similars of varying size. So the one-foot and ten-thousand-foot triangle[3] and square are equal

and the oily are "small." With different sizes the shapes would no longer belong to the same savours. Elsewhere Theophrastus makes a different point (*On the Senses*, xviii), that Democritus relies on difference of size to explain light and heavy and soft and hard, but on shapes to explain hot and cold "and the rest."

[2] That is, the special powers that make the difference between one savour and another. This can be seen from Democritus' omission of size in his description of the acid and pungent savours (numbers 3 and 4, *CP* 6 1. 6).

[3] It does not matter which side of the triangle is measured, since the triangles are similar, and any side of the larger will be 10,000 times the corresponding side of the smaller. The triangle is the simplest (rectilinear plane) figure (*cf.* Plato, *Timaeus*, 53 C 7–D 1; Aristotle, *Nicomachean Ethics*, vi. 9 [1142 a 28–29]). Theophrastus adds the square because it yields at once the irrational ratio of diagonal to side.

τετράγωνον [τέτρασιν].[1] καὶ ταῦτα μέν, <ὄντα>[2] καί, κατὰ τὸ ποσόν, διάφορα·[3] τὰ δὲ λόγῳ, [μὲν ὄντα][2] ταὐτόν,[4] οἷον ἡ ἀσυμμετρία τῆς διαμέτρου, τὸ δὲ ποσόν, διάφορον. ὥστ᾽ ἐπεὶ[5] ἐκ τούτων (ἢ ἀπὸ τούτων) ἐγίνετο, τὸ μὲν ποσόν,[6] διάφορον ἂν ἦν, τῷ δ᾽ εἴδει, ἀδιάφορον.

ἀλλὰ γὰρ αἱ μεν κατὰ <τὰ>[7] σχήματα καὶ τὰς μορφὰς δυνάμεις[8] πλείω, καὶ περὶ πλειόνων, ἔχουσιν ἀπορίαν.

3.1 ὅτι δ᾽ ἐναπομείξει[9] πως ὁ χυλὸς καὶ ἡ ὀδμή, καὶ ἐκ τῶνδε φανερόν· οὐδὲν γὰρ φαίνεται τῶν ἁπλῶν ἔχειν χυλὸν οὐδ᾽ ὀδμήν, ἐὰν μὴ λάβῃ μῖξειν, οἷον ὁ ἀήρ, ἢ πῦρ, ἢ ὕδωρ (ἄχυλον γὰρ καὶ τὸ ὕδωρ καθ᾽ αὑτό, διὸ καὶ οἱ παλαιοί φασι δι᾽ οἵας

§ 3.1: *Cf.* Theophrastus, *On Odours*, i. 1.

[1] ego : -σι a : τέτταρσι N P.
[2] ego. [3] u : διαφορᾶ U. [4] ego : αὐτα U.
[5] U : εἰ Schneider. [6] u : -ῶν U.
[7] Wimmer. [8] u : -ει U. [9] U : ἐν ἀπομίξει u.

---

[1] That is, the two similar triangles have their corresponding angles equal.

[2] A similar figure is one with proportional sides and equal angles (Aristotle, *Posterior Analytics*, ii. 17 [99 a 11–15]); *cf.* also Euclid, *Elements* vi, definition 1: "Similar rectilinear figures are all that have their angles, taken

in every angle.[1] All this they are, although they are also different, in size. In matters of ratio,[2] on the other hand (like the incommensurability of the diagonal), they are the same thing, although, in size, different.[3] Hence, since the savour-body was made to consist of a figure[4] or of a combination of them,[5] it would have been a thing with differences in the matter of size, but none in form.

But we do not pursue this further: when one attaches powers to figures and shapes the difficulties increase in magnitude and number.[6]

Evidence that Flavour and Odour are Due to Intermixture

3.1 That flavour and odour are due in some way to intermixture is further evident from the following: none of the simple bodies is observed to have flavour or odour unless it has acquired an admixture, as air or fire or water (water too being flavourless when pure, which is why the ancients say that water has the character of whatever sort of earth it flows

singly, equal and the sides about the equal angles proportional." [3] The incommensurability of the diagonal is expressed by the irrational ratio of diagonal to side, the ratio being the same in all squares, whatever the length of the side. But the quantities entering into the ratio will differ with the size of the square.

[4] As with Democritus' sweet (number 1, *CP* 6 1. 6).

[5] As with numbers 2–7 in *CP* 6 1. 6.

[6] *Cf. CP* 6 2. 3, beginning of second paragraph.

ἂν γῆς ῥέῃ, τοιοῦτο καὶ εἶναι· καὶ ἡ θάλαττα δὲ καὶ τὰ νιτρώδη καὶ σαπρὰ καὶ ὀξέα τῶν ὑδάτων ἔχει τινὰ μῖξιν, ὧν ἐνίοις[1] καὶ ἡ ὀσμή, μάλιστα <δὲ>[2] τῇ θαλάττῃ, συνακολουθεῖ)· γῆ δὲ καὶ λίθος ἄχυλα, πλὴν ἐάν τινα τοιαύτην λάβῃ μῖξιν.

3.2 καίτοι φαίνεταί γε ταῦτα μᾶλλον ἔχειν, οἷον ὅσαι θ' ἁλμώδεις καὶ ὅσα[3] καὶ (ὡς ἡ τέφρα) πικρά,[4] καὶ ἔνιαι γλυκεῖαι δοκοῦσιν εἶναι (καθάπερ καὶ ἡ ἄργιλος)· ἅμα δὲ καὶ οὐκ ἄλογον, εἴπερ τῇ τοῦ γεώδους ἐναπομίξει γίνεται χυλός· τὰ δὲ δὴ

[1] u : ἐνιους U (-ί- N aP).
[2] Schneider.
[3] ego : ὅσοι U : ὅσαι u.
[4] U : -αί u.

---

[1] *Cf.* Aristotle, *On Sense*, iv (441 a 30–b 7): "As many savours as are in pericarpia are also observed to be present in the earth. This is why many of the old natural philosophers assert that water has the character of the earth through which it passes. And this is clear in the case of saline waters especially, for salt is a species of earth. And water that sifts through ashes, which are bitter, produces a bitter savour. And there are many springs, some bitter, some acid, and some possessing other savours of all kinds."

[2] *Cf.* Aristotle, *On Sense*, v (443 a 8–15): That the affec-

through; sea-water too and waters that are soda-like and decomposed and acid have a certain admixture,[1] and in some these flavours are accompanied by odour, especially in sea-water); earth and stone are flavourless, except when they have acquired some mixture of this sort.[2] Still these last appear to have flavour more than the other simple bodies do: so with the earths that are salty and bodies that (like ashes) are even bitter; and some earths are considered sweet (as clay). Then too it is not unreasonable that this should be so, inasmuch as flavour arises from intermixing the earthy.[3] Earths that

3.2

tion (*sc.* that characterises odour) is from what is savorous is clear from the things that have and do not have odour: the elements are odourless (as fire, air, earth, water) because both the dry and the fluid among them are savourless, except when something mixed with them produces it. This is why sea-water has odour (since it has savour and dryness), and lumps of salt have more odour than soda (this is shown by the oil that exudes from them), whereas soda is more exclusively composed of earth. Again stone is odourless (being savourless), whereas timber has odour (being savorous), and of timber the watery kinds have less." *Cf.* Theophrastus, *On Odours*, i. 1: "Odours as a class come from mixture, like flavours; for everything unmixed is odourless, just as it has no flavour. This is why the simple bodies have no odour, such as water, air, fire. Earth mainly or only has odour, and hence is most mixed."

[3] *Cf. CP* 6 1. 1 (flavour is the intermixture in the liquid of the dry and earthy).

μεταλλευόμενα καὶ λίθων ἔνια γένη καὶ ὀσμὰς ἔχει πρὸς τοῖς χυλοῖς. ἅπαντα δ' οὖν ταῦτα μίξει τινὶ καὶ ἀλλοιώσει καὶ ἔγχυλα[1] φαίνεται καὶ ὀσμώδη· καὶ γὰρ ἡ τέφρα διὰ τὴν κατάκαυσιν ἐξηλλοιωμένη,[2] καὶ ὅσα πυροῦται, λαμβάνει τινὰ χυλόν, τὰ μὲν ἁπλᾶ, τὰ δὲ καὶ μιγνύμενα τοῖς ὑγροῖς.

ἡ μὲν οὖν φύσις ὅτι τοιαύτη καὶ πρότερον εἴρηται καὶ νῦν.

3.3 ἐπεὶ δ' ἐν τρισὶν οἱ χυμοί—φυτοῖς τε <γὰρ>[3]

§ 3.3: *Cf.* Theophrastus, *On Odours*, ii. 4, 6.

[1] u : ἔνχυλα U.
[2] ego : ἐξαλλοιωμένη U N P^{ac}(?) : ἐξαλλοιουμένη aP^{c}.
[3] aP.

---

[1] *Cf.* Aristotle, *On Sense*, v (443 a 16–21): "Further of earths that are mined gold is odourless, since it is savourless, but copper and iron are odorous. But when their fluid is burnt off, the slag of all becomes less odorous. Silver and tin are more odorous than the slag, less so than copper and iron, since they are watery."

[2] Contrast *ibid.* iv (443 a 15): "Further stone is odourless, since it is savourless . . ."

[3] *Cf.* Theophrastus, *On Stones*, viii. 50: "As for the differences that produce the savours of these (*sc.* of

are mined[1] and certain kinds of stone[2] have not only flavour[3] but odour.[4] But all these earths and earthy substances are seen to possess flavour and odour in virtue of a certain mixture and alteration: so ashes (which have received their alteration by the burning), and all substances that are exposed to fire, acquire a certain flavour, some in the pure state, some when mixed with liquids.

That the nature of flavour is as described was said before[5] and has been said now.

We Treat First the Natural Savours Arising in (1) *Inanimates and* (2) *Plants*

Savours occur in three things[6]: (2) plants and (3) 3.3

earths), they have a certain nature of their own, just as those that produce the savours of plants."

[4] *Cf. ibid.* vii. 39: "The natures of stones furthermore that are mined are also numerous; some such natures contain both gold and silver, the silver alone being apparent to the eye, and these are heavier both in weight and smell . . ."

[5] *CP* 6 1. 1.

[6] For the threefold division (inanimates, plants, animals) *cf.* Theophrastus, *On Odours*, i. 3: "The stink of decomposition is found in all things, plants, animals and inanimates . . ."; ii. 4: "Each of the animals, plants and inanimates that are odorous has an odour of its own . . ."; ii. 6: "Since of odours some are in plants and their parts . . ., some, according to our distinction, in animals and inanimates . . ."

καὶ ζῴοις εἰσί τινες καὶ ὀσμαὶ καὶ χυμοὶ κατὰ τὰς κράσεις, ἔτι δ' ἐν τοῖς κατὰ τέχνην[1] παρασκευῇ τινι[2] μιγνυμένοις, ἢ καὶ αὐτομάτως ἀλλοιουμένοις ὁτὲ μὲν ἐπὶ τὸ βέλτιον, ὁτὲ δὲ ἐπὶ τὸ χεῖρον (ὥσπερ τῶν σηπομένων)—τὸ μὲν ὑπὲρ ἁπάντων τούτων εἰπεῖν κοινότερόν τε καὶ καθόλου μᾶλλόν ἐστιν· ὑπὲρ δὲ τῶν φυσικῶν πρῶτον λεκτέον (ἀρχὴ γὰρ ἐν ἅπασιν ἡ φύσις) καὶ τούτων ὅσοι κατά τε τὰ ἁπλᾶ φαίνονται καὶ ὅλως τὰ ἄψυχα, καὶ ὅσοι κατὰ τὰ φυτὰ καὶ τοὺς καρπούς· καὶ γὰρ ταῦτα πρότερά τις τίθεται τῶν ζῴων, ἅμα δὲ καὶ ἡμῖν ἡ σκέψις ὑπὲρ τούτων, οἷον συνεχὴς οὖσα τοῖς πρότερον.

[1] U : τέχνης u.
[2] ego : παρασκευήν τι (τινα u) U.

---

[1] *Cf. ibid.* iii. 7: "As for the odours that arise by art and ingenuity, we must endeavour to discuss them as we did flavours [*sc.* in *CP* VII]. In both it is evident that what governs the procedure is the production of what is better for man, this being the aim of all art. Now even in unmixed bodies there are certain odours that men endeavour to contribute to by their procedures, just as they endeavour to promote the agreeableness of flavours. Nevertheless odours are for the most part in mixture . . ." For

animals have certain odours as well as savours depending on the tempering of their qualities; again savours are found in (1A) things mixed by some procedure of art,[1] or else in (1B) things that alter spontaneously, sometimes (1Ba) for the better, sometimes, as in decomposition, (1Bb) for the worse. To deal with all of these would spread our discussion too wide and make it too general.[2] We must first[3] discuss the natural savours (1Ba,[4] 2, 3), since nature everywhere comes first, and of these natural savours all that appear in the simple bodies or in inanimates in general (1Ba) and all that appear in plants and fruits (2). For not only does one account (2) plants prior to (3) animals,[5] but our enquiry is meanwhile concerned with them, being continuous (one may say) with what precedes.[6]

the lost seventh book of the *CP* see p. 459 ff. below.

[2] Perhaps these five divisions were treated in the lost five books "*On Flavours*," for which see p. 460 below.

[3] This discussion ends with book VI; in the lost book VII artificial mixture of flavours was dealt with.

[4] Spontaneous savours altered for the worse (3Bb) are not counted as natural, since art and nature always aim at what is better (*cf.* *CP* 1 16. 11–12).

[5] *Cf.* Aristotle, *History of Animals*, viii. 1 (588 b 4–11): "Thus nature passes from inanimates to animals gradually . . . Thus after the class of inanimates comes first that of plants . . . And the transition from them to animals is a continuous one . . ."

[6] *CP* I–V and the whole *HP*.

3.4 οἱ μὲν οὖν ἐν τῇ γῇ ξυνιστάμενοι χυμοὶ τῶν ὑγρῶν (οὗτοι γὰρ ἐμφανέστατοι· λέγω δ' οἷον ὀξεῖς) καὶ[1] γινόμενοι[2] μὲν δῆλον ὅτι διὰ τὰς αὐτάς πως ἢ παραπλησίας ἀνάγκας δι' ἃς καὶ ἐν τοῖς φυτοῖς, οὐ μὴν οὔτε ἴσοι τὸ πλῆθος οὔθ'[3] ὅμοιοι τούτοις, ἀλλὰ θολερώτεροι διὰ τὸ μὴ τὴν αὐτὴν πέψιν μηδ' ὁμοίαν ἔχειν, τὸ δὲ πλῆθος οὐ συμβλητοί· πολλοὶ γὰρ οἱ τῶν καρπῶν καὶ ἀνόμοιοι χυλοί, καὶ οὐδαμῶς ἔνιοί γε ἐμφαινόμενοι τοῖς ἐκ τῆς γῆς (οἷον ὁ αὐστηρός, καὶ δριμύς, καὶ ἄλλοι δὲ πλείους).

3.5 εἰ μή τις λέγοι τῇ μίξει καὶ τῇ κράσει τῶν ἄλλων γίνεσθαι τούτους, τοῖς δὲ γένεσιν εἶναι τοὺς

§3.5: Menestor, Fragment 7 (Diels-Kranz, *Die Fragmente der Vorsokratiker*, vol. i[10], p. 376.

[1] ὀξεῖς· καὶ U : *acuti et reliqui* Gaza : οἱ ὀξεῖς καὶ οἱ λοιποὶ Schneider : ὀξεῖς καὶ ἁλμυροὶ Wimmer.
[2] U[r] aP : -ον U[ar] N.
[3] Schneider : οὔ U : οὔδ' u.

---

[1] That is, causes due to the matter and its movements, as opposed to a final cause. Plants also have a final cause, absent in these inanimates, which aims at perfect development of the fruit and accounts for its better concoction.

[2] A correction of Aristotle, *On Sense*, iv (441 a 30–b 7): "As many savours as are in pericarpia are also observed to be present in the earth" (the whole passage is cited in

Naturally Produced Savours:
(1) *In Inanimates (Fluids in the Earth);*
Their Connexion with Fruit Savours

Now the savours forming in the earth that belong to fluids, these being the most noticeable savours there (I mean for instance the acid savours), and arising evidently from the same necessary causes[1] as in plants, are nevertheless equal neither in number to plant savours nor yet in quality, but are muddier because they do not have the same concoction as theirs nor anything approaching it, and in number they cannot be paired with them, fruits having savours that are numerous and of superior quality, and some at least that do not appear at all in fluids coming from the earth (as the dry-wine savour, the pungent and several more).[2] 3.4

Unless one should say that these savours of fruits arise from mixing and tempering the others, but are the same with them in their general kinds.[3] This 3.5

note 1 on *CP* 6 3. 1). Aristotle here mentions three savours: the salty, the bitter and the acid; he concludes "with savours of all kinds," not mentioning explicitly the sweet, oily, pungent, dry-wine and astringent. Theophrastus knew about fattiness or "oil" in waters coming from the earth (Frag. 159, p. 209. 22–27 Wimmer); by "several more" he is therefore referring, in all probability, to the sweet and the astringent.

[3] Evidently what Aristotle meant; see the citation in note 1 on *CP* 6 3. 1.

αὐτούς. οὕτως δ' ἂν εἴησαν ἄπειροι, ἄπειροι γὰρ οἱ λόγοι τῶν μίξεων· καὶ ἔτι δὴ κατὰ τὸ μᾶλλον καὶ ἧττον αἱ διαφοραί, διὸ καὶ οἱ ὁμογενεῖς πλείους (οἷον αὐστηροί, λιπαροί, πικροί, γλυκεῖς), ὅθεν καὶ οἱ παλαιοὶ τῶν φυσιολόγων ἀπείρους ἐτίθεντο τοὺς χυλούς, ὥσπερ καὶ Μενέστωρ·[1] ὁποία γὰρ ἄν τις ἡ μῖξις καὶ ἡ σῆψις γένηται τοῦ ὑγροῦ [φυτοῦ],[2] τοιοῦτον εἶναι καὶ τὸν χυμόν.

οὐ μὴν ἀλλὰ βέλτιόν γε οἱ ὡρισμένους λέγοντες, ἐκ τούτων δ' ἤδη κατὰ τὰς μίξεις ποιοῦντες[3] τὰς διαφοράς. ἴδιος γοῦν τῶν ἐκ τῆς γῆς ὁ ἁλμυρός· οὐδενὸς γὰρ καρποῦ τοιοῦτος, ἀλλ' εἴπερ, φύλλου καὶ καυλοῦ[4] καὶ κελύφους. ἀλλὰ περὶ μὲν

[1] μενέστῶρ u : εστωρ U : λεεστῶν N (-ὼν P) : μικτῶν a (after Gaza).

[2] Gaza : <τοῦ Schneider> ἐμφύτου Heinsius.

[3] Wimmer : ποιοῦνται U.

[4] Schneider : καρποῦ U.

---

[1] Aristotle produces the savours from the extremes sweet (which is positive) and bitter (which is its privation); the rest are mixtures of these in various proportions and approximations to them [Aristotle does not here admit irrational ratios] (*cf. On Sense*, iv [442 a 12–27]).

would make them infinite in number,[1] since the ratios of mixture are infinite (and distinctions are moreover due to differences of degree, which is why there are groups of savours of the same kind, as dry-wine savours, oily savours, bitter ones and sweet). Hence the old natural philosophers accounted flavours infinite, like Menestor, who held that the character of the savour went with the character of the mixture and of the decomposition[2] of the fluid.

Those do better nevertheless who speak of a definite number of savours[3] and only then resort to mixture to produce the varieties of each. At all events the salty savour is restricted to fluids coming from the earth: no fruit has such a savour (if anywhere it is in the leaf or the stem or the husk).[4] The

Theophrastus so qualifies the distinction between positive and privative as to make it useless for Aristotle's purpose and drops the notion that the primary savours are due to mixture of other savours; like the rest they are all due to concoction of various degrees and kinds. Like Aristotle he believes that the number of primary savours is limited, but since he does not make them mixtures he ignores Aristotle's proof (*On Sense*, vi [445 b 20–446 a 20]) that we do not perceive actually all the steps that intervene between the extremes.

[2] "Decomposition," as Diels remarks, is Menestor's term for concoction (*cf. CP* 6 8. 4 with note 1).

[3] As Aristotle; *cf. On Sense*, iii (440 b 23–25), iv (442 b 21–22).

[4] *Cf. CP* 6 10. 6–8.

τούτου τὴν αἰτίαν ὕστερον λεκτέον.

4.1 αἱ δὲ ἰδέαι τῶν χυλῶν ἑπτὰ δοκοῦσιν εἶναι (καθάπερ καὶ τῶν ὀσμῶν καὶ τῶν χρωμάτων)· τοῦτο δ᾽ ἐάν[1] τις τὸν ἁλμυρὸν οὐχ ἕτερον τιθῇ τοῦ πικροῦ, καθάπερ καὶ τὸ φαιὸν τοῦ μέλανος, ἐὰν δὲ χωρίζῃ, συμβαίνει τοῦτον[2] ὄγδοον εἶναι. γλυκὺς γὰρ καὶ λιπαρὸς καὶ πικρὸς καὶ αὐστηρὸς καὶ δριμὺς καὶ ὀξὺς καὶ στρυφνὸς ἀριθμοῦνται, προστίθεται δὲ καὶ ὁ ἁλμυρὸς ὄγδοος. οἴονται δέ τινες καὶ τὸν οἰνώδη δεῖν· ἔν τε γὰρ τοῖς καρποῖς ὑπάρχει πλείοσιν, καὶ ἐκ τῆς γῆς ἀναπιδεύει τις ἐνιαχοῦ

§4.1: Pliny, *N. H.* 15. 106–8; Galen, *De Simpl.* i. 38 (vol. ix, p. 451. 11 Kühn).

[1] N aP : δε ἀν U : δὲ, ἄν u.
[2] u aP : τοῦτο U N.

---

[1] *CP* 6 10. 1–10.

[2] *Cf.* Aristotle, *On Sense*, v (442 a 12–28): "Just as the colours come from a mixture of white and black, so savours come from one of sweet and bitter . . . So the oily is a savour belonging to the sweet, whereas the bitter and salty are about the same, and the pungent, dry-wine, astringent and acid come in between. For the species of savour are just about equal in number to those of colour, there being seven species of each if one counts (as is reasonable)

reason for this will be discussed later.[1]

(2) In Plants and Fruits: Their Number

4.1 The kinds of flavours, as of odours and colours, are held to be seven in number. This holds if one counts the salty as not distinct from the bitter (just as one counts grey as not distinct from black); if one keeps them separate, the salty becomes the eighth.[2] For the list is (1) sweet (2) oily (3) bitter (4) dry-wine (5) pungent (6) acid (7) astringent; and the salty[3] is also added to these as an eighth. Some think that the vinous flavour should also be added, for it is not only present in several fruits but a savour of this sort also wells up from the earth in some

grey as a sort of black, for it remains that yellow should belong to white (as the oily belonged to the sweet), and red, violet, green and blue should come between white and black, the rest being mixtures of these." The number seven was perhaps suggested by the theory of music, there being seven different notes in the octave. Plato mentions seven savours (*Timaeus* 65 B–66 C): astringent, dry-wine, bitter, salty, pungent, acid, sweet. For odours *cf. CP* 6 14. 12 and Aristotle, *On Sense*, v (443 b 7–11): "Further the odours must be analogous to the savours. But this is the fact with some: thus there are pungent, sweet, dry-wine, astringent and oily odours, and one could call the decomposing smells analogous to the bitter savours." See list in Appendix II.

[3] *Cf.* Aristotle, *On the Soul*, ii. 10 (422 b 10–14), cited in note 2 on *CP* 6 1. 2.

τοιοῦτος χυμός· ἔτι δ' οὐδ', ὥσπερ ὁ γαλακτώδης εἰς τὸν γλυκὺν ἂν τεθείη (καὶ γὰρ οὗτος ἐν ἐνίοις ὡς εἶδος ἄν τι τιθοῖτο[1] τοῦ γλυκέος[2]), οὕτω τὸν οἰνώδη τῶν ἄλλων τινὶ προσνεῖμαι ῥᾴδιον, ἀλλὰ ἰδία τις ἡ φύσις καθ' αὑτήν, ἐπιδεχομένη καὶ τὸ

4.2 γλυκὺ καὶ τὸ στρυφνὸν καὶ αὐστηρόν. ἀλλὰ γὰρ τοῦτο μὲν οὐδὲν <ἂν>[3] ἴσως διαφέροι πρὸς τὴν τῶν ἄλλων θεωρίαν·[4] ὁ δὲ ἀριθμὸς ὁ τῶν ἑπτὰ καιριώτατος καὶ φυσικώτατος.

πότερον δὲ τοὺς μὲν ὥσπερ ἀρχάς, τοὺς δὲ ὡς στερήσεις δεῖ λέγειν, οἷον τὸν μὲν γλυκὺν καὶ λιπαρὸν καὶ ὅσοι σύστοιχοι τούτοις, ἀρχάς, ἐκ τούτων γὰρ αἱ γενέσεις καὶ αἱ τροφαί, τοὺς δ' ἄλλους

§ 4.2: *Cf.* Galen, *De Simpl.* iv. 9 (vol. xi, p. 650. 1–15 Kühn).

[1] ἄν τι τ. Wimmer : ἀντιτιθοῖτο U.
[2] u P : -κέως a : -κύος U N.
[3] Wimmer after Schneider.
[4] Gaza (*contemplationem*), Itali : θεραπείαν U.

---

[1] *Cf.* Theophrastus *On Waters* (Fragment 159 ed. Wimmer p. 209. 18–21): "In many places there are springs

places[1]; furthermore it is not easy to assign it to some other savour, as the milky might be put under the sweet (this too being a flavour that in some cases could be reckoned as a species of the sweet); its nature instead is one that stands apart by itself, admitting sweetness, astringency and the dry-wine quality. But we drop the point; this would perhaps make no difference for the understanding of the rest of the subject, and the number seven is most apposite and natural.[2]

4.2

The Distinction Between Natural and Privative Flavours

Are we to speak of some savours as principles (as it were), and of the rest as privations? For instance are we to speak of the sweet and the oily and all of the same order as principles, since generation and nutrition proceed from them, but of the rest as

which are in some cases of a more potable and vinous character, as the one in Paphlagonia where the natives come for their potations . . ."

[2] *Cf.* Alexander, *Comm. on Aristotle's Met.* (p. 38. 16–20 ed. Hayduck): "Again they (*sc.* the Pythagoreans) called the number seven 'appositeness' (*kairós*), for natural occurrences are held to achieve the fullness of time of both their birth and maturity in hebdomads, as with man. So children are brought forth in seven months, and teethe in seven months and reach puberty in the second hebdomad and get beards in the third."

στερήσεις, ἢ πάντας φυσικούς; ἔοικε γάρ, εἰ μὲν εἰς τὸ ἄριστον δεῖ τάττειν τὴν φύσιν, ἐν τούτοις εἶναι μόνοις· ἅμα δὲ καὶ τὸ τῆς τροφῆς καὶ γενέσεως ἀληθινόν, οὐδὲν γὰρ (ὡς εἰπεῖν) οὔτε τρέφεσθαι δοκεῖ διὰ τῶν ἑτέρων οὔτε γίνεσθαι· εἰ δ' εἰς τὸ πλεῖον αὖ δεῖν αὐτῆς,[1] ἐκεῖνό γε μᾶλλον, ἢ οὐχ ἧττον, κατὰ φύσιν, ὀλίγοι γὰρ ἐν τούτῳ τῷ μέρει.

4.3 σχεδὸν δέ τι παρόμοιόν ἐστι τὸ ζητούμενον καὶ ἐπὶ τῶν ἄλλων ζῴων πρὸς τὸν ἄνθρωπον, πότερα κατὰ φύσιν ἢ κατὰ στέρησιν τἆλλα· καθάπερ γὰρ

[1] U (*plura naturae tribui quam pauciora dignius sit* Gaza : πλείονα εἴδη εἶναι Wimmer) : πλείονα ἰδεῖν αὐτῆς u.

---

[1] Aristotle's view: *cf. On Sense*, iv (441 b 23–442 a 27): "That the savours are not an effect, or a privation, of every sort of dryness, we must establish from the point that neither the dry without the wet nor the wet without the dry is nutritive, since not just one of them alone is food for animals, but only the mixture of the two . . . But the food taken feeds insofar as it is gustable. For all animals are fed by the sweet, either pure or mixed . . . Heat causes increment and prepares the food. And it attracts what is light, but leaves the salty and bitter behind because of their weight. What the external heat does in external bodies, this the internal heat does in the nature of animals and plants, which is why they are fed by the sweet. The other savours are mixed with the food in the same way as the salty and the sharp: they serve as seasoning. They do so

privations[1]? Or are we to speak of all as natural? For it appears that if we are to put nature under the head of what is best,[2] nature is found in the first group alone; then too the point[3] about nutrition and generation has the aspect of truth, since practically nothing is held either to be fed or to be generated through the rest. If on the other hand we are to determine nature by the criterion that it must be with the majority,[4] the second group is more natural (or at least not less so), there being few flavours in the first.

4.3 Of a similar sort (one might say) is the question raised about the other animals compared to man, whether they are such by nature or by privation[5]

by counteraction, since the sweet is all too nutritive and rises in the stomach. And just as colours come from a mixture of white and black, so the savours come from a mixture of sweet and bitter . . . And as black is the privation of white in the transparent, so salty and bitter are a privation of sweet in the nutritive wet."

[2] *Cf. CP* 1 16. 11 and note *b*.

[3] Made two sentences earlier.

[4] *Cf. CP* 2 17. 3 (*ad fin.*).

[5] Plato (*Timaeus*, 91 D 5–92 C 3) and Aristotle after him (*On the Parts of Animals*, iv. 10–14) speak as if the lesser animals were produced by subtracting from the mental and physical traits of man. *Cf.* Aristotle, *History of Animals*, xi. 1 (608 b 6–8): ". . . for this animal (*sc.* man) has his nature brought to completion . . ."

καὶ τούτων ἴδιαί τινες αἱ οὐσίαι, καὶ τῶν φυτῶν ὁμοίως. εἰ οὖν ἡ μὲν στέρησις ἐν ἀτελείᾳ τινὶ καὶ ἐνδείᾳ, ταῦτα δὲ καὶ γεννᾷ καὶ ἐκτρέφει τὰ ὅμοια, καὶ ἰσχυρότερα πολὺ τῶν ἑτέρων, οὐκ ἂν εἴησαν κατὰ στέρησιν.

ἔτι δὲ ἔν γε τοῖς φυτοῖς ἅμα συμβαίνει μὴ τελεοκαρπεῖν, καὶ ἧττον εἶναι πικρὰ τὰ ἐν τούτῳ τῷ γένει (καθάπερ ἐπὶ τοῦ κενταυρίου καὶ ἀψινθίου καὶ τῶν τοιούτων), ὡς ἐκεῖσε τῆς φύσεως φερομένης, καὶ τὴν γλυκύτητα καθάπερ ἀπεψίαν τινὰ οὖσαν ἢ ἔλλειψιν πέψεως. 4.4 ἀεὶ γὰρ ἡ φύσις ἐξομοιοῦν ἐθέλει τῷ ὑποκειμένῳ, καὶ τοῦτο[1] καὶ ἐν τοῖς ζῴοις καὶ ἐν τοῖς φυτοῖς· ποιεῖται[2] γάρ, καὶ τροφὴ γίνεται πᾶσιν, ὅταν κρατῇ[3] τὸ λαμβανόμενον.

[1] U^r from -ω.
[2] N aP : ποιεὶ τε U.
[3] u (-εῖ U) : κρατηθῇ Scaliger.

---

[1] Some do not; see note 4.

[2] *Cf.* Aristotle, *On the Generation of Animals*, ii. 1 (732 a 25–733 a 1): "Of animals some produce completed offspring and bring forth something similar to themselves . . . whereas others bring forth something that is not fully formed and has not received its proper shape . . . And just as the animal brought forth is a completed thing, whereas

(for just as the others have certain distinctive natures of their own, so equally have plants). In consequence, if privation involves a certain incompleteness and deficiency, whereas these animals also[1] generate and rear offspring like themselves,[2] and are much stronger[3] than the rest,[4] they would not be by privation.

Moreover in the plants it happens that two things go together: the failure to bear perfect fruit is accompanied by less bitterness in those that belong to the bitter kind, as in centaury and wormwood and the like,[5] which would come from this: that their nature moves toward being bitter, and sweetness is in them an absence (as it were) or inadequacy of concoction.[6] For the nature of a living thing always aims at assimilating the intake to its own goal,[7] in plants as well as in animals; for the creature is being made, and food comes about for all, when the creature masters what it takes. 4.4

the grub or egg is incomplete, so the perfect offspring is of a nature to be produced from the more perfect animal . . ."

[3] *Cf. ibid.* ii. 1 (732 a 17–18): ". . . the higher animals are also more self-sufficient in their nature, and so have size."

[4] The layers of eggs and grubs.

[5] *Cf. CP* 3 1. 3.

[6] Against Aristotle, *On the Generation of Animals*, iii. 1 (750 b 25–26): ". . . for in all the concocted is sweeter"; v. 6 (786 a 16–17): ". . . for the concoction makes them sweet . . ."

[7] *Cf. CP* 5 2. 1; 5 5. 7; 6 17. 2.

εἰ δὲ δὴ καὶ ἔνια τρέφεται διεφθαρμένοις τισίν (ὥσπερ καὶ τῶν ζῴων ὧν καὶ ἡ γένεσις[1] ἐκ τούτων), οὐδ' ἂν ἡ τῆς τροφῆς ἱκανὴ πίστις[2] εἴη πρὸς τὸ κατὰ φύσιν, οὐδ' ἂν πρώτη[3] σύστασις, ἀλλ' ἑκάστοις μεμερισμέναι κατὰ τὰς μίξεις.

4.5 ἐπεὶ οὐδ' οἱ γλυκεῖς χυμοὶ τρόφιμοι πάντες ἡμῖν, ἀλλ' οἱ μὲν ἐκστατικοί (καθάπερ ἡ ὁμοία τῷ σκολύμῳ ῥίζα καὶ ἄλλαι[4] τινές), οἱ δ' ὑπνωτικοί, πλείους δὲ διδόμενοι[5] καὶ θανατηφόροι (καθάπερ ὁ μανδραγόρας), ἔνιοι δὲ ὁμολογουμένως ἐπιθάνα-

[1] N aP : γένησις U. [2] u : πίστ U.
[3] ἡ πρώτη Schneider.
[4] u : ἀλλα U. [5] u : διδομενα U.

---

[1] The grubs produced by a plant feed on the corruption that produced them (*CP* 2 9. 5; 2 9. 6; 3 22. 4; 5 10. 5).

[2] Feeding and generation (initial formation) are due to the nutritive part: *cf. CP* 1 12. 5. *Cf.* also Aristotle, *On the Generation of Animals*, ii. 4 (740 b 25–741 a 2): "And just as the products of art are produced by means of the tools . . . , so the power of the nutritive soul, just as later in the actual animals and plants it produces growth out of the food . . . , so at the start it forms whatever creature is naturally produced. For the matter by which the creature grows is the same as that out of which it was initially formed, and therefore the acting power does the same now as it did at the start, the later power, however, being greater. So if this power is the nutritive soul, it is also the generative soul; and this is the nature of every creature, present in all plants and animals, whereas the

If moreover some plants feed on certain corrupt substances (as do those animals too whose generation is from these substances)[1] then also the appeal to the character of a flavour as food, or yet its character of having formed the creature at the start,[2] would not suffice to establish its character as natural.[3] Instead, for different types of creatures there are different types of food, depending on the mixtures in each.

4.5 In fact it is not even true that all sweet savours are nutritious for us. Some instead cause derangement, like the root resembling golden thistle and some other roots, some are soporific and when taken in large quantities even fatal, like mandragora, and a few are admittedly deadly,[4] for many persons in

other parts of the soul are present in some living things but not in others."

[3] *Cf. CP* 6 4. 2 (its nutritive character shows the sweet as natural).

[4] *Cf. HP* 7 9. 4: "Some [that is, roots of herbaceous plants] are sweet but deadly or cause disease . . ."; *ibid.* 7 15. 4 (the *strýchnos* is surely a case of homonymy; one is edible, but there are two other plants of the name) "the one able to produce madness, the other sleep and in large doses death"; *ibid.* 9 13. 4: "Some of the sweet roots cause derangement, like the one at Tegea resembling golden thistle, which Pandius the sculptor went out of his mind after eating when he was working in the temple; others are deadly, as the root growing near the mines at the works in Thrace. It is light and very agreeable to the taste, and causes a sleepy and easy death."

τοι· πολλοὶ γὰρ ἤδη καὶ πολλαχοῦ ῥίζας φαγόντες ἃς ἠγνόουν, γλυκείας τε τῇ γεύσει καὶ ἡδείας, ἀπέθανον· καὶ ἄλλα τοιαῦτα πλείω τυγχάνει, τὰ μὲν βλάπτοντα, τὰ δὲ καὶ ὅλως ἀναιροῦντα, τῶν κατὰ τὴν προσφορὰν ἡδέων ἢ ἀλύπων.

4.6 ἔνια δὲ πάλιν ὠφελοῦντα <τῶν>[1] ἀηδῶν καὶ πικρῶν, οἷον καὶ τὰ ἄρτι λεχθέντα, κενταύριον καὶ ἀψίνθιον, καὶ ὅσα δὴ φαρμακωδεστέρους ἔτι τοὺς χυμοὺς ἔχει· πολλὰ γὰρ καὶ τούτων ὀνίνησιν. τῶν δὲ καὶ ὅλως ὁ χυλὸς ἄτροφος ὢν ὥσπερ[2] ἥδυσμα τῆς τροφῆς ἐστιν, οἷον ὁ ἁλμυρὸς καὶ <ὁ>[1] ὀξύς· οὐ[3] γὰρ δυνάμεθα κρατεῖν ἀκράτου, καθάπερ οὐδὲ τῶν ἄλλων[4] ἔνια ζῴων, διὸ καὶ τούτοις παρέχομεν τοὺς ἅλας·[5] ἔνια δὲ καὶ αὐτὰ ἑαυτοῖς εὑρίσκει τὰ πρόσφορα πρὸς τὴν τοιαύτην ἐπικουρίαν, ὥσπερ καὶ οἱ ὄρνιθες.

[1] Schneider.
[2] ego : ὡσπερεὶ U ar (ὡσ- U r).
[3] U N : οὐδὲ aP.
[4] U : ἁλῶν u.
[5] u (ἅλους N) : ἄλλους U aP.

---

[1] *CP* 6 4. 3.
[2] *Cf. HP* 7 9. 4–5: "Not all the fragrant or sweet or agreeable roots (*sc.* of herbaceous plants) are likewise edible, nor yet the bitter ones inedible; only those are edible

many different places have eaten unfamiliar roots that were sweet and pleasant eating and died of it. And a number of other parts that make pleasant or painless eating either injure or kill outright.

4.6 On the other hand there are some evil-tasting and bitter plants that are good for us, as the centaury and wormwood just mentioned[1]; again those with a still more medicinal taste, many of these too being beneficial.[2] Elsewhere the flavour, although quite without nutritive value, is spice (so to speak) for our food, as the salty and the acid,[3] since we cannot assimilate the food when it has not been tempered, no more than can some animals, which is why we feed them salt[4] (a few, like birds,[5] even discover for themselves the substances that are helpful here).

that after consumption bring no harm to the body; for some sweet roots are deadly or cause disease, and others are bitter or evil-smelling but beneficial. Similarly with the leaves too and the stalks, as with wormwood and centaury."

[3] *Cf.* Aristotle, *On Sense*, iv (442 a 8–10), translated in note 1 on *CP* 6 4. 2.

[4] *Cf.* Aristotle, *History of Animals*, viii. 10 (596 a 18) [lumps of salt are fed to sheep and goats].

[5] *Cf.* *ibid.* ix. 7 (613 a 2–5): "When the nestlings are hatched he (*sc.* the male swallow) chews some of the most salty earth and opens their mouths and spits it into them, preparing the young to take their food."

ἤδη καὶ φανερὸν ἂν δόξειεν ὡς οὐκ ἔστιν ἁπλῶς μεμερισμένα, τὸ μὲν ὥσπερ φυσικόν, τὸ δ' ὡς ἐν

4.7 στερήσει καὶ παρὰ φύσιν. ἀλλὰ καὶ τὸ ἡμῖν ἄτροφον, ἑτέροις τρόφιμον, καὶ τὸ ἄλλοις τρόφιμον, ἄλλοις ἄτροφον. ὡς γὰρ ἂν αἱ φύσεις ἔχωσιν κατὰ τὰς κράσεις, οὕτως καὶ αἱ τροφαὶ καθ' ἕκαστον ἁρμόσουσιν, ὁμοίως δὲ καὶ αἱ ἡδοναὶ καὶ αἱ λῦπαι[1] καὶ αἱ βοήθειαι πρὸς τὰ πάθη καὶ τὰς διαθέσεις, ὃ καὶ φανερὰ ποιοῦντα πολλὰ τῶν ζῴων ἐστίν, οὐ πρὸς τὰ συμβαίνοντα πάθη μόνον αὐτομάτως, ἀλλὰ καὶ κατ' αὐτὰς τὰς ἐδωδάς, ὅταν ἄλλο φάγωσιν, ἕτερον ἐπεσθίοντα (καθάπερ οἱ ἔχεις τὸ πήγανον, ὅταν τὸ σκόρδον[2]).

[1] u : λοιπαὶ U.
[2] U : τὸν σκορπίον G. R. Thompson.

---

[1] *Cf.* Aristotle, *History of Animals*, ix. 5 (611 a 17–19) [of the deer]: "when it has given birth, it eats the afterbirth first. And it also runs for hartwort, and after eating it returns to its young"; 611 b 20–22: "When deer have been bitten by the venomous spider or the like, they collect crabs and eat them"; ix. 6 (611 b 33–612 a 7) [of bears]: "When they leave their lair after hibernating they first eat arum . . . and chew wood . . . And many other quadrupeds show intelligence in the remedies they seek; so they say that the wild goats in Crete seek out dittany when they have been shot . . . And when bitches have a certain complaint they eat a certain grass as an emetic";

It would by now appear evident that there is no simple division of savours into a class that is "natural" and another that is "privative" and so unnatural. Instead what is non-nutritive for man is nutritive for others; and what is nutritive for one set is non-nutritive for another. For the nutritive character varies in each case with the different types of natures in the matter of the tempering of the qualities. (Similarly with the pleasantness or distastefulness of what is taken and its character as a remedy for the animal's ailments and dispositions. Many animals are observed to take food in this way, not only as a remedy for accidental ailments[1] but also in the plain business of eating, following the consumption of one thing with that of another,[2] as vipers take rue after eating garlic.[3]) 4.7

612 a 31–34: "When bitches get worms they eat the standing cereal in the fields."

[2] Theophrastus seems to hint that the animals are here guided by the taste; in Aristotle the purpose is remedial: *cf. History of Animals*, ix. 6 (612 a 7–8): "When the panther has eaten the poison called 'panther-choker,' it looks for human dung"; 612 a 24–31: "When a tortoise has eaten viper, it follows this by eating marjoram . . . When the snake eats fruit, it swallows the juice of ox-tongue [*pikrís*, literally 'bitter-wort': the animal must remain venomous in spite of the sweet fruit] . . ."

[3] Not mentioned elsewhere. Dioscorides (*M. M.* iii. 45. 4) says that rue remedies the odours arising from garlic and onion.

5.1 ἔχει δὲ καὶ κατὰ τὰς ὀσμὰς[1] ὁμοίως· ἄλλαι γὰρ ἄλλοις[2] ἐναντίαι καὶ οὐ πρόσφοροι, καὶ οὐ μόνον εἰς τὸ[3] μὴ δεῖσθαι, μηδὲ ζητεῖν, ἀλλ' ὑπὸ τῶν ἡδίστων ἡμῖν ἀναιρεῖσθαι, καθάπερ οἱ γῦπες ὑπὸ τῶν μύρων, καὶ οἱ κάνθαροι ὑπὸ τῶν ῥόδων· πολεμοῦσι δὲ δὴ σφόδρα καὶ αἱ μέλιτται τοῖς μεμυρισμένοις.

ὅλως μὲν γὰρ ἢ οὐδέν, ἢ βραχύ τι πάμπαν, ἐστὶν ἐν τοῖς ἄλλοις ζῴοις τὸ τὴν εὐωδίαν διῶκον αὐτῆς χάριν, ἀλλ' εἴπερ, κατὰ συμβεβηκός, ὧν ἡ τροφὴ τοιαύτη· πρὸς ταύτην γὰρ ἦν ἡ ἐπιθυμία 5.2 καὶ ταύτης. εἰ[4] δ' ἄρα καὶ τοιοῦτόν ἐστιν, ἡμῖν γε οὐ σαφές. ἐπεὶ καὶ τὴν πάρδαλιν τοῖς μὲν ἄλλοις ἡδύ φασιν δοκεῖν αὐτὴν <ὄζειν>[5] (διὸ καὶ

§5.2: *Cf.* Aristotle, *Historia Animalium*, ix. 6 (612 a 12–15); [Aristotle], *Problems*, xiii. 4 (907 b 35–37); Pliny, *N. H.* 8. 62.

[1] τὰς ὀσμὰς Gaza, Scaliger : ταύτας U. [2] aP : -ως U N.
[3] εἰς τὸ U : ὥστε Schneider after Gaza.
[4] u : ἡ U. [5] αὐτὴν ὄζειν ego : ὄζειν Gaza, Schneider.

---

[1] *Cf.* Theophrastus, *On Odours*, ii. 4: "Some animals appear to suffer from odours and fragrances, if what is said of vultures and beetles is true"; *cf.* [Aristotle], *On Amazing Reports*, cxlvii (845 a 35–b 2); Pliny, *N. H.* 10. 279; Plutarch, *Mor.* 87 E, 710 E; Aelian, *On the Nature*

The Application to Odours

So too with odours: different odours are bad or unsuited to different animals, not only to the point that the animal does not want them or seek them out, but to the point that it is killed by odours most delightful to man, as vultures are killed by perfumes and beetles by roses[1]; and bees are violently hostile to persons wearing perfume.[2] 5.1

In fact among animals other than man there are either none at all or very few indeed that seek out good odour for its own sake; if animals do seek it out, they do so only incidentally, when their food has that odour,[3] for their desire is relative to food and it is the food that it wants. In the second place, if there is in fact such an animal, the matter is not evident, at least to us. So the panther is reported to have an agreeable smell for other animals in its own 5.2

of Animals, iii. 7, iv. 18; Clement, *Paed.* p. 197. 5–9 Stählin; [Eustathius], *Commentary on the Hexaemeron*, chap. xviii (736 Migne).

[2] *Cf.* Aristotle, *History of Animals*, ix. 40 (626 a 26–28) [*of bees*]: "They are offended . . . by bad smells and by the smell of perfumes; this is why they sting persons who use them."

[3] *Cf.* Aristotle, *On Sense*, 5 (443 b 17–444 a 8) [cited in note 1 on *CP* 6 5. 3]; Theophrastus, *On Odours*, ii. 4: "No animal (so to speak) appears to delight in good odour by itself, but only in things that lead to feeding and consumption."

θηρεύειν αὐτὰ[1] *[ὅταν]*[2] *κατακειμένην,*[3] *ταύτῃ*[4] *προσαγομένην)· ἡμῖν δ' οὐδεμίαν εὐωδίαν ἐμφαίνει. εἰ μὴ κἀκεῖνο ἀληθές (ὥσπερ ἐστὶν ἀληθές), ὅτι χειρίστην ἔχοντας*[5] *πάντων τὴν ὄσφρησιν (ὡς*[6] *εἰπεῖν) πολλαὶ λανθάνοιεν ἂν ἡμᾶς ὀσμαί, καὶ τὰ ἡδέα καὶ τὰ λυπηρὰ τὰ κατὰ ταύτας.*

5.3 *ὃ δ' ἄν τῳ*[7] *δόξειεν ἄτοπον εἶναι, τὸ τὰ ἄλλα ζῷα, σαφεστέραν ἔχοντα, μηδ' αἰσθάνεσθαι, μηδὲ*

§ 5.3: Aristotle, *On Sense*, chap. iv (442 a 17).

[1] ego : *αὐτὴν* U.
[2] Wimmer.
[3] Wimmer (*κατακέηται* Schneider) : *κατακαειν ἡ* U.
[4] Heinsius (*gratia sui odoris* Gaza : *καὶ ταύτῃ* Wimmer) : *ταύτην* U.
[5] ego : *ἔχομεν* U.
[6] *καὶ διὰ τοῦτο ὡς* aP.
[7] u : *ἄντο* U.

---

[1] *Cf. CP* 6 17. 9 and Theophrastus, *On Odours*, ii. 4: "The odours even of animals considered fragrant escape us." *Cf.* Aristotle, *History of Animals*, ix. 6 (612 a 12–15): "It is said that the panther, noticing that the beasts delight in its odour, hunts them by concealing itself, since they draw near, and in this way catches even deer"; *cf.*

right (this being why it hunts them by lying down, luring them by this smell)[1]; but it has no noticeable fragrance to us. (Unless another point, which is true, is true here, that we have the worst sense of smell of all animals,[2] so to speak, and it would follow that many odours escape us, together with any attendant pleasantness or painfulness.)

An Apparent Difficulty

5.3 What some might take to be odd, that the other animals have a more exact sense of smell than man but do not perceive fragrances and do not discrimi-

also [Aristotle], *Problems*, xiii. 4 (907 b 35–37); Pliny, *N. H.* viii. 62; Plutarch, *De Sollertia Animalium*, chap. xxiv (976 D); Aelian, *On Animals*, v. 10; for still later versions see M. Wellmann in *Hermes*, vol. li (1916), pp. 17–18 and *Philologos*, Suppl. xxii (1931), pp. 25–26, 114–115. The story (with others) was no doubt suggested by the apparent connexion of the name *pordalis* (panther) with *pordē* ("breaking wind"). Since the animal was invisible, the prey must have been attracted by the odour.

[2] *Cf.* Theophrastus, *On Odours*, ii. 4: "Each animal, plant and inanimate among those that have odour has a peculiar odour of its own. But many cases are not apparent to man, because man has the worst sense of smell (so to speak)"; Aristotle, *On Sense*, iv (440 b 31–441 a 2): ". . . we have the worst sense of smell among animals and among our own senses . . ."; *On the Soul*, ii. 9 (421 a 9–10): ". . . we do not have this sense (*sc.* that of smell) exact, but have it inferior to that of many animals . . ."

κρίνειν τὰς εὐωδίας, οὐκ ἔστιν ἄτοπον.

τάχα μὲν γὰρ καὶ αἰσθάνονται, πρὸς δὲ τὴν φύσιν αὐτῶν οὐχ ἁρμόττουσιν, ἀλλὰ καὶ ἐναντίαι, καθάπερ ἐπὶ τῶν γυπῶν ἐλέχθη, καὶ ὡς τὰ ἔντομα πάντα ὑπὸ τοῦ ἐλαίου βαρύνεται, φεύγει γὰρ αὐτὴν[1] *τὴν ὀσμήν· ἕτερα δ' ὑφ' ἑτέρας τινὸς ἴσως. (ἄλογον δ' ἐνταῦθα δόξειεν ἂν συμβαίνειν πρὸς ἐκείνην τὴν ὑπόθεσιν, <ὡς> ὁ*[2] *λιπαρὸς τοῦ γλυκέος χυλός· ἔνια δὲ προσφιλέστατα τῷ γλυκεῖ, καθάπερ οἱ κνῖπες· οὐδὲν γὰρ τῶν οἰκείων*
5.4 *φθαρτικόν, εἰ μὴ ταῖς ὑπερβολαῖς. ἀλλὰ τούτου*[3] *μὲν τὴν αἰτίαν ἐν τῇ δριμύτητι ληπτέον, ὥσπερ καὶ τῆς ὀριγάνου καὶ τῶν τοιούτων· ἅπαντα γὰρ*

[1] U : *αὐτὰ* u.
[2] ego (*ὅτι* Itali : *ὅτι ὁ* Schneider : *εἰ ὁ* Wimmer) : *ὁ* U.
[3] u aP : *τοῦτο* U N.

---

[1] *Cf.* Aristotle, *On Sense*, v (443 b 17–444 a 8): "There are two species of the odorable . . . One corresponds to the savours . . . and these odours have a pleasant or unpleasant character incidentally (for since they are affections of what is nutritive, the smells of these savours are pleasant when one desires food, but not pleasant for all those for whom the food that has the odours is not also pleasant). These odours then . . . have pleasantness and unpleasantness incidentally . . ., and this is why they are

nate between them,[1] is not odd.

(1) Perhaps they do perceive them, but the fragrances are not suited to their natures but even injurious to them, just as we said[2] of vultures, and just as all insects find olive oil oppressive, for they avoid the mere smell of it, and other animals are perhaps distressed by some other fragrance. (Here a difficulty might appear to arise for the hypothesis that "the oily is a flavour belonging to the sweet,"[3] some insects being greatly attached to the sweet, like the fig-insects,[4] since nothing suited to an animal is destructive of it except by excess. But the cause in the case of olive oil is to be found in its pungency,[5] just as with marjoram and the like, for

5.4

shared by all animals. Whereas some odours are pleasant intrinsically . . . This odorable is peculiar to man . . .; whereas the other species, because its character of being pleasant is incidental, is divided into types answering to those of savours. But such a division will not apply to this one, since its nature is intrinsically pleasant or unpleasant."

[2] *CP* 6 5. 1.

[3] Aristotle, *On Sense*, iv (442 a 17): ". . . the oily is a savour belonging to the sweet . . ."; *cf. CP* 6 4. 2.

[4] *Cf.* Aristotle, *History of Animals*, iv. 8 (534 b 18–25) [cited in note 1 on *CP* 6 5. 4] for the fondness of this insect for honey.

[5] The pungency of olive oil is not elsewhere mentioned. But *cf. CP* 6 8. 3 and especially *HP* 5 4. 4 (where the immunity of olive and wild olive to being eaten by the wood-worm is ascribed to their bitterness).

φεύγει.)

τὸ δὲ μὴ αἰσθάνεσθαι καθάπερ λέγουσιν, οὐ σαφὲς ἡμῖν. εἰ μὴ ἄρα καὶ τοὺς πόρους ἐνταῦθα ἄν τις αἰτιάσαιτο, καθάπερ ἔνιοι τῶν φυσικῶν, τῷ εἰσαρμόττειν[1] ᾒ[2] μή· διὸ καὶ διαιροῦσιν τοῖς μεγέθεσιν, τὰ μὲν μικρὰ φάσκοντες τῶν τοιῶνδε αἰσθάνεσθαι, τὰ δὲ μείζω τῶν τοιῶνδε μᾶλλον· οὐχ ἱκανὴν δ',[3] οὐδὲ οἰκείαν ἴσως διαίρεσιν διαιροῦντες, οὐδ' ἐν[4] τοῖς πόροις ἡ αἴσθησις, ἀλλὰ μᾶλλον πρὸς τὴν διάθεσιν ἄν τις καὶ τὴν κρᾶσιν τό τε πάθος ἀποδοίη[5] καὶ τὴν ἀπάθειαν, ἐν οἷς τὸ ἡδὺ καὶ τὸ λυπηρόν.

5.5 ὅπερ συμβαίνει καὶ ἐπὶ τῶν ἀνθρώπων· οὐ γὰρ πᾶσιν αἱ αὐταὶ προσφιλεῖς, ἀλλὰ μᾶλλον ἐν ταῖς

[1] U^r (-ά- U^ar) : μὴ ἁρμόττειν N aP : ἁρμόττειν εἰς αὐτοὺς Schneider : ἐναρμόττειν Diels (*Doxographi* p. 502. 5, note).
[2] N aP : ἢ U. [3] [δ'] Wimmer.
[4] Wimmer : οὐ δεῖ U N : οὐδ' εἰ u : οὐ γὰρ ἐν aP.
[5] aP : ἀποδεοι U : ἀποδώῃ u N.

[1] *Cf.* Aristotle, *History of Animals*, iv. 8 (534 b 18–25): "For insects are aware of things at a distance . . . , as bees and fig-insects notice honey from afar, which implies that they know it by its smell, and many are killed by the smell of sulphur. Further, ants are made to abandon their ant-hills by sprinkling ground marjoram and sulphur about, and most such animals take to flight when deer horn is

insects avoid them all.[1])

(2) As for the statement[2] that animals do not perceive the fragrances, the matter is not evident to us. Unless here too one should (like some of the natural philosophers) make the sense-passages responsible, because the fragrances fit them or do not fit them. This is why they divide the animals by size, and say that small animals perceive small perceptibles, whereas the larger animals perceive the larger better.[3] But the division is inadequate[4] and perhaps inappropriate[5] as well, and the perception is not to be found in the passages; instead one would rather account for both the sensitivity and the non-sensitivity (and on this depends the pleasurableness or painfulness of the sensation) by the relation of the perceptible to the disposition and tempering of qualities of the percipient.[6]

5.5 This occurs also in man, for the same odours are not agreeable to all of us. Uniformity occurs rather

burnt and most of all when storax is burnt."

[2] *Cf.* Aristotle, *On Sense*, v (443 b 16–444 a 8), cited in note 1 on *CP* 6 5. 3.

[3] *Cf.* Theophrastus, *On the Senses*, xxix–xxx, xxxiii–xxxv.

[4] Because some animals have larger passages than man, some smaller, and the size of the passages must account for insensibility to fragrance in all.

[5] Since we are dealing not with a quantity but with a quality.

[6] *Cf.* Theophrastus, *On the Senses*, xxxii, xxxv, lxiv.

κακώδεσι καὶ βαρείαις, ὧν ἔνιαί γε κοιναὶ πᾶσιν, οἷον αἱ ἔκ τινων ἐκπνέουσαι χασμάτων καὶ ἄντρων, θανατηφόροι τοῖς προσπελάζουσιν. εἰ δ' ἄρα καὶ αὗται[1] τοῖς ἀναπνευστικοῖς μόνοις, ἀλλὰ τοῦτό γε φανερόν, ὥσπερ ἐπὶ τῶν χυλῶν, ὅτι καὶ ὀσμαί τινες ἑκάστοις εἰσὶν οἰκεῖαι, καὶ ὅτι τὸ εὐῶδες τὸ μὴ κατὰ τὴν τροφὴν ὀλίγοις ἢ οὐδενὶ προσφιλές· εἰ δὲ δή τινι καὶ συμφέρον, ἀδηλότερόν ἐστιν.

ἀλλὰ γὰρ ταῦτα μὲν ἴσως ἐπὶ πλέον εἴρηται τῶν ὑποκειμένων· ὅθεν δ' ὁ λόγος, ἐκεῖσε ἐπανιτέον, ὡς[2] ἑκάστου τῶν χυλῶν καὶ τῶν ὀσμῶν ἐστίν τις[3] φύσις, ὥσπερ καὶ τῶν ζῴων, ἣ πρὸς ἑκάστην διάθεσιν ἁρμόττει καὶ κρᾶσιν.

5.6

οὐ μὴν ἀλλ' ὥς γε[4] καθόλου καὶ κοινῶς εἰπεῖν

[1] Gaza (*illi*), Schneider : αυταὶ U.
[2] ego : ὧν U : u erases. [3] u : τισιν ἡ U.
[4] ego (*ceterum quantum* Gaza : ἀλλ', ὥστε Moreliana) : ἄλλως τε U.

---

[1] *Cf.* Aristotle, *On Sense*, v (444 b 7–28), on the possession of the sense of smell by animals that do not breathe.
[2] *CP* 6 4. 4.
[3] Contrast Aristotle, *On Sense*, v (444 a 14–15): ". . . the class of smell described (*i.e.* smell pleasant in its own

in evil and oppressive smells, some of which affect all alike, as the odours exhaling from chasms and caves that are lethal to all who draw near. Even if these odours affect only the animals that breathe,[1] yet this point is clear, as it was with flavours,[2] that with odours too there are some that are appropriate to different kinds of animal, and that fragrance not associated with food is welcome to few or none. Whether it is also beneficial to some animal[3] is still more uncertain.

The Discussion of the Distinction Between Natural and Privative Savours Concluded

But we have perhaps gone farther in discussing these matters than our plan warrants, and we must return to where we left off,[4] making the point that each flavour and each odour possesses a certain nature, as the animals[5] do too, and that this nature of the odour and flavour fits the different disposition and tempering of qualities of the different animals.[6] 5.6

Nevertheless, to speak generally and broadly,

right) has come about as a remedy for health . . ."

[4] The discussion whether there are privative savours CP 6 4. 2–6).

[5] The percipients.

[6] For flavours *cf.* CP 6 4. 4 *ad fin.*; 6 4. 7; for odours CP 6 5. 1; 6 5. 4 *ad fin.*; 6 5. 5.

οἱ γλυκεῖς καὶ οἱ κατὰ ταύτην τὴν συστοιχίαν τροφιμώτεροι καὶ μᾶλλόν εἰσι κατὰ φύσιν. ἐναντίωμα δέ τι (τάχα δὲ οὐκ ἀκόλουθον) ἔσται τῷ[1] ἀνὰ λόγον, ὅτι χυλὸς μὲν ἅπασι προσφιλὴς ὁ γλυκύς, ὀσμὴν δὲ οὐκ ἔστιν οὕτω λαβεῖν, οὕτω γε[2] τῷ γένει λαμβάνοντας,[3] εἰ μή τις ἄρα λανθάνει διὰ τὴν ἡμετέραν ἀσθένειαν τῆς αἰσθήσεως· ἴσως δ' οὐδὲ τὸ περὶ τοῦ γλυκέος εἰρημένον ἁπλῶς ἀληθές.

ὑπὲρ μὲν οὖν τούτων ἀρκείτω τὰ εἰρημένα.

6.1 τῶν δὲ φυτῶν (μᾶλλον δὲ πάντων ἐν οἷς οἱ χυλοί) διὰ τί μὲν ἕκαστα τούτους ἔχει (λέγω δ' οἷον γλυκεῖς ἢ πικροὺς ἢ λιπαρούς) τὴν πρώτην αἰτιατέον σύστασιν (ὑπὲρ ἧς οἱ μὲν τοῖς σχήμασιν διορίζοντες, ὥσπερ Δημόκριτος, οἴονται λέγειν τινὰς

[1] U : τὸ u.
[2] Schneider : δὲ U N : aP omit.
[3] Schneider : λανθάν- U.

---

[1] That is, aside from the accidental character of being associated with pleasant food.
[2] *Cf. CP* 6 5. 2.

sweet flavours and those of that order are more nutritive than the rest and more natural. But in the analogous statement about odours a certain inconsistency (or perhaps lack of correspondence) will arise, because whereas the sweet flavour is agreeable to all, it is impossible to find an answering odour that we can similarly formulate as constituting a class of its own,[1] unless some odour escapes us owing to the weakness of our sense of smell.[2] But perhaps the statement about the sweet flavour[3] is also not unqualifiedly true.

Let the present discussion suffice for these matters.

A Difficulty in Explaining the "Privative" as Unconcocted

6.1 Now in explaining why in plants (or rather everything where flavours are found) each kind of plant or other thing has the flavour we observe (I mean for instance sweet or bitter or oily) one must go back to the way in which the flavour was first formed (and those who, like Democritus, use shapes to distinguish the kinds of flavour believe that they are speaking of a set of causes[4] that bears on the origi-

[3] The statement "the sweet flavour is agreeable to all," made a few lines before.

[4] The shapes.

αἰτίας, ὁμοίως δὲ καὶ εἴ τις κατ᾽ αὐτὴν ἔχει τὸν ἴδιον ἀποδοῦναι περὶ ἑκάστου λόγον)· ὡς[1] δὲ κατὰ πέψιν καὶ ἀπεψίαν διαιροῦντες τὰ γλυκέα καὶ τὰ πικρὰ καὶ τὰ ἄλλα, τάχα δ᾽[2] ἂν δόξειεν[3] κωλύειν τὰ μικρῷ πρότερον ἡμῖν εἰρημένα· πέψει γὰρ πρότερόν τινι φαίνεται πάντ᾽ ἐξομοιοῦσθαι καὶ τοῖς φυτοῖς καὶ τοῖς ζῴοις.

6.2 ἀλλὰ τούτοις ὁμοίως λεκτέον ὅτι καὶ οἱ γλυκεῖς (καὶ οἱ ἁπλῶς τρόφιμοι) χυλοί, καὶ αἱ ὀσμαὶ αἱ[4] εὐώδεις, πέψει τινὶ γίνονται καὶ κατεργασίᾳ,[5] καὶ ὅτι μᾶλλον (ὡς τῷ γένει λαβεῖν) πεπεμμένα[6] τῶν ἑτέρων (δηλοῖ δὲ καὶ ἐπ᾽ αὐτῶν τῶν ὁμογενῶν· καὶ γὰρ εὐχυλότερα, καὶ εὐοσμότερα, τὰ πε-

[1] U N : οἱ aP. [2] U N : τάχ᾽ aP.
[3] U : -αν Schneider. [4] u : καὶ U.
[5] κατεργασίᾳ U r : -αι U ar.
[6] u (-εμέ- U) : -οι Schneider.

---

[1] Aristotle: *cf. On Sense*, iv (442 a 12–19): "Just as the colours come from a mixture of white and black, so the savours come from sweet and bitter, and each set of them is in a ratio or else due to difference of degree (either following certain numbers measuring the mixture and the movement imparted [*i.e.* the ratio is between integers, like those representing the concords of music; in the savours the numbers would represent the quantity or the movement imparted by the components of the mixture] or else [*i.e.* where the difference was not in numbers, but

nal formation, and so too does anyone else[1] who is able to come forward with the special ratio,[2] resting on the formation, for each flavour); but when people rely on concoction and the want of it to distinguish things sweet and things bitter (and the rest), the point that we made a short while before[3] might seem to stand in the way, since it appears that in both plants and animals all assimilation has been previously brought about by a type of concoction.[4]

We must nevertheless say with these last[5] that both the sweet (and simply nutritive) flavours and the fragrant odours are produced by a certain concoction and processing, and that their possessors (taken as a class) are better concocted than the other set (this can be seen directly in plants of the same kind, the concocted ones being of better flavour and of better odour than the unconcocted), 6.2

only in degree] numerically indeterminate—the savours which when mixed produce pleasure being the only ones expressible in numbers [*i.e.* they correspond to the musical concords]). The oily savour then is a savour belonging to the sweet, the salty and bitter are about the same, and the pungent, dry-wine, astringent and acid are intermediate."

[2] In Aristotle the unmixed savours, sweet and bitter, would not be expressible by a ratio.

[3] *CP* 6 4. 4 *init.*

[4] The "mastering" of *CP* 6 4. 4. In this sense all flavours are concocted and therefore sweet.

[5] Those who rely on concoction.

πεμμένα τῶν ἀπέπτων), ἑκατέρων πεπεμμένων κατὰ τὴν οἰκείαν πέψιν.

ὑπὲρ μὲν οὖν τούτων ἐν τοῖς ἑπομένοις ἔσται φανερώτερον.

6.3 *τῶν δ' ἄλλων ἐκεῖνο δεῖ λαβεῖν πρῶτον ὃ κοινόν ἐστιν τῆς τῶν φυτῶν οὐσίας· πάντα γὰρ ἐν τῇ γενέσει τῶν χυλῶν μεταβάλλει κατὰ*[1] *τὴν πέψιν ἐξ ἄλλων εἰς ἄλλους χυλούς, ὡς μὲν ἁπλῶς εἰπεῖν, ἐκ τῶν στερητικῶν εἰς τοὺς κατ' εἶδος (οἷον ἐκ πικρῶν καὶ στρυφνῶν εἰς γλυκεῖς καὶ λιπαροὺς καὶ εἴ τις ἄλλος ὁμόστοιχος), ὡς δὲ καθ' ἕκαστον, ὡς ἂν ἡ διάκρισις ἔχῃ τῆς φύσεως, καὶ τὰ μὲν εἰς πλείους, τὰ δ' εἰς ἐλάττους· τὸ*[2] *μὲν γὰρ ἐκ τοῦ πικροῦ (καθάπερ ἐλαία, συνεμφαίνουσα πρὸς τὸ στρυφνόν), τὸ δ'*[3] *ἐκ τοῦ στρυφνοῦ (καθάπερ ἄπιοι καὶ τἆλλα· καὶ γὰρ τὰ ὀξέα πρότερον στρυφνά), τὸ δ' ἐκ τοῦ στρυφνοῦ πρῶτον εἰς τὸν*

§ 6.3: Galen, *De Simpl.* iv. 12 (vol. xi, p. 660. 5–10 Kühn); Plutarch, *Quaest. Nat.* v (913 B).

[1] Heinsius (*per* Gaza) : *μετὰ* U.
[2] ego : *τὰ* U.
[3] U : *τάδ'* u.

although the two groups are each concocted in accordance with its own type of concoction.

This will become clearer in the following discussion.[1]

Concoction Involves Change From One Flavour to Another

6.3 Passing to the other related points we begin with one that is common to all plants as such[2]: in producing their flavour all change in the process of concoction from one flavour to another. To speak broadly,[3] the passage is from the privative flavours to those having a form (as from bitter and astringent to sweet and oily and any other of like order); but to enter into particulars, the flavours depend on the distinction that sets off the nature of one plant from that of another, and some plants pass through a greater number of flavours, some through a smaller. So one fruit passes to its final taste from bitter[4] (like the olive, which has also a hint of the astringent), another from astringent (as pears and the others that belong here; in fact acid fruits start by being astringent), and still another passes first

[1] *CP* 6 6. 3–6 7. 4.

[2] Literally "to the essence of plants."

[3] *Cf. CP* 6 5. 6 "to speak generally and broadly," that is, to pass over the comparatively few exceptions.

[4] Supply "to its fully concocted flavour."

ὀξύν, εἶτ' ἐκ τοῦ ὀξέος εἰς τὸν γλυκὺν (ὥσπερ βότρυς).

6.4 ὡς γὰρ ὅλως εἰπεῖν, πλείστας ἀλλοιώσεις οὗτος[1] λαμβάνει, καὶ ὅλως οἱ οἰνώδεις[2] χυλοί· τὸ μὲν γὰρ πρῶτον ὥσπερ ὑδατώδης γίνεται, μετὰ δὲ ταῦτα στρυφνός, εἶτ' ὀξύς, εἶτ' ἔσχατον γλυκύς. ὡσαύτως δὲ καὶ τὸ συκάμινον· ἐκ στρυφνοῦ γὰρ ὀξύ, καὶ ἐξ ὀξέος γλυκύ· καὶ ἐπ' ἄλλων δὲ τοῦτο συμβαίνει, τῆς γὰρ οἰνώδους γλυκύτητος ἐγγυτάτω κεῖται τὸ ὀξύ. διὸ καὶ οἱ ἀποροῦντες ὅτι[3] τὸ συκάμινον, ἐρυθρὸν ὄν, ὀξύτερόν ἐστιν ἢ λευκόν, ἐγγυτέρω τῆς ὄψεως[4] ὄν, οὐκ ὀρθῶς ἀποροῦσιν· τότε γὰρ οἷον γένεσίς ἐστιν αὐτοῦ τοῦ οἰκείου χυμοῦ, λευκοῦ δ' ὄντος ἡ στρυφνότης πλέον ἀπηρτημένη καὶ κοινοτέρα. διὰ τοῦτο γὰρ καὶ ἐνταῦθα ὅταν ᾖ, ξηρότερόν ἐστιν· ἐρυθραινόμενον δὲ

[1] U^{r} : οὗτως U^{ar} : οὕτω N aP.
[2] οἱ οἰνώδεις Schneider : οἰνώδεις οἱ U.
[3] U : διότι Gaza (*cur*), Heinsius.
[4] U : πέψεως Gaza, Itali.

---

[1] That is, colour. *Cf.* the story of Polyidus: Minos, seeking to find his little son, who had fallen into a great jar of honey and drowned, consulted an oracle, and was

from astringent to acid, then from acid to sweet (like the grape).

6.4 For generally speaking this flavour of the grape (and indeed all vinous flavours) undergoes the greatest number of qualitative changes. So at first the flavour produced is watery (as it were), next it becomes astringent, and finally sweet. So too with the mulberry: it passes from astringent to acid and from acid to sweet, and this passage occurs in other vinous fruits as well, acidity lying nearest to the vinous sweetness. This is why those who raise the difficulty that the mulberry when red is more acid than when it is white, although when red it is closer to its final appearance,[1] are mistaken in seeing a difficulty here, for when the mulberry is red the very savour that belongs to it is (as it were) being produced, whereas when the mulberry is white its astringency is remoter from the proper flavour and less peculiar to the plant. This is also why the fruit is drier when white, but gets more

told that whoever could explain or find a likeness for a prodigy that had occurred among his cattle would find the boy. The prodigy was a cow (or calf) that changed its colour daily from white to red to black. Polyidus compared the prodigy to the *móros* (blackberry or mulberry): *cf.* Sophocles, frag. 395 Pearson: "first you will see the ear flowering white, then the round mulberry turned red; then Egyptian eld takes hold of it"; Aeschylus, *Cressae*, frag. 116 Nauck[2], 54 Smyth: "for at the same time it is heavy with mulberries white and swart and brightly red."

ἐξυγραίνεται, καθάπερ ἔγχυλον γινόμενον.

6.5 ὅλως γὰρ πᾶν [τὸ][1] περικάρπιον, ξηρὸν[2] τὸ πρῶτον, ἀνυγραίνεται, καὶ ἔστιν γένεσις αὕτη τῶν χυλῶν, ἐπιρρέοντος καὶ ὥσπερ διηθουμένου[3] πλείονος ἀεὶ τοῦ ὑγροῦ καὶ ἀεὶ συναύξοντος· ᾗ[4] καὶ στρυφνὰ τὰ πολλὰ κατ᾽ ἀρχὰς οὐκ ἀλόγως, ἅτε καὶ ξηρὰ ὄντα (ξηρότατα δὲ[5] δοκεῖ τῶν ὑγρῶν ὧν ὀπώδης ἡ ὑγρότης, ὥσπερ τῆς συκῆς, κατὰ τὸ πάχος).

τούτοις μὲν οὖν ὥσπερ ὑλικοί τινές εἰσιν οἱ κατὰ τὰς στερήσεις, ἐξ ὧν ἡ μετάβασις εἰς τὴν πέψιν· τοῖς δ᾽ οἷον οὐσία ταῦτα καὶ τέλη, πολλοὶ γὰρ καὶ ψυχροὶ[6] καὶ στρυφνοὶ κατὰ φύσιν καὶ ὀξεῖς, ὧν ἐξ ἀρχῆς ἧττον ἔγχυλος ἡ ὑγρότης.

6.6 ἅπαντες[7] δέ πως ἐμφαίνουσιν οἱ καρποὶ καὶ

§6.5: Galen, *De Simpl.* iv. 7 (vol. xi, pp. 637. 17–638. 3 Kühn).

§6.6: Galen, *De Simpl.* iv. 7 (vol. xi, pp. 636. 16–637. 2 and 637. 17–638. 3 Kühn); iv. 4 (vol. xi, p. 633. 7–13 Kühn).

[1] ego.

[2] ξηρὸν ὂν Wimmer.

[3] Palmerius : διηρθρομένου U.

[4] ᾗ aP : η U : ἤ u N.

[5] Schneider (*sed* Gaza) : τε U.

fluid as it reddens, as if becoming filled with its flavour-juice.

For in general every pericarpion, dry at first, becomes fluid, and this is how the flavours are produced: more and more fluid flows in and strains (as it were) through the plant, and increases the size of the pericarpion more and more. Hence, reasonably enough, most pericarpia are astringent to begin with, since they are dry.[1] (Those fluids are considered driest whose fluidity, like that of the fig-tree, has the thickness of fig-sap.) 6.5

In these fruits, then, the privative flavours have the character of matter (as it were), and the passage is from these flavours to concoction. In some, however, these privative flavours are essence (so to speak) and final goals of the plants, for many flavours are naturally[2] cold or astringent or acid in plants whose fluid is from the start less juicy than that of the rest.

All fruit has a certain taste of the flavour of the 6.6

[1] Plato, *Timaeus* 66 C–D (cited in note 3 on *CP* 6 1. 4; *cf.* *CP* 6 1. 3), says that the astringent dries out the vessels of the tongue. To do this, one may suppose, it must itself be dry.

[2] In their fully developed form, and not as a stage in their progress to it.

---

[6] U : πικροὶ Gaza, Moreliana.

[7] Gaza, Scaliger : ἃ πάντως U.

τὸν τοῦ φυτοῦ χυλόν, ὅπερ ἴσως ἀναγκαῖον, ἐξ ἐκείνου γε ὄντας,[1] ὠμοὶ μὲν ὄντες μᾶλλον, πεπαινόμενοι δ' ἧττον, ἅτε καὶ τῆς ἐκκρίσεως ἀεὶ καθαρωτέρας γινομένης.

καὶ τῶν μὲν τοιαύτη τις σύστασις· ἐνίων δὲ ὥσπερ ἄχυλος ἡ πρώτη καὶ ὑδατώδης, οἷον τῶν σιτωδῶν (ὥσπερ πυροῦ καὶ κριθῆς καὶ τῶν ὁμοίων), ὧν[2] δὴ καὶ τελείωσις[3] εὐθὺς ἐκ ταύτης, οὐ λαμβάνουσα πλείους μεταβολάς, οὐδὲ γὰρ οὐδ' ἡ γλυκύτης πόρρω ταύτης, ὥστε μὴ δεῖσθαι πλειόνων.

6.7 διὸ καὶ φαίνεται διττή τις ἡ γένεσις εἶναι τῶν χυλῶν, ἑκατέρα κατὰ τὰς ὑποκειμένας[4] φύσεις·

[1] Scaliger (*cum ... suam sibi trahant originem* Gaza : ὄντας Itali : γενομένους Heinsius) : γνῶντας U.
[2] u : ὧι (?) U.
[3] ἡ τελείωσις Schneider.
[4] Schneider : συγκειμένας U.

---

[1] *Cf. HP* 1 12. 2: "Speaking broadly, all (*sc.* saps or plant fluids) accord with the distinctive nature of each kind of tree and plant in general; for every kind has its

plant,[1] and this is perhaps necessary, since the fruit has come from the plant. In unripe fruits this taste is stronger, but it becomes fainter as they ripen, since the secretion into them keeps getting purer.

Such then is the formation of the flavour in some. But in others the first formation is flavourless (so to speak) and watery, as in cereals (as wheat, barley and the like), and these are the plants where the finished flavour comes directly from this initial formation and does not pass through several changes; indeed its sweetness too is not far removed from this formative stage, which means that no more stages are required.

The Two Types of Flavour Production

6.7 This is why we find the process of producing flavours to be of two types, each depending on the natures to be achieved. One process changes from

own special tempering of qualities and mixture. The fluid is evidently the one appropriate to the fruit that is the goal, and in most fruits a certain resemblance to the fluid of the plant is found, only not exact or distinct (but it is in the fruit as pericarpion [*sc.* and not in the fruit as seed]). Hence the nature of the flavour-juice undergoes more elaboration (*sc.* than that of the sap) and more concoction into a state that is pure and unmixed (for we must take the one thing [*sc.* the sap] to be as it were matter, the other [*sc.* the flavour] to be form and shape).

ἡ μὲν γὰρ ἐξ ἀπέπτων καὶ πλειόνων μεταβάλλουσα πρὸς τὴν τελέωσιν, ἡ δέ, ὥσπερ ἄχυλος καὶ ἀειδής,[1] ἐν ἁπλῇ τινι γενέσει καὶ ἀλλοιώσει, διὰ τὴν ὑποκειμένην φύσιν. ὧν ἑκάτερον (ὡς ἁπλῶς εἰπεῖν) τὸ μὲν ἐν τοῖς ἐπετείοις μᾶλλον, τὸ δ' ἐν τοῖς δένδροις ἐστίν, ὡς πλείονος δεομένων[2] πέψεως.

6.8 τάχα δ' ἀληθέστερον ἐκείνως[3] ἀποδοῦναι, καὶ μὴ ἐπετείοις καὶ δένδροις, ἀλλ' ὅλως τῇ κράσει διορίζοντας· ὁποία γὰρ ἄν τις αὕτη τυγχάνῃ, αἱ μὲν[4] μεταβολαὶ συνακολουθήσουσιν κατὰ λόγον. ἐπεὶ καὶ ἡ ἀμυγδάλη[5] τὸ πρῶτον ὑδατώδης, ἕως ἂν ᾖ χλωρά, ξηραινομένη δὲ λίπος λαμβάνει, καὶ ἄλλα δέ τινα τῶν καρυωδῶν,[6] ὧν καὶ τὰ κελύφη στρυφνά (τῶν δὲ καὶ ὀξέα, καθάπερ τῶν ἀμυγδαλῶν), ὡς ἐνταῦθα παντὸς ἐκκρινομένου τοῦ περιτ-
6.9 τωματικοῦ καὶ γεώδους. ἡ δ' οὖν φύσις ἔοικεν ὥσπερ ἐκ μεμιγμένων τινῶν χυλῶν ἐκκρίνειν τῇ πέψει τὸν ἁπλοῦν.

[1] U : ἀηδὴς u.
[2] U : δεομένοις Heinsius.
[3] u : ἐκεῖνος U.
[4] [μὲν] Schneider.
[5] u : -αλῇ U.
[6] Scaliger : κυρσιωδῶν U.

flavours that are unconcocted[1] and passes through several flavours to achieve the perfected nature; whereas the other begins with no flavour (as it were) and no form and lies in a simple sort of production and qualitative change, owing to the nature in view. Of the two, broadly speaking, the second type is found more in annuals, the first in trees, since tree fruit needs more concoction.

6.8 But perhaps it is truer to allot the two types in the earlier way,[2] and not by distinguishing by annuals and trees, but simply by the tempering of qualities, for the character of the tempering will account for the changes that lead up to it. Thus the almond fruit too[3] is at first watery (so long as it is green), but gets drier and acquires oiliness, and so too do certain nuts, which moreover have astringent shells (even acid in some, as the almond), which would come from the secretion into the shells of all that is excremental and earthy. 6.9 Here at all events the nature of the tree appears in the concoction to single out the simple flavour from a group of mixed ones, as it were.

[1] "Unconcocted" is here used of strong flavours of the privative type, as the astringent, bitter and acid of *CP* 6 6. 3.

[2] *CP* 6 6. 7: "depending on the nature to be achieved"; "owing to the nature in view."

[3] Like the cereals.

ᾗ[1] καὶ δόξειεν ἂν ἐναντίως καὶ[2] ἐπὶ τῶν ἄλλων· ἐν[3] μὲν γὰρ τοῖς ἄλλοις ἐκ μίξεως ἡ γένεσις, ἐνταῦθα δ' ὥσπερ ἐξ ἀφαιρέσεως καὶ χωρισμοῦ· καὶ ἔνθα μὲν ἐκ τῶν ἀρχῶν, ἔνθα δ' αὐτῆς τῆς ἀρχῆς (ἀρχὴ γὰρ τὸ γλυκύ, τούτου δ' ἡ πέψις ὡς ἁπλῶς εἰπεῖν).

6.10 ἀλλὰ ταῦτα μὲν ἴσως[4] ἄλλην τινὰ ἔχει θεωρίαν· ἐπεὶ[5] κἀκεῖνο ἄτοπον, εἰ οἱ χυλοὶ πάντες ἐκ τοῦ πικροῦ καὶ γλυκέος μιγνυμένων, ὥσπερ ἐκ τῶν στοιχείων τὰ σώματα· πάλιν δ' αὖ θάτερον,[6] εἰ ἐκ μὲν τούτων μηδέν, ταῦτα δ' ἐκ τῶν ἄλλων, οὐ γὰρ ἀρχῶν ἡ τοιαύτη φύσις.

εἰ μὴ ἄρα πλεοναχῶς αἱ ἀρχαὶ καὶ κατὰ πλείους τρόπους, ὅπερ ἔοικεν, ὥσπερ κατὰ συστοιχίαν ἐνίων λεγομένων.

[1] ᾗ aP : ἡ U (ἢ N).
[2] ᾗ καὶ Schneider : ᾗ Moreliana.
[3] u aP^c : ἐκ U : εἰ N P^ac (?). [4] Coray : ὡς U.
[5] ego : ἔπειτα U. [6] ego : αὐθ' ἕτερον U.

---

[1] *Cf.* Aristotle, *On Sense*, iv (42 a 12–13): ". . . savours come from sweet and bitter . . ."

[2] "In a row" because no member of the row is prior to any other. *Cf.* Aristotle, *Metaphysics* A 5 (986 a 22–26): "Another group among the same persons (*sc.* among the so-called Pythagoreans) say that the primary things are

Difficulties in This View

In so doing it would seem to act in an opposite way to what it does in the rest: in the rest the flavour is produced by admixture, whereas here it is due to subtraction and separation; and there the primary flavours are the sources, whereas here the primary flavour is the result (sweet being primary, and concoction producing the sweet, broadly speaking).

6.10 But perhaps these matters are to be studied from another approach. Indeed the one view is absurd, that all flavours come from mixing the bitter and the sweet,[1] as bodies come from the elements; and so again is the other, that nothing is produced from these two, but these two come from the rest, since this is not the nature of primary things.

Solution

Unless "primary things" is an expression with more than one meaning, and there is more than one way of being "primary," which appears to be the case (some "primary things" being presented in a row, as it were[2]).

;en, the ones presented in a row: limited indeterminate, odd even, one many, right left, male female, at rest in motion, straight curved, light darkness, good evil, square oblong."

ἀλλ' οὖν δὴ τούτων μὲν πέρι λόγος ἕτερος.

7.1 αἱ δὲ γενέσεις τῶν χυλῶν ἐν τοῖς εἰρημένοις τρόποις, ἢ ἐκ τοῦ ἀειδοῦς[1] εἰς εἶδος, ἢ ἐκ τῶν ἐναντίων εἰς τοὺς ἐναντίους ἀλλοιουμένων. κοινὴ δ' ἡ ὕλη πάντων τὸ ὑγρόν· ἀλλοιοῖ δὲ καὶ ποιεῖ τὸ θερμὸν τὸ[2] ἐν αὐτῷ, καὶ τὸ[2] τοῦ ἡλίου, τοῦτο γὰρ (ὥσπερ εἴρηται) μάλιστα τῶν φυτῶν οἰκεῖον, οὐχ ὥσπερ τῶν ζῴων τὸ ἐν αὐτοῖς, ἐπεὶ[3] μεταβολαί γε καὶ ἀλλοιώσεις καὶ ἐν ἐκείνοις εἰσὶ τῶν χυλῶν.

7.2 Δημοκρίτῳ μέν γε πῶς ποτε ἐξ ἀλλήλων ἡ γένεσις ἀπορήσειεν ἄν τις. ἀνάγκη γάρ, ἢ τὰ σχήματα μεταρρυθμίζεσθαι, καὶ ἐκ σκαληνῶν καὶ ὀξυγωνίων περιφερῆ γίνεσθαι· <ἢ>[4] πάντων ἐνυπαρχόντων, οἷον τῶν τε τοῦ στρυφνοῦ καὶ ὀξέος καὶ γλυκέος, τὰ μὲν[5] ἐκκρίνεσθαι, τὰ τῶν

§7.1: Galen, *De Simpl.* iv. 7 (vol. xi, p. 637. 10–14 Kühn); iv. 11 (vol. xi, p. 654. 6–10 Kühn).

[1] ego : εἴδους U. [2] u : τῶ U.
[3] u : ἐπι U.
[4] Wimmer. [5] τὰ μὲν U in [ ] below : ἡ U here.

---

[1] *CP* 6 6. 7. [2] The final or formal cause.
[3] The material cause. [4] The efficient cause.

But this belongs to a different discussion.

Flavour Production: Matter and Agent

Flavours arise in the ways mentioned,[1] by alteration from what lacks a form to what possesses one or from flavours of one type to flavours of the opposite type.[2] The matter common to all is fluidity,[3] and the agent[4] that alters the quality and produces the flavours is the internal heat of the plant and the heat of the sun,[5] this last (as was said)[6] being the heat most appropriate to plants, and not the internal heat as in animals (there being changes of kind and quality of flavours in animals as well). 7.1

Flavour Production: The Difficulties of Democritus with Matter and Agent

One might raise the problem of how the production from one another is to occur for Democritus. For his figures must either assume a new shape and become curved instead of irregular and sharp-angled; or all the figures must be present (as those of the astringent, acid and sweet), and the figures 7.2

[5] Aristotle in this connexion speaks only of the internal heat: *cf. On Sense*, iv (441 b 15–19), cited on p. 202 on *CP* 6 1. 1, and iv (442 a 4–8), cited in note 1 on *CP* 6 4. 2.
[6] *CP* 2 6. 2.

πρότερον ἀεί, θάτερα δ' ὑπομένειν· ἢ τρίτον, τὰ μὲν ἐξιέναι, τὰ δ' ἐπεισιέναι. ἐπεὶ δ' ἀδύνατον μετασχηματίζεσθαι (τὸ γὰρ ἄτομον[1] ἀπαθές), λοιπὸν τὰ μὲν εἰσιέναι, τὰ δ' ἐξιέναι,[2] ἢ <τὰ μὲν ἐξιέναι>, [πάντων ἐνυπαρχόντων, οἷον τοῦ τε στρυφνοῦ καὶ ὀξέος καὶ γλυκέος τὰ μὲν ἐκρινεσθαι τὰ τῶν πρότερον ἀεὶ ·] τὰ δ' οἰκεῖα καθ' ἕκαστον [θάτερα δ'] ὑπομένειν. [ἢ τρίτον, τὰ μὲν ἐξιεναι, τὰ δ' ἐπεισιεναι].[2] ἄμφω δὲ καὶ ταῦτα ἄλογα· προσαποδοῦναι γὰρ δεῖ καὶ τί τὸ ἐργαζόμενον ταῦτα καὶ ποιοῦν.

ἀλλὰ γὰρ τούτῳ[3] μὲν ὑπὲρ πλειόνων ἴσως ὁ λόγος, ἁπάντων γὰρ οὕτω ποιεῖ τὰς γενέσεις, οἷον οὐσίας, πάθους,[4] ποσότητος.[5]

7.3 τῶν [δ'][6] ἄλλων δ'[7] ὑπὸ τοῦ θερμοῦ πέψις γινομένη τὰ μὲν ἐκκρίνει καὶ διατμίζει, τὰ δὲ παχύνει καὶ συνίστησιν, τὰ δὲ λεπτύνει, τὰ δὲ ὡς ἁπλῶς εἰπεῖν ἀλλοιοῖ, καθάπερ καὶ ἐν τοῖς πυρουμένοις. ὅλως δὲ τοῦτο ληπτέον, ὅτι πάντες

[1] Scaliger : ἄτοπον U.

[2–2] ego : ἢ τὰ μὲν ἐξιέναι, τὰ δὲ ὑπομένειν Schneider : ἢ τὰ μὲν ὑπομένειν τὰ δ' ἐξιέναι Wimmer : aP omit.

[3] U : τούτου u : τοῦτο N aP.

[4] U : ποσοῦ Gaza (*quantitatis*), Heinsius.

belonging to the earlier stages must be separated[1] out while the figures belonging to the final flavour remain; or thirdly some figures must leave and new ones enter. Since it is impossible for the figures to be reshaped, for nothing can affect the atom, what is left is that new figures should enter while the old depart, or that the old depart while the figures of the new flavour remain. But these two suppositions are also both unreasonable, since a further explanation is required: what brings all this about and produces the finished savour?

But we do not pursue the question; Democritus must explain much more perhaps than this, since this is how he produces all change, as of substance, affections and quantity.

How the Agent Operates

7.3 Concoction, operated by heat, separates some things from the rest and vaporizes them, thickens and sets others, reduces others to fine particles, and changes others (to speak broadly) in quality, just as it does in cooking. We must take it as a general

[1] For an argument against separating out some constituents, *cf.* Aristotle, *On the Heaven*, iii. 7 (305 a 35–b 28).

---

[5] ego : ποιότητος U.
[6] ego.
[7] U N P (ἡ Wimmer) : a deletes.

ἐνυπάρχουσιν οἱ χυμοὶ δυνάμει πρὸς οὓς ἡ μεταβολή,[1] τότε δὲ[2] γίνεσθαι καὶ εἶναι κατ' ἐνέργειαν. πεφθέντων δ', οἱ μὲν εὐθὺς φανεροὶ τῇ αἰσθήσει, καθάπερ ὁ γλυκὺς καὶ ὁ λιπαρός, καὶ γὰρ ἀποπιεζόμενος οὗτός γε ἐξ ἐνίων ῥεῖ, καὶ χωρι-
7.4 σμὸς ἔκ γε[3] τῶν πλείστων οὕτως[4] [αὐτοῦ].[5] οἱ δ' οὐχ[6] ὁμοίως ἔνδηλοι, καθάπερ ὁ οἰνώδης ἐν τῷ μύρτῳ, ἔτι δ' ἧττον ὁ λιπαρός (ἔχει γὰρ καὶ τοιοῦτον, ὃς ἐφίσταται τῷ οἴνῳ σακκιζομένων,[7] καὶ ἀφαιροῦσιν αὐτὸν καὶ κάουσιν ἐπὶ τῶν λύχνων).

ἀκρατέστεροι δὲ πάντες γίνονται χωρισθέντες τῶν περικαρπίων, καὶ ὅσοι δὲ χρονίζονται, μᾶλλον. τότε γὰρ ἥ τε τοῦ ὑδατώδους ἀποπνοή,[8] καὶ ἡ τοῦ γεώδους ὑπόστασις, καὶ τὸ ὅλον ἡ τοῦ θερμοῦ τοῦ ἐν ἑαυτῷ[9] δύναμις, ἐπεὶ καὶ ἀφαιρεθέντων τῶν καρπῶν ἐν αὐτοῖς τούτοις ἀλλοιοῦνται·

§ 7.4: *Cf.* Pliny, *N. H.* 15. 118; 124.

[1] u (*in mutatione quae fuerit* Gaza : προσούσης [προσιούσης Schneider] δὲ μεταβολῆς Heinsius) : προσούσηι μεταβολῆι U : προσούσῃ μεταβολῇ N aP.
[2] ego (τότε Schneider) : το δε U. [3] Schneider : τε U.
[4] U : οὗτος Schneider. [5] ego.
[6] οὐχ' u : ὀχ' U. [7] ego : -νωι U.
[8] N aP : ἀποπνοιὴ U. [9] u (ε- U : ἑαυτῶ N P) : αὐτῷ a.

premise that all the savours to which the change is directed are present potentially and in the process are brought into being and exist actually. When concoction is complete some of the savours are immediately evident to perception, as the sweet and the oily (the latter indeed will flow from certain fruits under pressure, and this is how the flavour is separated from most fruit); other savours are not so noticeable, 7.4 as the vinous savour in the myrtle berry, still less its oily savour (the berry having also an oil that remains on the surface of the wine into which the berries are strained through a cloth,[1] the oil being then removed and burned in lamps).

Changes After Extraction

All flavour-juices become stronger in taste on extraction from the pericarpia, and do so more when kept for some time. For then the watery part evaporates and the earthy part settles, and in a word the heat within the juice does its work.[2] In fact the flavours change in quality in the fruit itself after its

[1] *Cf.* Palladius, *On Agriculture*, iii. 31. 1: "The Greeks also have this recipe for mixing myrtle wine: put ripe myrtle berries that have been dried in the shade and then crushed, to the amount of eight *unciae*, in a linen bag which you hang in a container of wine. Then cover the container and seal it air-tight . . ."

[2] *Cf.* Galen, *De Simpl.* iv. 11, translated p. 462 below.

καὶ τά γε κάρυα, καὶ ὅσα γε τοῦτον τὸν τρόπον ἐλαιώδη, χρονιζόμενα μᾶλλον ἐξελαιοῦται διὰ τὰς εἰρημένας αἰτίας. ᾗ[1] καὶ φανερὸν (ὥσπερ ἐλέχθη) διότι δυνάμει πάντες ἐνυπάρχουσιν οὗτοι· τὰς δὲ ἐνεργείας αἱ ἀλλοιώσεις ποιοῦσιν.

7.5 εἰσὶ δ', ὥσπερ ἐν αὐτοῖς τοῖς περικαρπίοις, ὡρισμέναι μεταβολαὶ χωρισθέντων, ὁμοίως εἴς[2] τε τοὺς κατὰ τὰς πέψεις, καὶ εἰς τοὺς κατὰ τὰς φθοράς. μάλιστα δὲ τοῦτ' ἔνδηλον ἐπὶ τοῦ οἴνου· καὶ γὰρ παριστάμενος, καὶ ἐιστάμενος καὶ οἷον γηράσκων, ἐν ὡρισμένοις τισὶ μεταβάλλει χυλοῖς· ἀλλοιωθεὶς[3] γάρ, εἰ[4] μὲν ὥσπερ κατὰ φύσιν, ὅταν[5] παλαιούμενος, ἐκπεπίκρωται (τοῦτό γε[6] συμβαίνει διότι τὸ πότιμον ὁ ἀὴρ ἐξάγει καὶ τὸ περιέχον· ἀπιόντος γὰρ τούτου, καταλείπεται τὸ γεῶδες καὶ πικρόν)· ὁ δ'[7] ὥσπερ βίᾳ καὶ παρὰ φύσιν, εἰς τὸ

[1] a after Gaza : ἡ U (ἦ u N P).
[2] u : ὁμοιώσεις U.
[3] u : ἀλλοιωθεισης U. [4] U : ὁ Wimmer.
[5] ὅτ' ἂν U : ὁ Wimmer : Schneider deletes.
[6] U : δὲ Wimmer.
[7] U (ὃ δ' a) : εἰ δ' Gaza (*Sed si*), Schneider : ὅδ' u N P.

---

[1] *Cf.* Aristotle, *On Sense*, iv (441 a 11–17): "For we observe that savours are changed by the heat when the

removal from the tree,[1] and nuts and all fruits that are oily in this way get oilier in storage for the reasons mentioned.[2] This also makes it evident that these flavours are all present potentially (as we said[3]), the actualizations being produced by the qualitative changes.[4]

7.5

The Changes in Flavour are Fixed

Just as when the flavour is left in the pericarpia, so after extraction from them, the change, whether into a flavour of concoction or one of corruption, is restricted. This is most noticeable in wine: both when it becomes fit to drink and when it turns and (as it were) gets old, the change is confined to determinate flavours. So when the qualitative change is complete, if the change has been (so to speak) natural, by aging, the wine has become bitter (this happens because the air and the container extract the potable portion, for when this is gone the earthy and bitter component remains); but when the wine suffers alteration forcibly (as it were) and

pericarpia are taken from the tree and set out in the sun and also when they are exposed to fire . . . , and they lose their fluid both when they are kept . . . and when they are boiled (*i.e.* fermented) . . ."

[2] In the preceding sentence; *cf. CP* 6 8. 8.

[3] *CP* 6 7. 3.

[4] That is, flavours were not imported into the fruit ready-made.

ὀξύ (τοῦτο γὰρ ἐναντίον). ἔχει δὲ καὶ ἐπὶ τῶν ἄλλων παραπλησίως.

7.6 συμβαίνει <δὲ>[1] τῷ οἴνῳ τὴν ἐκστατικὴν ταύτην ποιεῖσθαι φθορὰν ἐξ οὗπερ καὶ ἡ φυσικὴ γένεσις· ἐκ γὰρ τοῦ ὀξέος, καὶ εἰς τὸ ὀξὺ καθάπερ εἰς τὴν ὕλην ἀναλυόμενον. μεταβάλλειν δὲ καὶ[2] ἀποκαθίστασθαι πάλιν[3] συμβαίνει μὲν ὡσαύτως, σπανίως <δέ>,[1] καὶ μάλισθ' ὅταν ἢ θαλαττωθῇ, ἢ ἐπὶ τῆς τοῦ ἄστρου ἐπιτολῆς παρακινήσῃ (καὶ γὰρ τότε παραπλήσιον τὸ πάθος, ἂν μὴ γένηται σφοδρόν).

ἀλλ' ὑπὲρ μὲν τῶν τοιούτων ἐν τοῖς ἑπομένοις οἰκειότερον ἐπελθεῖν.

§ 7.6: *Cf.* Pliny, *N. H.* 14. 118 and 2. 107.

[1] Schneider.
[2] U : μεταβάλλει. Schneider : μεταβάλλει. καὶ Wimmer.
[3] δὲ πάλιν Schneider.

---

[1] The opposite of sweet. *Cf.* Galen, *On Simple Medicaments*, iv. 12 (vol. xi, p. 656. 4–15 Kühn): "But since in the present discussion we have treated the natural change of wines in the course of time, let us also say something about the unnatural change. So just as astringent wine in time becomes first sweeter, next more pungent and bitterer, and finally turns bitter, in the same way wine that has suffered a chill at once becomes more acid, and if completely

unnaturally, it passes to acid, acid being the opposite.[1] And the case is similar with other separated fruit-juices as well.

7.6 It so happens that this corruption of wine results in the very same flavour from which the production of the juice began in nature: the juice began with acidity,[2] and the wine ends with acidity, as if resolved into its matter. And it happens that it changes to acid and is restored again to the uncorrupted state, but seldom, and especially when sea water is added or the lees get disturbed at the rising of the dog-star[3] (for then too the result, if the wine is not too violently disturbed, is much the same).

But such matters are more properly treated in what follows.[4]

chilled becomes completely acid. And wine that has been filled with foreign fluid from the very start of harvesting, owing to prevailing rain, and wine that in some other way has acquired a component of water, all such wines readily turn acid on the slightest occasion."

[2] The grape passes from astringent to acid to sweet (*CP* 6 6. 3), or from watery (as it were) to astringent to acid to sweet (*CP* 6 6. 4); in both cases sweetness comes immediately from acidity.

[3] Presumably both the sea water and the shaking in the hot season result in more heat, which reverses the change brought about by cold: *cf.* note 1 on *CP* 6 7. 5.

[4] In the lost seventh book, which dealt with flavours produced by man (G. R. Thompson).

7.7 ἡ δὲ φθορὰ καὶ ἡ ἀλλοίωσις ἐν τεταγμένοις πάθεσιν· ἐπεὶ καὶ τὸ μέν, διαφθειρομένων,[1] ἐξυγραίνεται (πολλάκις δὲ ἐξοξύνεται), γηράσκον δὲ καὶ χρονιζόμενον ἀποξηραίνεται, καταλειπομένου καὶ ἐνταῦθα τοῦ γεώδους.

ὅλως δ' αἱ φθοραὶ πάντων ἢ καταμίξει τοῦ ἀλλοτρίου, ἢ ἐκλείψει τοῦ οἰκείου διὰ χρόνον. αἱ δὲ καταμίξεις ὁτὲ μὲν τῶν ἔξω [τινος][2] μιχθέντων, ὁτὲ δὲ τῶν αὐτοῦ, νοσήσαντός τινος καὶ πλεονάσαντος· τάχα δ' ἀληθέστερον εἰπεῖν ὅτι κινηθέν-

7.8 τος τινὸς καὶ διαφθαρέντος ὑπὸ τῶν ἐκτός, οὕτως γὰρ καὶ ἡ τῶν περικαρπίων φθορά, καὶ αὐτῶν τῶν ὑγρῶν (οἷον οἴνου γάλακτος). οὐδὲν δ' ἧττον τεταγμένη καὶ ὑπὸ[3] τούτων ἡ μετάβασις· ἐναντία γὰρ πᾶσιν [τῶν],[4] αἱ δὲ ἐναντιότητες πλείους.

ὥστε τοῦτο μὲν φανερόν.

[1] ego : -όμενον U.

[2] ego : τινος U : -ὸς u : -ῶν N aP.

[3] U : *in* (*his*) Gaza : Schneider deletes : ἐπὶ Wimmer.

[4] Schneider : τῶν U N : τούτων aP. (Relic of a variar πάντων.)

Changes in the Fruit On Removal

The corruption and qualitative change is fixed in its bodily effects. So, as the pericarpia spoil, a part turns fluid (and often gets acid); but as it ages and is kept long, the fluid dries out, the earthy portion here too being left behind.[1] 7.7

In general the corruption in all is either brought about by intermixture of what is alien or disappearance through time of what belongs to the fruit. The alien intermixtures are sometimes due to admixture of things external, sometimes to admixture of things internal (when some part gets diseased and spreads). But here it is perhaps truer to say that a part has been affected and corrupted by external things, for this is how the corruption not 7.8 only of pericarpia but also of free fluids occurs (as of wine and milk). But the transition brought about by these external things too is no less fixed than the others,[2] since in all these cases the transition is to an opposite (there being several oppositions[3]).

This point is then clear.

[1] As with wine: *cf. CP* 6 7. 5 *ad fin.*

[2] *Cf. CP* 6 6. 3–6.6. 10 (of the living flavours); *CP* 6 7. 5 (of flavours in the separated juice).

[3] Such as formless (watery)—with form (*CP* 6 6. 6; 6 7. 1); "privative" (strong and unpleasant)—"with form" (sweet or oily or the like) (*CP* 6 6. 3; 6 7. 1); vapoury—thickened (*CP* 6 7. 4).

ἡ δὲ πέψις πάντων τῶν καρπῶν γίνεται μὲν ὑπὸ τοῦ θερμοῦ (καθάπερ εἴρηται), δοκεῖ δὲ τῶν ὀψικάρπων ὑπὸ τοῦ ψύχους, διὰ τὴν ὥραν. πέττει γάρ, ὥσπερ ἀεὶ λέγομεν, τὸ θερμὸν ἀντιπεριστάμενον.

8.1 ἴδιον δ' ἐπὶ τῆς ἐλάας τὸ συμβαῖνον (εἴπερ ἀληθές)· λέγουσιν γὰρ ὡς οὐδὲν πλεῖον ἴσχει τοὔλαιον[1] [ἢ][2] μετ' Ἀρκτοῦρον ἢ ὅσον λαμβάνει τοῦ θέρους, ἅμα δὲ καὶ ὁ πυρὴν τότε γίνεται σκληρός, μεθ' ὃν οὐκέτι δύνανται τὸ ὑγρὸν ἐξελαιοῦν· ὥστ' εἰ μὲν τῷ φυσικῷ δεῖ πάθει τῷ[3] συμπτώματι τὸν ὅρον λαμβάνειν, τῷ πυρῆνι[4] ληπτέον, εἰ <δὲ>[2] τῇ ὥρᾳ, τῇ τοῦ ἄστρου [δύσει].[5] τάχα <δ'>[2] ἄμφω συμβαίνει διὰ τὴν αὐτὴν αἰτίαν, τοῦ μὲν οἷον πεττομένου,[6] <τοῦ δ' ἐλαιουμένου τῇ πλείονι θερμότητι.

[1] u : τοῦ λαιον U : τοῦ ἐλαιῶν N : τοῦ ἐλαιοῦν aP.
[2] aP.
[3] τῷ ego (καὶ μὴ Schneider) : και U (καὶ N) : καὶ τῶ aP.
[4] Wimmer (*qualitati aeris ambientis* Gaza : ἀέρι Schneider) : πυρὶ U. [5] ego.
[6] u : πετομενου U.

---

[1] *CP* 6 7. 1.
[2] *Cf. CP* 2 8. 1.

An Error About the Agent

Although the concoction of all fruits is brought about by heat (as we said),[1] it is believed to be brought about in late-fruiting trees by the cold, owing to the lateness of the season.[2] The explanation is that the heat concocts them (as we keep saying[3]) —when it is concentrated by the cold.

A Disputed Peculiarity of Olive Oil Production

8.1 What occurs in the olive, if true, is isolated: it is said that after the rising of Arcturus[4] the fruit gets no increase of oil beyond the amount it received in summer,[5] and that at this time the stone also gets hard, after which date the trees are no longer able to turn their fluid to oil; so that if we are to get the terminus to the event by the natural[6] affection, we must set it by the hardening of the stone, if by the season, by the season of the star. But perhaps both occurrences[7] are due to the same cause, the stone being (as it were) concocted, and the fluid changed to oil, by the greater heat.

[3] *CP* 2 8. 1; *cf. CP* 1 12. 3; 2 6. 1; 6 8. 8.
[4] The end of summer and beginning of autumn: *cf. CP* 1 10. 5; 1 13. 3; *HP* 1 9. 7; 1 14. 1; 9 8. 2.
[5] *Cf. CP* 1 19. 3–4. [6] That is, of internal origin.
[7] That is, the cessation of production and the hardening of the stone.

8.2 τοῦ>το[1] διαμφισβητοῦσί τινες, καὶ τῷ εἰκότι προσάγοντες (ὡς ἄτοπον τὸ[2] μηδὲν πληθύεσθαι τῆς τροφῆς καὶ ὑδάτων ἐπιγινομένων) καὶ τῇ αἰσθήσει (φανερὸν γὰρ εἶναι καὶ τῇ πείρᾳ· λευκὰς γὰρ τριβομένας ἔλαττον ἀφιέναι, πασῶν δὲ ἄριστα ῥεῖν τὰς ἐσχάτας συλλεγομένας, κεχειμασμένας[3] τε μάλιστα καὶ τετελεωμένας, διὰ μὲν[4] γὰρ τὴν τελέωσιν <πλεῖον>,[5] διὰ <δὲ>[6] τὸ κεχειμάσθαι, τοῦ ὑδατώδους ἀφῃρημένου, τὸ λοιπὸν ἐκπέττεσθαι μᾶλλον).

ὁ μὲν οὖν λόγος ὁ ἀντιλεγόμενος οὗτος.

8.3 ἐνδέχεσθαι δὲ δοκεῖ καὶ φαίνεσθαι πλεῖον διὰ τὸ[7] ὑδατῶδες καὶ τὴν[8] ἀμόργην, ἐπεὶ ὅτι γ' ἔχουσι τὸ ἔλαιον πρὸ τοῦ μελανθῆναι φανερόν, καὶ ὅτι καθαρώτερον καὶ λευκότερον (ἐκ γὰρ τῶν

[1] ego : <. . .> Wimmer : τως U.
[2] u : τω U.
[3] ego (καὶ κεχειμασμένας Schneider) : καὶ ἀχειμασμένας U.
[4] [μὲν] Wimmer.
[5] ego.
[6] ego : καὶ διὰ Schneider.
[7] u aP : τῷ U : τοῦ N.
[8] u aP : τὸν U N.

Objections

Some dispute the fact not only with an appeal to likelihood but with an appeal to sense. The appeal to likelihood is that it is strange that no food should be left over[1] at a time when there is also rain[2]; and the appeal to sense is that the fact is clear from experience: olives pressed when green yield less oil, and the best yield of all is from the olives gathered last, which have been most exposed to cold and have reached their full development, the greater amount being due to their development, and the removal of the watery portion and consequent better concoction of the rest being due to exposure to cold. 8.2

This then is the argument presented in rebuttal.

A Reply

But it is held that it is possible that the greater yield may be only apparent, owing to the presence of the watery part and the dregs, since the fact that olives contain the oil before they turn dark is evident, and that the oil is purer and lighter in colour (so the white oil comes from coarse olives when they 8.3

[1] From feeding the flesh.

[2] At *CP* 1 10. 3 Theophrastus speaks of the first rains as coming "after Arcturus."

φαυλίων ὠμῶν[1] τὸ λευκόν), τὸ δ' ἐμφαίνειν τινὰ πικρότητα τῶν λευκῶν θλιβομένων οὐδὲν ἄτοπον, ᾗ[2] ὕστερον ἐκπεφθεῖσιν[3] παύεται· τοῦτο γὰρ οὐκ ὀλιγότητος καὶ πλήθους σημεῖον, ἀλλ' ὅτι συναπολαύει[4] τι τοῦ πέριξ ὁ χυλός, ὥσπερ καὶ τῶν ἀγγείων. ὅτι δὲ οὐκ ἀναμένει τὴν τοῦ περικαρπίου πέψιν φανερόν, ῥεῖ γὰρ ἐκ τῶν λευκῶν· ὥστε οὐδὲ δυνάμει καὶ ἐνεργείᾳ διαιρεῖν ἔστιν ὅτι τὴν ὕλην ἐξ ἧς μέλλει πρότερον ἔχει.

8.4 τίς οὖν ἡ αἰτία, καὶ τίς ὁ τρόπος, εἴπερ ἀληθές, ἢ ἁπλῶς ἢ ὡς βραχεῖ τινι πλέον; ἀμφότερα γὰρ ἄλογα, μεῖζον δὲ θάτερον.

ἢ πρῶτον μὲν οὐχ ἡ αὐτὴ πέψις τοῦ τε χυλοῦ, καὶ τοῦ περικαρπίου πρὸς ἐδωδήν; τὸ μὲν[5] γὰρ

[1] Schneider : ὠμὸν U.
[2] u : ἦ U : ἢ N : ἢ εἰ aP.
[3] u : -εμφ- U.
[4] Gaza (*affici*), Schneider : συναπὄλλυσι U.
[5] u : τὸ με U : τὸ N : τοῦτο aP.

---

[1] *Cf.* Theophrastus, *On Odours*, iv. 15 (on the oils used as a base for perfumes): ". . . indeed the olive oil most used is that which is pressed from coarse olives when they

are unripe).[1] That a certain bitterness should be noticeable in the oil when green olives are pressed, a bitterness which is not found later in the fully ripened fruit, is not at all strange: this is no proof that the amount of oil increases later, but merely shows that the flavour-juice is tainted by the surrounding flesh, as by a container. It is evident that the oil does not wait for the pericarpion to be concocted, since oil is pressed from green olives. In consequence one cannot distinguish here between potentially and actually containing oil, and say the olive at this earlier period has the matter from which it is going to produce the oil.

Solution

What then is the cause, and in what way does this happen, supposing that the report is either true as it stands or true with the qualification that only a slight increase of oil occurs after Arcturus? For no increase or only a little are both unreasonable, the first more so than the second. 8.4

Is the answer as follows? In the first place the concoction that produces the flavour-juice[2] and the concoction that makes the pericarpion fit to eat are not the same: the pericarpion must be made agree-

are unripe, for the coarse olive is considered to have the oil that is least fatty and has the finest particles."

[2] The oil.

δεῖ προσφιλὲς εἶναι τῇ γεύσει, τοῦτο δ' ἐν ἀλλοιώσει τοῦ γεώδους (ὡς δ' ἔνιοί φασι σήψει·[1] καὶ γὰρ ἡ δρυπεπὴς ἐν σήψει)· τὸ δ' ἔλαιον αὐτῇ τῇ τοῦ χυλοῦ μεταβολῇ. τοῦτο δ' ὅσῳ[2] χαλεπώτερον, οὐκ ἄλογον ὑπὸ τοῦ πλείονος θερμοῦ δημιουργεῖσθαι, καὶ εὐθὺς ἐν ἀνεπιμίκτῳ[3] χυλῷ τῷ γεώδει

8.5 καὶ ὑδατώδει. πρὸς Ἀρκτοῦρον μὲν οὖν ἄμφω ταῦτ' ἐστίν· καὶ γὰρ ἡ ὥρα θερμοτέρα καὶ ὁ χυλὸς ἀμιγέστερος. μετ' Ἀρκτοῦρον δ' ἐναντίως· αὔξεται γὰρ ἡ σὰρξ τότε καὶ τελεοῦται.

ὅτι δ' οὐκ ἐν πλήθει τροφῆς, οὐδ' ἐν εὐσαρκίᾳ τῶν ἐλαῶν τὸ τοῦ ἐλαίου πλῆθος, ἐκ πολλῶν φανερόν· αἵ τε γὰρ ἐπομβρίαι ποιοῦσιν ἔλαττον, αἵ τ' ἀρδόμεναι χεῖρον ῥέουσιν, ὧν δὲ ἡ σὰρξ πολλή,

[1] u : σηψιν U.
[2] ὅσῳ u aP : ὅσα U N.
[3] ἐν ἀνεπιμίκτῳ ego : ἐνεπιμίκτῳ Gaza (*minus ... permistus est*), Schneider : ἐναπομίκτω U : ἐν ἀπομίκτω u : ἀναπομίκτω N aP.

---

1 "Decomposition" is an old term for concoction (or qualitative change): *cf. CP* 2 9. 14; 6 3. 5; for the verb *cf. CP* 3 11. 6 and Empedocles' saying that wine is "water decomposed in a piece of wood" (Frag. B 81, Diels-Kranz, *Die Fragmente der Vorsokratiker*, vol. i^{10}, p. 340. 32). Theophrastus does not like to use "decomposition" of a change for the

able to the taste, and this means that the earthy component must undergo qualitative change (some term this "decomposition";[1] so the tree-ripened olive is said to be in a state of "decomposition"[2]); whereas the oil is due to the change of the juice and of nothing else. To the extent that this last is the harder task, it is not unreasonable that the operation should be performed by the greater heat and on juice before it is intermixed with the earthy and watery.[3] Now as the season draws on towards the rising of Arcturus both these conditions for producing oil exist: the season is hotter than it is later and the juice freer of admixture. But after the appearance of Arcturus the opposite is the case, for then the flesh grows and is matured.

8.5 That abundance of oil does not depend on abundance of food nor yet on fleshiness in the olives is evident from many considerations: rainy weather reduces the production of oil and the trees that are watered yield less to the oil-press; and olives with much flesh in proportion to the stone contain little

better. Here however the change involves the cessation of the production of oil.

[2] *Cf. CP* 2 8. 2; *HP* 4 14. 10: "Worms also occur in tree-ripened olives (the very ones that are inferior in their yield of oil), and are held in general to be products of decomposition."

[3] The earthy and watery component is increased by the rains, both those of summer and of autumn.

ὁ δὲ πυρὴν μικρός, ὀλιγοέλαιοι, καθάπερ αἱ φαύλιαι,[1] ὡς ἐνταῦθα τῆς φύσεως κεκμηκυίας. οὐδὲ

8.6 δὴ ἐν τοῖς ψυχροῖς ἴσχουσιν,[2] ἀλλὰ σάρκα πολλήν.

ὃ καὶ ἀποροῦσιν· διὰ τί ἡ μὲν ἄμπελος ἐν τοῖς χειμεριωτάτοις χυλόν,[3] ἡ ἐλάα δ' οὔ;

αἴτιον δὲ τό τε νῦν εἰρημένον, ὅτι εἰς τὴν σάρκα ἡ δύναμις, τοῦ βότρυος δ' ὁ χυλὸς ἐν αὐτῇ τῇ σαρκί· καὶ ἔτι πρότερον ἴσως καὶ κυριώτερον, ὅτι ἡ θερμότης ἡ ποιοῦσα τοὔλαιον[4] ἀσθενής· ἦρι μὲν γὰρ οὐ γίνεται, τὸ δὲ θέρος οὐ καλόν, ἀλλὰ τὸ μετόπωρον, οὐκέτι δύναται <δ'>[5] ἐν τούτῳ κατακρατεῖν ὁ ἥλιος. ὃ καὶ σημεῖον[6] ἄν τις λάβοι διὰ τί τοῦ θέρους γίνεται· τὸν μὲν γὰρ οἰνώδη καὶ ὑδατώδη χυλὸν δύναται τὸ ψῦχος ἐκπέττειν, τὸν ἐλαιώδη δ' ἀδύνατον.

8.7 τὰ ὕδατα δέ, τὰ ἐκ Διὸς γινόμενα καὶ τὰ ἐκ τῶν ὀχετῶν ἄρδοντα, τὴν σάρκα πληθύει (καθάπερ εἴρηται), τὸ δ' ἔλαιον ἄμικτον, ὥσπερ καὶ ὅταν τοῦ περικαρπίου χωρισθῇ.

[1] ego : φαυλίαι U. [2] Schneider : ἰσχύουσιν U.
[3] χυλὸν ἔχει Schneider (*sapore non caret* Gaza) : ἔχει χυλὸν Wimmer. [4] Schneider : το ὔλαιον U : τὸ ἔλαιον u.
[5] ego : δὲ δύναται Wimmer. [6] u : σημει|U.

---

[1] *CP* 6 8. 5.

oil, like the coarse olives, which would come from the fact that the nature had spent so much of its force on the flesh that it was too exhausted to produce much stone or oil. Again in cold countries too the olives get little oil and much flesh. 8.6

About this a difficulty is raised: why does the grape produce juice in the most winter-like countries, but not the olive?

The cause is (1) the one just mentioned,[1] that the power is spent on the flesh, and the juice of the grape is bound up in its very flesh[2]; and (2) further a cause that is perhaps prior and more decisive: the heat, which produces the oil, is too weak for this in a winter-like climate.[3] For in spring no oil is produced, and it is not the summer but the autumn that is sunny, and by then the sun is not strong enough to master the food. One might moreover take this as proof of why the oil is produced in summer: whereas cool weather is capable of bringing the vinous and watery juice to full concoction, it is incapable of doing this with the oily juice.

Rain and irrigation make the flesh abundant (as we said[4]); but the water does not mix with the oil any more than it does when the oil has been extracted from the pericarpion. 8.7

[2] *Cf. CP* 6 8. 7: flesh and oil are not so intimately united in the olive as flesh and juice are in the grape.

[3] *Cf. CP* 1 13. 8. [4] *CP* 6 8. 5; *cf.* 1 19. 5.

ἐν Αἰγύπτῳ δὲ τὸ μὲν τῶν Αἰγυπτίων καλουμένων ἐλαῶν γένος τῶν πολυσάρκων τε καὶ μεγάλων ἀνέλαιόν[1] ἐστι, θάτερον δ' ἐλαιῶδες (πολλὰ γάρ εἰσιν, ὥσπερ ἐν ταῖς ἱστορίαις εἴρηται).

περὶ μὲν οὖν τῆς τοῦ ἐλαίου γενέσεως τοιαύτην τιν' ὑποληπτέον τὴν αἰτίαν.

ὑπὲρ δὲ τῶν ἄλλων τῶν λίπος,[2] ὥσπερ αὗται,[3] συμφωνοῖτ'[4] ἂν ἐκ πάντων. 8.8 εἰ δ' ἔνια κατ' ἄλλην ὥραν, διαιρετέον ἢ τὰς φύσεις ἑκάστων, ἢ ποῖόν τι τὸ λιπαρόν, ἢ τὸν τρόπον τῆς γενέσεως· πλὴν[5] χρονιζόμενα, καθάπερ τὰ καρυώδη. ταῦτα[6] δύναμιν μὲν ἔχει (καθάπερ ἐλέχθη), τὴν δ' ἐνέργειαν ὁ χρόνος ἀποδίδωσι, τοῦ μὲν ὑδατώδους ἀπογινομένου, τοῦ λιπαροῦ δὲ ξυνισταμένου καὶ πεττομένου· τῶν δὲ καὶ κομιδῇ ξηρά τις ἡ λιπαρότης, εἰ μὴ ἄρα καὶ ἀντιπεριιστάμενον τὸ θερμὸν ἔν τισιν ἐνδέχεται ταὐτὸ τοῦτο ποιεῖν.

[1] Scaliger : ἔλαιον U.
[2] U N (*pinguibus* Gaza : λιπαρῶν Wimmer) : λῖπος ἐχόντων aP. [3] Schneider : αὐταὶ U.
[4] aP : -εῖτ' U N. [5] πλὴν εἰ Wimmer.
[6] ταῦτα γὰρ aP.

---

[1] *HP* 4 2. 9 (of the Thebaic nome): "For the olive too is grown in this region, but is not watered by the river

As for Egypt, the one kind, the so-called "Egyptian" olive which is fleshy and large, has no oil; but the other kind (the trees being plentiful there as was said in the History[1]) produces it.

So we are to suppose that the cause of the production of oil is much as described.

Other Oily Trees

On the subject of other trees that (like the olive) produce fattiness, there would be agreement on all sides. If some of the trees differ from the olive in the season of producing the oil, we must distinguish in each group either the nature of the trees, or the character of the oil or the manner of production, except for the cases where the fruit gets oily after some time in storage, as with nuts. These have the potentiality of being oily (as we said[2]), whereas time gives the nuts the actuality, the watery part disappearing while the fattiness becomes set and concocted (in some this fattiness is in fact a very dry one). Unless it is also possible that heat[3] should produce the same result in some by counter-displacement.

8.8

(which is more than 300 stades distant), but by ground water, for there are many springs. Its oil is not at all inferior to ours, but is more evil-smelling . . ."

[2] *CP* 6 7. 4–5.

[3] Here external heat.

9.1 ἁπάντων δ' ὅσα χυλὸν ἔχει, τὰ μὲν εὐθὺ καὶ ὀσμώδη τυγχάνει, τὰ δὲ πολλὰ συνεμφαίνει τινὰ γευομένοις ὀσμήν, ἔνια δὲ καὶ θλιβόμενα μόνον καὶ κινούμενα· καὶ πάλιν τὰ ὀσμώδη, διαμασωμένοις καὶ γευομένοις, χυλόν, σύνεγγυς[1] τῶν αἰσθήσεων κειμένων, ἀλλὰ τρόπον τινὰ καὶ τῶν αἰσθητῶν. ᾗ[2] καὶ οὐ κακῶς ἂν δόξειε λέγεσθαι τὸ κατὰ τοὺς χυλοὺς τελεῖν τὰς ὀσμάς· αὕτη γὰρ εἰς ἐκείνην φέρει τὴν δόξαν, ὡς ἐχόντων τινὰ συγγένειαν καὶ ἀπὸ ταὐτοῦ πως γινομένων. ἃ δὴ καὶ δεῖ διελεῖν· εἴ τι ἑκάτερον, ἤ πως ταὐτὸ διαφέρον.

§ 9.1–2: Theophrastus, *On Odours*, i. 1.

[1] aP : σύγγυς U N.
[2] ᾗ aP : ἡ U : ἧ u N.

---

[1] *Cf.* Theophrastus *On Odours*, iii. 9: "For the senses (*i.e.* sense organs of odour and taste), which lie close to one another, produce a certain tainting of the one by the other, which is why people try to make the objects of taste themselves fragrant"; xiv. 67: "Perfume is also held to improve the agreeableness of wine . . . It is not unreasonable that since the senses lie close together they should also go partners to some extent when their object is the same, since on the whole no flavour is without odour or odour without flavour. This last is so because no odour comes from an object possessing no flavour."

[2] *Cf. CP* 6 14. 12.

The Association of Flavours with Odours

Of all things possessing flavour some are also fragrant just as they are, but most convey a certain odour only when tasted, although a few do so also when merely pressed and handled. Things possessing odour, on their part, show flavour on being chewed and tasted, since the sensoria are close together,[1] and not only the sensoria, but in a way the two sensibles as well.[2] Hence the saying that "odours rate with flavours"[3] is, it would appear, not badly put, since this view leads us to the other, that odours and flavours have a certain kinship and are in a way derived from the same origin. These are matters that we must proceed to determine: are the two of them two distinct things or the same thing with a difference?[4] 9.1

[3] The saying is not mentioned elsewhere, although it may well have suggested Aristotle's phrase (*On Sense*, v [443 b 19–20, 444 a 4]) about the class of odours that are "stationed with the savours". "Rate" is from the language of politics and refers to a person's ranking or counting or paying taxes with one of the classes into which the community was divided.

[4] The answer (limited to "good odour") appears in its most general form in *CP* 6 16. 8: good flavour lies in the flavour-juice, good odour proceeds from it; with good odour, the flavour is less perfect or is privative; and for both good odour and good savour the flavour must not be too faint or too extreme.

9.2 οὐ μὴν ἴσως κατά γε τὰς προσηγορίας ἀποδοθήσονται πᾶσαι· πικρὰν γὰρ ὀδμὴν καὶ ἁλμυρὰν καὶ λιπαρὰν καὶ στρυφνὴν[1] οὐκ ἄν <τις> ἐθέλοι[2] λέγειν· οὐδὲ γὰρ οἱ χυλοὶ πάντες κατὰ τὰς ὀσμάς· ὡς[3] οὐδὲν ἧττόν ἐστίν πως κατάλληλα καὶ συνακολουθεῖ καὶ θατέρῳ θάτερον.

ἑκάστου δὲ τῶν χυλῶν ἰδέαι πλείους, οἷον γλυκέος πικροῦ τῶν ἄλλων· καὶ γὰρ μελιτώδης καὶ οἰνώδης καὶ γαλακτώδης καὶ ὑδατώδης ἐστίν, τάχα δὲ καὶ τῷ μᾶλλον καὶ ἧττον διαφέρουσαι, κυριωτάτως δὲ τῇ καταμίξει τῶν ὑποκειμένων· ὁμοίως δὲ καὶ ἐπὶ τῶν ἄλλων.

ταῦτα μὲν οὖν σχεδὸν συμφωνοῦσι πάντες.

[1] a : στρυφνὰν U N P.
[2] ego (*cf.* Aristotle, *On Sense*, v (443 b 11) ἄν τις . . . εἴποι): *quispiam velit* Gaza : ἐθέλοι τις Heinsius : ἐθέλοις Wimmer : ἐθέλει U a : ἐθέλοι u : ἐθέλῃ N P.
[3] U : *tamen* Gaza : ἀλλὰ Schneider.

---

[1] A correction of Aristotle, *On Sense*, v (443 b 6–11): ". . . the odours must be analogous to the savours. But this is the case with some: thus there are 'pungent' and 'sweet' odours, and again 'dry-wine,' 'astringent' and 'oily' ones, and one would call the odours of decomposition analogous to the bitter savours." *Cf. CP* 6 14. 12 *ad fin.* and also Aristotle, *On the Soul*, ii. 9 (421 a 28–b 9). *Cf.* also

9.2 Perhaps, however, not all odours will be found to answer to the names (at any rate) of the flavours, since one would hardly be willing to speak of a "bitter" or "salty" or "oily" or "astringent" odour[1]; in fact not all the flavours answer to the names of odours either.[2] But the two are in some way correspondent and the one accompanies the other.

Each flavour (such as sweet, bitter and the rest) has several varieties; so we have[3] the varieties honeyed, vinous, milky and watery. It may be that the difference of these varieties is also a difference of degree, but in the strictest sense it is due to the admixture of the underlying substances (and this holds of the varieties of the other flavours as well).

On these points, then, everybody (one may say) agrees.

Galen, *On Simple Medicaments*, iv. 22 (vol. xi, pp. 698. 17–699. 8 Kühn): [The savours most appropriate to the tongue are sweet, whereas those which are not appropriate to it have many differences; so with vapours (*i.e.* odours): the ones appropriate to the *pneuma* in the brain are agreeable and pleasant, whereas the rest have many differences among themselves]: ". . . and not all of them have accepted names as do the savours. For we say that this or that has an 'acid' or a 'pungent' odour, but we do not go on to say that it has a 'dry-wine' or 'astringent' or 'salty' or 'bitter' odour . . ." See list in Appendix II.

[2] Perhaps because the flavours (at least those distinguished by man) are far more numerous.

[3] For the sweet.

9.3 ἡ δὲ τῇ γεύσει τῶν ὀσμῶν αἴσθησις οὐκ ἄλογος, ᾗπερ μάλιστα ἔνδηλος[1] ἐπὶ τῶν εὐστόμων λεγομένων τῶν τε λαχανηρῶν (ὥσπερ ἀνήθου μαράθου μυρρίδος,[2] ἐνίων δ' οὐδ' ἐχόντων ὅλως ὀσμήν) καὶ ἔτι μᾶλλον ἐπὶ τῶν ξηρῶν, ἀόσμων δὲ τελέως (οἷον φακοῦ κνήκου[3] τῶν τοιούτων)· διαθραυόμενα γὰρ ἅμα τῇ μασήσει καὶ διαθερμαινόμενα, ποιεῖ τινα ἀτμόν, ὃς ἀναπέμπεται λεπτὸς ὢν διὰ τῶν πόρων εἰς τὴν ὄσφρησιν. ἐμφαίνεται δὲ καὶ 9.4 ἐνίοις[4] διαμασωμένων χυλός, ὡς ἐπίπαν δὲ τά γ' εὔοσμα πάντα πικρά. τούτου μὲν οὖν τὴν αἰτίαν ὕστερον λεκτέον.

ἔοικε δέ, δυοῖν ὄντοιν[5] ἐναντίων, οἷον τοῦ τε γλυκέος καὶ πικροῦ, τὸ μὲν οἷον εὐχυλίας ἀρχήν, τὸ δ' εὐοσμίας εἶναι, καὶ τρόπον τινὰ μᾶλλον τὸ πικρὸν τῆς εὐοσμίας. εὔοσμον μὲν γὰρ ἔργον λα-

[1] N aP : ἔνδηλως U.
[2] ego : μύριδος U[c] (ρ from δ) : ἴριδος Itali.
[3] U[ar] : κνίκου U[r] N aP.
[4] ego (ἐνίων Wimmer) : εν and a blank of 4–5 letters U.
[5] aP : ὄντι U N : ὄντων u.

---

[1] *Cf. CP* 6 9. 1: (most flavoured things) "convey a certain odour only when tasted."

Smells Noticed by Tasting

That we should perceive odours by tasting[1] is not unreasonable. This is especially noticeable in the so-called "good tasting"[2] plants, (1) the vegetables (as dill, fennel, sweet cicely, some even with no odour at all),[3] and (2) still more in dried products that are completely odourless (as lentils, safflower and the like). For when crushed and heated by the chewing they produce a certain vapour that owing to its fineness is sent up through the passages to the seat of smell. A flavour[4] too is noticeable in certain fragrant things when one chews them, but by and large all fragrant substances are bitter. We shall deal with the reason for this later.[5] 9.3 9.4

The Bitter Flavour and Fragrance

It seems that of the two opposites, namely sweet and bitter, the sweet is the origin (as it were) of good flavour, whereas the bitter is the origin of fragrance; and in a way the bitter is to a greater extent the origin of fragrance,[6] since it is hard to find any

[2] Apparently used of flavours that are pleasant without being sweet.

[3] That is, none perceptible without tasting the substance.

[4] That is, agreeable flavour. [5] *CP* 6 16. 8.

[6] For the difficulties of taking sweet as the only "good" or "concocted" flavour *cf.* *CP* 6 4. 2–6.

βεῖν μὴ πικρόν, εὔχυλα δὲ πολλὰ καὶ μὴ γλυκέα, σχεδὸν δὲ ταῦτα καὶ ὀσμώδη κατὰ τὴν γεῦσιν καὶ τὴν προσφοράν,[1] ἡ δὲ γλυκύτης σπανίως, καὶ ἥκιστα εὔοσμον, ὡς οὐ μιγνυμένων ἅμα τοῦ γλυκέος καὶ εὐόσμου· καίτοι ἄμφω γε διὰ πέψεως. ἀλλὰ περὶ μὲν τούτων ὕστερον.

10.1 ἐπεὶ δ' οἱ χυλοὶ πλείους, ἀπορήσειεν ἄν τις διὰ τί ποθ' οἱ μὲν ἄλλοι πάντες ἐν τοῖς φυτοῖς καὶ καρποῖς γίνονται, καὶ γὰρ πικρὸς καὶ δριμὺς καὶ ὀξύς, ὁ <δὲ>[2] ἁλμυρὸς οὐκέτι· οὐδὲν γὰρ τῶν φυομένων ἁλυκὸν ὥστε καὶ ἐν ἑαυτῷ τοιοῦτον ἔχειν τὸν χυλόν, ἀλλ' ἐν τοῖς περὶ τὰ ἔξω γίνεταί τις ἁλμυρίς,[3] οἷον καὶ τοῖς ἐρεβίνθοις, αὐτοὶ δὲ γλυκεῖς.

αἴτιον δ', ὅτι ἄτροφον καὶ ὥσπερ ἀγέννητον[4] τὸ ἁλμυρόν. σημεῖον δ', ὅτι οὐδὲ φύεται οὐδὲν (ὡς εἰπεῖν) ἐν ταῖς τοιαύταις χώραις· διεσθίει γὰρ καὶ ἐξαιρεῖται τὰς δυνάμεις ὥστε κωλύειν τὴν

§10.1: *Cf.* Plutarch, *Quaest. Nat.* v (913 A–B).

[1] u : προφ- U. [2] aP.
[3] Gaza (*salsugo*), Schneider : ἁλμυρός U.

fragrant thing that is not bitter, but many non-sweet things have excellent flavour (these being more or less the substances that are also fragrant when tasted and eaten).[1] Sweetness on the other hand has rarely any odour at all, least of all a good one. All this suggests that the sweet and the fragrant do not mix; and yet both are products of concoction. But of this problem later.[2]

Absence of the Salty Flavour in Plants

10.1 There being a number of flavours one might raise this problem: why do all the rest occur in plants and fruits (so the bitter, the pungent and the acid) whereas the salty does not? For no plant is salty to the extent of having this flavour internally; it is only on the surface that any saltiness occurs, as in chickpea, the chickpea itself being sweet.

The General Reason

The reason is that the salty does not feed and (as it were) does not procreate. Here is proof: virtually no plant will grow on salty land, since the salt eats through it and takes away its powers and so

[1] *Cf. CP* 6 9. 3.
[2] *CP* 6 16. 1–8.

[4] a : ἀγενητον U (-γένη- u N P).

10.2 σύστασιν. ὃ δὴ καὶ τοῖς ἄλλοις τούτου αἴτιον, εὔλογον μηδὲ καθ' αὑτὸ γεννᾶν· ἐπεὶ καὶ τὰ ἐν τῇ θαλάττῃ φυόμενα γλυκύτητί τινι καὶ ἑτέροις χυλοῖς φύεται καὶ συνίσταται (καθάπερ ἰχθῦς [1] καὶ τἆλλα ζῷα τὰ ἐν αὐτῇ).

καθόλου μὲν οὖν τοιαύτη τις ἡ αἰτία. δεῖ γὰρ ἐξ οὗ τι μέλλει [2] γίνεσθαι μεταβλητικὸν εἶναι· τὸ δ' ἁλμυρὸν ἀσαπὲς καὶ ἀμετάβλητον, διόπερ οὔτε φύεται οὐδὲν ἐξ αὐτοῦ οὔτε αὐτοτελὲς οὐδέν.

10.3 ὅλως δὲ ἐν τοῖς καθ' ἕκαστα καὶ ἐν τοῖς παρακολουθοῦσιν καὶ τὰ τοιαῦτα δόξειεν ἂν συμφωνεῖν· [3] οἷον ὅτι ὁ ἥλιος καὶ τὸ ἐν ἑκάστῳ θερμὸν ἕλκει τὸ κουφότατον καὶ τὸ τροφιμώτατον, τὸ δ' ἁλμυρὸν βαρὺ φύσει καὶ ἄτροφον (ἐπεὶ τἀσαπὲς [4]

§10.2: Aelian, *H. A.* ix. 64.

[1] οἱ ἰχθῦς Schneider.
[2] u : μελλειν U.
[3] u : -φανεὶν U.
[4] ego (ἔπειτα ὂν ἀσαπὲς Schneider): ἔπειτα ἀσαπὲς U.

---

[1] Growth depends on the formation (or "procreation") of new parts.

[2] *Cf.* Aristotle, *History of Animals*, viii. 2 (590 a 18–22): "Thus among testacea those that are stationary feed on

prevents formation.[1] Now it is reasonable that what prevents other things from generating will also do no generating itself. In fact even plants growing in the sea grow and are formed by sweetness of a certain kind and by other flavours than the salty, just as fish and the other marine animals.[2] 10.2

In its general formulation, then, the cause is as described, since that from which a thing is to be produced must be capable of change, whereas the salty is immune to decomposition[3] and change, which is why nothing grows from it and why it has no independent power of production.

The General Formulation Apparently Confirmed

Indeed among more particular matters and their consequences the following points would appear to be in agreement with this general formulation. For example 10.3

(1) The sun and the internal heat of each plant draw to themselves what is lightest and most nutritive; but the salty is naturally heavy and non-nutritive (since what will not decompose will not

the potable water (for it percolates through the close-textured parts because it has finer parts than the sea-water, which undergoes concoction), just as they acquire their original generation from it"; *cf.* Aristotle, *Meteorologica*, ii. 2 (355 b 4–11).

[3] That is, concoction: *cf. CP* 6 8. 4, note 1.

καὶ ἀναλλοίωτον)· καταλειπόμενον οὖν καὶ οὐ συνελκόμενον ὑπὸ τῶν ῥιζῶν οὐκ ἀναμίγνυται τοῖς φυτοῖς. ἔτι δὲ ἐπείπερ ἀπερίττωτον <τὸ>[1] *φυτόν, οὐδ' ἐπισπᾶσθαι καὶ ἕλκειν εἰκὸς τὸ ἄτροφον, ἔδει γὰρ καὶ ἔκκρισίν τινα γίνεσθαι.*[2] *καὶ διὰ τοῦτο· ἥκιστά τε ὑπὸ τοῦ ἡλίου ἀνάγεσθαι, καὶ ἐπιπολάζειν, πανταχοῦ γὰρ πλατέα καὶ μεγάλα τοῖς ὑγροῖς ἐπιφέρεσθαι, ἀσύμπλεκτα δὲ καὶ ἄκολλα*[3] *διὰ τὸ μηδὲν ἔχειν σκαληνές, ἀλλὰ γωνοειδῆ*[4] *τε εἶναι καὶ πολυκαμπῆ.*

ταῦτα μὲν εἰ κωλύει πρὸς τὴν τῶν φυτῶν κατάμιξιν ἔξεστι σκοπεῖν.

[1] aP. [2] *γίνεσθαι* <. . .> Wimmer.
[3] U : *inconglutinata(que) democrito placet* Gaza : *ἄκολλά φησι Δημόκριτος* Schneider.
[4] U (-*εῖ* N) P : *γωνιοειδῆ* a.

---

[1] From Aristotle, *On Sense*, iv (442 a 4–8), translated in note 1 on *CP* 6 4. 2. Theophrastus suggests that the internal heat of the plant does not attract the salty in part (at least) because the salty is non-nutritive. Attraction is not merely mechanical, but also a matter of desire. So Plato assigns to plants the desiderative part of the soul and speaks of their sensations and desires (*Timaeus*, 77 B). So when Theophrastus speaks of plants as "taking delight" or "seeking" or "liking," it may be more than a mere manner of speaking; at *CP* 2 7. 2 he speaks of "appetition" and at *CP* 2 18. 4 of the vine as sensitive to smell.

undergo qualitative change). It is therefore left behind and not attracted with the rest of the food by the roots and so does not become mingled with the plant.[1]

(2) Furthermore, since a plant has no excrement,[2] it is not likely to attract to itself and draw in what is non-nutritive, since this would then have to be somehow excreted.

(3) Another reason: salty shapes are least of all drawn up by the sun and remain on the surface,[3] since they everywhere, being flat and large, float on liquids and do not intertwine or adhere because they have no interlocking irregularities, but instead are angular and wrinkled.[4]

We may consider whether these characters prevent the salty from combining with plants.

[2] *Cf.* Aristotle, *On the Parts of Animals*, ii. 3 (650 a 20–23): ". . . plants take their food, already processed, by their roots from the earth (which is why plants have no excrement, since they use the earth and the heat in it in lieu of a stomach) . . ."; ii. 10 (655 b 32–36); iv. 5 (681 a 32–34); *History of Animals*, iv. 6 (531 b 8–10). The argument that this would explain the failure of plants to attract the salty is apparently due to Theophrastus.

[3] To "remain on the surface" is also a medical term: *cf.* Aristotle, *On Sense*, iv (442 a 11–12) and *Posterior Analytics*, ii. 11 (94 b 14–16): the salty "rises in the stomach" and is not digested.

[4] *Cf.* Democritus in *CP* 6 1. 6 (with note 5).

10.4 πρὸς δὲ τὰ πρότερον εἰρημένα, ζητήσειεν ἄν τις περὶ τὰ καθόλου λεχθέντα διὰ τί ποτε ἐνίοις ἐγγίνεται (ἢ ἐπιγίνεταί γε) καὶ πόθεν ἡ ἁλμυρίς. εἰ μὲν γὰρ ἐν αὐτοῖς ὑπάρχει, δῆλον ὡς οἰκεῖον ἄν τι τῆς τροφῆς εἴη καὶ τῆς φύσεως· εἰ δ' ἔξωθεν ἐπιγίνεται, καὶ[1] τοῦτο μὲν ἧττον, ἐκεῖνο δ' ἂν[2] ὁμοίως ἀπορήσειε, πόθεν καὶ ὑπὸ τίνος· ἀνάγκη γὰρ ἐκ τοῦ ἀέρος, ἢ ἐκ τῆς ἀτμίδος τῆς ἀναφερομένης, ἢ κατὰ τὰς ῥίζας ἑλκυσθὲν ἐξανθεῖν, οἷον περίττωμά τι, φαίνεται δ' ἡ ἅλμη, καὶ ὅλως τὸ ἁλμῶδες, ἐπιπολάζειν.

10.5 εἰ μὲν οὖν οὕτω, φανερὸν ὅτι ἕλκοιεν ἄν· εἰ δ' ἐκείνως, ἄτοπον διὰ τί μόνοις ἐπικαθίζει τούτοις, οἷόνπερ ἐρεβίνθῳ καὶ ἁλίμῳ[3] καὶ τοῖς τοιούτοις (ὅσα γὰρ ἁλμᾷ νοσηματικῶς,[4] ὥσπερ ἡ ῥοδωνιὰ καὶ ἄλλ' ἄττα, περὶ τούτων ἕτερος λόγος).

[1] [καὶ] Schneider. [2] U N : ἄν τις aP.
[3] καὶ ἁλίμῳ Wimmer (*cf.* *CP* 6 8. 10) : καταλαμβάνει U.
[4] Wimmer : ἄλμαι νοσηματικῶι U.

---

[1] *CP* 6 10. 1–3.
[2] *CP* 6 10. 1 (last paragraph), 2 (last paragraph).
[3] The received term for the salty coat of chickpea: *CP* 3 22. 3; 3 24. 3; 4 8. 4; 5 9. 6; 5 10. 1.
[4] Democritus' observation: *cf.* Theophrastus, *On the Senses*, lxvi (cited in note 5 on *CP* 6 1. 6) and *CP* 6 10. 3.
[5] The "brine" is here perhaps a powdery mildew. *Cf.*

A Question: How to Explain the Saltiness that in fact Occurs?

10.4 Confronted with the preceding discussion[1] one might enquire with reference to the general statement[2] (1) why it is that in some plants (or at least *on* them) saltiness does in fact occur, and (2) where it comes from. For if it exists within the plant itself, the inference is clear: it is something properly belonging to the food and nature of the plant; if on the other hand it is an accession from outside, one would wonder less about the reason for its presence, but would wonder none the less about the source and the agent, since it must necessarily come (a) from the air, or (b) from the vapour arising from the ground or else (c) be drawn in at the roots and crop out, like some rejected residuum (and the brine,[3] and indeed everything that is similarly salty, tends to come to the surface).[4]

10.5 Now if it comes from the plant, it is evident that the roots must attract it; but if it comes from the air or the vapour, it is odd that it is deposited only on these briny plants (such as chickpea, purslane and the like). (As for plants where the brine is due to disease, as the rose-bush and some others,[5] that is another matter.)

HP 7 5. 4: "In the dog days . . . coriander gets 'briny'"; *HP* 8 10. 1 (of diseases of grains): "Some also get scab and 'brine,' as cummin."

τῷ δ' ἐρεβίνθῳ καὶ οἰκεῖον φαίνεται καὶ χρήσιμον· ἀποπλυθέντος[1] *γοῦν ὅταν ἀνθοῦσιν*[2] *ἐφύσῃ*[3] *καὶ ἄρτι συνισταμένοις, ἀπόλλυνται καὶ διαφθείρονται σφακελίσαντες, ὥστε πρὸς σωτηρίαν ἡ φύσις ἐπάγοιτ' ἂν τὰ τοιαῦτα <ὡς>*[4] *συγγενῆ.*

10.6 *φαίνονται δὲ καὶ ὅλως τινὰ ἔχειν τοιοῦτον χυλὸν ἔν τε τοῖς φύλλοις καὶ τοῖς κλωσίν, ὃς καὶ ἀποπλυθέντων ὅλως*[5] *ἔνδηλός ἐστιν κατὰ τὴν γεῦσιν, οὐ μόνον ἐν τούτοις, ἀλλὰ καὶ ἐν αὐτῷ τῷ καρπῷ. δῆλον δὲ τοῦτο ἐν τῇ γεύσει γίνεται ἐάν τις ἐπὶ τὴν γλῶτταν ἐπιθῇ μὴ διαμασησάμενος ὅλως· ἐν τῷ κελύφει γὰρ ἡ ἁλμυρίς, οὐκ ἐν τῷ ἐντός, ἣ*[6] *καὶ συνδιατηρεῖ πρὸς τὸ ἄκοπον εἶναι. καὶ φαίνεται τὴν αὐτὴν ἔχειν τάξιν ἥπερ καὶ πεφυκό-*

[1] ego (*sc.* τοῦ ἁλμώδους : *abluta* [*sc. salsugine*] Gaza : ἀποπλυθέντων Wimmer) : ἀπολυθέντων U N : ἀπολουθέντων aP.

[2] u (ἀνθῶσιν N aP) : ἀνθούσηι U.

[3] Heinsius : ἐκφύσηι U. [4] Wimmer.

[5] U : ὅμως Schneider.

[6] u : ἡ U (ἦ N) : ἣ aP.

---

[1] *Cf. CP* 6 10. 4: "it is something properly belonging to the . . . nature of the plant."

[2] *Cf. CP* 6 10. 4: "why . . . saltiness does in fact occur."

[3] *Cf. CP* 3 22. 3; 3 24. 3; 4 2. 2; 4 8. 4; 4 10. 1; 4 13. 4; 4 14. 4; 5 9. 6.

[4] A combination of elements taken from the preceding

In Chickpea Salinity is Excreted to the Surface by the Nature of the Plant

But in chickpea it appears both to belong properly to the plant[1] and to have a purpose[2]: we see that if it is washed off by rain at flowering time and when the fruit is just forming, the plants are destroyed, dying of necrosis.[3] So that it is in the interests of preserving the plant that its nature brings in such substances, and brings them in not as foreign, but as her own intimate allies.[4]

10.6 We see moreover that chickpea has a general saline flavour in both the leaves and the twigs, a flavour which even after the plant has been washed clean is quite noticeable to the taste,[5] not only here but in the fruit itself. This becomes clear if in tasting the fruit you place it on the tongue and do not chew it at all, since the saltiness is in the skin and not in the interior; and it is this that keeps the fruit from getting worm-eaten in storage.[6] Here the saltiness appears to have the same station as in the

discussion (6 10. 4–5): the brine comes from the outside (*cf.* "brings in," used of calling in an ally from abroad), being in fact drawn in by the roots; but it is sought for by the nature of the plant, not rejected, since it is germane (and not alien or hostile) to that nature.

[5] This shows that the saltiness comes from the inside, and is not deposited on the surface by air or vapour.

[6] *Cf. CP* 4 2. 2; 4 15. 3.

10.7 τος, καὶ γὰρ ἐν ἀμφοῖν ἔξω καὶ ὥσπερ φυλακῆς χάριν. χλωροῦ μὲν οὖν ὄντος, ἐν τῷ καυλῷ καὶ λοβῷ (καθάπερ εἴρηται), ξηραινομένου δέ, καὶ τὸ κέλυφος λαμβάνει τοιοῦτον χυλὸν ὥσπερ ἐκκρινόμενον ἀπὸ τοῦ ἐντός· ἔξω γὰρ ἀφίστασθαι[1] τὰ τοιαῦτα εὔλογον ὡς ἂν ἀλλότρια, καθάπερ[2] καὶ ἐν τοῖς σικύοις[3] ἡ ἐν τῷ χροΐ[4] πιπρότης, καὶ ὡς ἐν Καρίᾳ φασὶν ἄπιόν τιν' ἔχειν χνοῦν ἁλμώδη[5] θαυμαστῶς, ὥστ' ἐὰν μὴ ἀποπλύνῃ τις μὴ δύνασθαι ἐσθίειν· οὐδὲν δὲ ἄτοπον οὐδ' εἰ πλείω τοιαῦτ' ἐστίν. ἐπεὶ οὐδὲ τοῦτο πόρρω τῶν εἰρημένων, οἷον τὸ ταῖς βαλάνοις ἐπὶ τῷ ἄκρῳ τὴν πικρίαν εἶναι, <καὶ>[6] τοῖς βολβοῖς ἐν τῇ καλουμένῃ κορυφῇ, καὶ τοῖς σκόρδοις τὴν δριμύτητα ἐν τῷ διήκοντι τῆς

§10.7: *Cf.* [Aristotle], *Problems*, xx. 25 (925 b 30–37); Pliny, *N. H.* 19. 97.

[1] u (no accent N) aP : ἀμφισταθαι U.
[2] u : καθερπερ U.
[3] u : συκίοις U N aP.
[4] Schneider : χαοΐ U : χνοΐ u aP : χλοΐ N.
[5] Gaza (*salsa*), Scaliger : ἀνιώδη U.
[6] Gaza, Wimmer.

---

[1] *CP* 6 3. 5; *cf.* 6 10. 6 ("in the . . . twigs").
[2] *Cf.* [Aristotle], *Problems*, xx. 25: "Why among pericar-

10.7 living plant: in both it is stationed outside and serves (so to speak) as a guard. So when the plant is green, the saltiness is in the stem and the pod (as we said[1]); but as the plant becomes dry, the skin too acquires a salty flavour which is (as it were) excreted from the interior. For it is reasonable that flavours of this sort, as foreign, should withdraw to the outside, just as the bitterness does in the skin of the cucumber,[2] and just as it is reported that in Caria a certain pear has a bloom on it so amazingly salty that one must wash it off before one can eat the pear. And there would be nothing odd if there are more such cases. (In fact what we find in the following group is not remote from the instances given: so acorns have their bitterness at the tip, purse-tassels in the so-called "crown,"[3] and garlic plants their pungency in the part emerging from the head

pia do some have the bitterer parts on the root side, as cucumber, whereas others have them at the upper extremity, as acorns? Is the answer this? That in the first group the food is unconcocted there because a constant influx of it s occurring at the root; whereas the others are dry by nature, and in consequence, as the sweet (*i.e.* sweet fluid) s drawn away from the tip and is already concocted, the ip becomes dry, and what is left behind is the bitter (just as lumps of salt). As the pericarpion gets drier it gets bitter, as olives and acorns do on being kept."

[3] Perhaps the base of the stem as it emerges from the bulb.

γέλγιθος·[1] πλὴν ταῦτα μὲν ὡς καθ' ὁμοιότητά τιν' εἰρήσθω.

10.8 ὅτι δὲ φαίνεται φυσικόν τι καὶ συγγενές, ἐκεῖθεν δῆλον· ὅπου γὰρ ἂν σπαρῇ καὶ φυῇ, πανταχοῦ λαμβάνει τὴν ἁλμυρίδα, κἂν μὴ τὸ ἔδαφος ᾖ τοιοῦτον.[2]

ἐπεὶ καὶ τά γ' ἐν τοῖς ἁλμώδεσι φυόμενα, τὸ[3] ἔχειν ἁλμυρίδα τινὰ οὐκ ἄλογον (ὥσπερ ἄλλα τε καὶ τὸ ἅλιμον), ἔνια <δὲ>[4] καὶ εὐχυλότερα καὶ βελτίω γίνεσθαι, καθάπερ τὴν ῥάφανον· ἐξεσθίει γὰρ αὕτη[5] τὴν δριμύτητα καὶ τὴν πικρότητα τὴν ἐνυπάρχουσαν, ἅμα δὲ καὶ εὐμέριστόν τινα ποιεῖ, παραιρουμένη τὴν ὑγρότητα τὴν πλείω (δεινοὶ 10.9 γὰρ οἱ ἅλες ἀφελεῖν). ἐμφανὲς δὲ τοῦτο καὶ ὠμῆς οὔσης (πίπτουσαν[6] γὰρ ἐπὶ τὴν γῆν ἔνια θραύεσθαί φασιν)· ὅταν οὖν τοιαύτη, καὶ οὕτως

[1] ego : γελγηθος U[cc] (θ for δ) : ι for η u (γελγιδος N aP).
[2] N aP : -το U.
[3] u N (τῶ aP) : τα U : Schneider deletes.
[4] Schneider.
[5] αὕτηι u : αὐτῆι U. [6] U (-ν dim) : -σα N aP.

---

[1] Presumably the base of the stem. *Cf.* perhaps [Aristotle], *Problems*, x. 30 (926 a 26–30): "Why does garlic smell more when it runs to stalk than when it is young'

of cloves.[1] But we mention these cases as presenting only a certain similarity.)

10.8 Our view that it appears[2] to be something belonging to the nature and bound to the plant by the closest ties is borne out by the following evidence: no matter where chickpea is sown and grows up it everywhere acquires this saltiness, even where the soil is not saline.

In Cabbage Internal Salinity is Natural

As for the plants that grow in saline ground, it is not unreasonable that they should possess a certain salinity (as purslane among others), and that some should even get a better flavour and improve in such ground, as cabbage[3]; for the salinity eats away the pungency and bitterness that is present in cabbage, and also makes it easier to divide the plant into pieces, since it takes away the excess fluid, salt being excellent at desiccating. 10.9 (This easy divisibility is noticeable even when the cabbage is raw, for it is said the pieces break off when you drop it, so that

Is the answer this? That when it is still young the presence of a good deal of alien fluid takes away its power. But when it is ripe, this fluid has already been excreted, and the garlic then has its proper smell; and this is by nature pungent."

[2] *Cf. CP* 6 10. 5: "it appears ... to belong ... to the plant."

[3] *Cf. CP* 2 5. 4; 2 16. 8.

ἔχουσα, παραδοθῇ τῷ πυρί, κατὰ λόγον[1] ἤδη τὸν χυλὸν[2] εἶναι[3] γλυκεῖαν καὶ ἁπαλήν. ἐπεὶ καὶ οἱ τὸ[4] λίτρον ἐμβάλλοντες[5] τοῦτο βούλονται ποιεῖν, ἀλλ' ἐξ ὀλίγου ποιοῦσιν·[6] ἡ δὲ φύσις ἐκ πολλοῦ καὶ κατὰ μικρὸν ποιήσασα τοιαύτην ἀπέδωκεν. ὅθεν οὐδὲ τοῦτο ἄλογον, τὸ περὶ τροπάς, καὶ περὶ τὸ ἄστρον ἐνιαχοῦ, καὶ ὅλως τοῦ θέρους, εἶναι βελτίω (καθάπερ ἐν Ἐρετρίᾳ)· τότε γὰρ μᾶλλον ἡ ἁλμυρὶς ἐργάζεται καὶ κρατεῖ, αὐτὴ[7] μὲν οὖσα πλείων, τῆς δ' ὑγρότητος ἐλάττονος γινομένης ἐν

10.10 ἐκείνῃ. κατὰ λόγον[8] δὲ καὶ τὸ ἐν Αἰγύπτῳ καὶ τὸ ἐν ἅπασι τοῖς τοιούτοις τόποις εἶναι χρηστήν.

ἀλλὰ γὰρ ταῦτα μὲν ἴσως ἐπὶ πλέον εἴρηται.

ἡ δ' ἁλμυρίς, εἴθ' ὑπὸ τῶν ῥιζῶν ἕλκεται, εἴθ' ὑπὸ τῆς τοῦ ἡλίου θερμότητος ἀναφέρεται, κατ'

[1] κατὰ λόγον u : all but -ν illegible in U.
[2] ego : του χυλὸν U : τοῦ χυλοῦ u.
[3] ego : εἶ καὶ U : εἶναι καὶ u : εἰ καὶ N aP.
[4] Schneider : τον U (-ν was once a variant for the following λ-).
[5] N aP : ἐμβαλόντες u : ἐκβαλόντες U.
[6] u : ποιεῖνσιν U.
[7] aP : αὕτη U N.
[8] Gaza (*ratio est*), Itali : κατολίγον U.

when with this character[1] and in this state[2] the cabbage is put on the fire, it is now reasonable that it should be sweet in its flavour and tender. In fact the persons who put soda in the cooking water[3] have this as their aim, but do not allow enough time, whereas the nature of the plant gives it this character by beginning early and proceeding gradually.) This is why another circumstance is also not unreasonable: in some countries cabbage improves at the summer solstice and during the dog days and in general in summer, as at Eretria; for then the salinity is more operative and masters the juice, since at that season the salinity is itself more abundant[4] and the amount of fluid is decreasing. It is also reasonable that cabbage should be excellent in Egypt and all hot countries.

10.10

But perhaps this discussion has been unduly prolonged.

Conclusion: Salinity is Raised from the Ground

Whether the salinity is attracted by the roots or is drawn up by the heat of the sun, in either event[5]

[1] The pungency and bitterness have been eaten out.
[2] The cabbage has been rendered easily divisible.
[3] *Cf. CP* 2 5. 3.
[4] None is washed away by rain.
[5] Both were denied by Aristotle: see the next note.

ἀμφοτέρους τοὺς τρόπους[1] οὐκ ἂν ἀκίνητος ἀπὸ τῆς γῆς εἴη, περὶ οὗπερ ἦν ὁ ἐξ ἀρχῆς λόγος.

ἀλλὰ περὶ μὲν τούτων ἅλις.

11.1 ἐκεῖνο δ' ὡς οἰκεῖον τῶν χυλῶν πειρᾶσθαι δεῖ διαιρεῖν, οἷον ποῖον μᾶλλον καθ' ἕκαστον γένος, ἢ τὸ ξηρὸν ἢ τὸ ὑγρόν· ὥσπερ ὁ ὀξὺς ὑγροῦ δοκεῖ μᾶλλον, καὶ ὁ αὐστηρός· ὁ δὲ δριμὺς ξηροῦ,[2] καὶ ὁ γλυκύς (παχυνόμενα δ' οὖν τὰ ὑγρὰ γλυκύτερα)· ὁ δὲ πικρὸς <. . .>

τάχα δὲ οὐθὲν ἂν ταῦτα[3] διαφέροι,[4] γινομένων γε[5] πάντων ἐξ ἀμφοῖν. εἰ μὴ αὐτῶν τούτων

11.2 θάτερον μᾶλλον ὑλικόν, οὕτω δὲ πάντες ἀπὸ τοῦ

[1] C^ac Gaza : τόπους U.
[2] u : ξυροῦ U.
[3] πικρός <. . .>—ταῦτα Wimmer (πικρὸς τάχ' οὐθὲν ἂν αὐτῶν Schneider) : πικρός |(the line is a letter shorter than any other on this folio [255^r]) ταχα δὲ οὐθεν ἀν ταυτα U.
[4] a : διαφερει U (dot over ει) : διαφέρει u (-η N P).
[5] Schneider : δε U.

---

[1] A correction of Aristotle, *On Sense*, iv (442 a 4–8), cited in note 1 on *CP* 6 10. 3 and note 1 on *CP* 6 4. 2.
[2] *CP* 6 10. 3.

it would not be immovable from the ground[1]; and this is the point with which the discussion began.[2]

But enough has been said about this matter.

Are Some Flavours Drier, Others Wetter?

11.1 Another point that we must endeavour to settle as relevant to flavours is this: which, the dry or the fluid, prevails more in each of the kinds of flavour? Thus the acid[3] is believed to belong more to the fluid, and so the dry-wine; whereas the pungent is believed to belong more to the dry, and so the sweet (at all events fluids get sweeter as they thicken[4]); the bitter . . .[5]

But perhaps this would make no difference, all flavours being produced from both.[6] Unless as between the dry and the fluid themselves the one 11.2 has more the character of matter,[7] which would make all flavours come from the same one of the

[3] In Greek the very word *óxos* ("acidity") means vinegar.

[4] *Cf. CP* 6 16. 2.

[5] Either Theophrastus broke off, or the account of the bitter, astringent, oily and salty has dropped out.

[6] *Cf. CP* 6 1. 1; also Aristotle, *On Sense*, iv (441 b 25–26): ". . . neither the dry without the fluid nor the fluid without the dry (*sc.* has savour)."

[7] *Cf. CP* 6 7. 1 (of flavours): "The matter common to all is fluidity . . ."

αὐτοῦ, καί εἰσιν (ὥσπερ ἐλέχθη) πάντες ἐν ξηροῖς.

ἀλλὰ τοὺς μὲν ἀποχωρίζομεν αὐτῶν (καθάπερ καὶ τοῦ βότρυος[1] καὶ τῆς[2] ἐλάας)· τοῦτο δ' αὖ[3] πρὸς τὴν χρείαν ὁρῶντες. ἐνίους δὲ καὶ ὕδωρ ἐπιχέοντες ἐλλαμβάνουσιν[4] (ὥσπερ τοὺς ἐπὶ[5] τῶν ἀκροδρύων καὶ σύκων[6]), τοὺς δὲ καὶ ἐξιστάντες τῆς φύσεως καὶ ὑποσήποντες εἰς χυλοὺς ἄγουσι ποτίμους (οἷον ὡς οἱ τοὺς οἴνους ποιοῦντες ἐκ τῶν κριθῶν καὶ τῶν πυρῶν, καὶ τὸ[7] ἐν Αἰγύπτῳ καλούμενον ζῦθος[8]).

ἁπάντων δὲ τούτων[9] αἱ μὲν ἀρχαὶ καὶ αἱ δυνάμεις φυσικαί, τὰ δὲ γινόμενα τέχνης μᾶλλον καὶ

[1] U (τοὺς τοῦ βότρυος Schneider) : τοὺς βότρυας u.
[2] Schneider : τας U.
[3] ego (Schneider deletes) : οὐ U : οὐ u.
[4] U : λαμβάνουσιν u.
[5] U : Schneider deletes : ἀπὸ Wimmer.
[6] Schneider : συκῶν U.
[7] u : τῶ U.
[8] Schneider : ζύθος U.
[9] N aP : τοῦτουτων U : τούτουτων u.

---

[1] *CP* 6 6. 5.

[2] That is, solid things (pericarpia or seeds mainly). Aristotle speaks as if the fluid were the matter (as we should expect from Plato, *Timaeus*, 59 E–60 A, cited n.2, p. 209 on *CP* 6 1. 1): *cf. On Sense*, iv (441 a 20–21): "...

two, and all (as we said)[1] are in dry things.[2]

But we separate some at least of the flavours from the dry things (as we separate them from the grape-cluster and from the olive). But we do this again with our own ends in view.[3] People also obtain a few flavours in water that they pour on the dry things, as with the flavours in tree fruit and figs. They even make some depart from their nature by inducing partial decomposition and thus turn them into juices that we can drink, as do the makers of wines from barley and wheat and of the so-called *zŷthos*[4] in Egypt.

In all these cases[5] the starting-points and the powers at work are it is true natural,[6] but the result is rather the achievement of art and of the intelli-

the water undergoes a certain affection . . ."; *ibid.* 441 b 8–9: "fluid is of a nature to be affected . . . by its opposite": *ibid.* 441 b 19–20: "And savour is this: the affection produced by the aforementioned dry in the fluid." The same no doubt holds of *ibid.* 441 b 23–25: ". . . the savours are either an affection or a privation not of every dry, but of the nutritive dry . . ." (*i.e.* an affection imposed on the fluid by the dry, or a privation of the affection so imposed).

[3] And not to develop the nature of the juice.

[4] Made from barley: Herodotus, ii. 77. 4; Diodorus, i. 34. 10.

[5] Of separation from the dry and of addition of water (sometimes with partial decomposition).

[6] In the fruit (or plant or part), not outside it.

συνέσεως. ἀλλὰ περὶ μὲν τῶν ἀπὸ διανοίας καὶ τέχνης γινομένων αὐτὰ καθ' αὑτὰ δεῖ θεωρεῖν· τῶν δὲ φυσικῶν χυλῶν τὰ πάθη καὶ τὰς γενέσεις ἐκ τῶν εἰρημένων θεωρητέον.

11.3 ὑποκειμένων δ' οὖν καὶ δεδειγμένων [καὶ δὲ δειγμένων][1] τούτων, ἀπορήσειεν ἄν τις διὰ τί ποτ' οὐκ ἐν τοῖς αὐτοῖς[2] μέρεσιν ἡ εὐχυλία καὶ ἡ εὐοσμία γίνεται πᾶσιν, ἀλλὰ τοῖς μὲν ἐν τοῖς ἄνω, τοῖς δὲ ἐν τοῖς κάτω καὶ περὶ τὰς ῥίζας· καὶ οὐδὲ τῶν ἄνω πάντων ἐν τοῖς αὐτοῖς, ἀλλὰ τοῖς[3] μὲν ἐν τοῖς περικαρπίοις, τῶν δὲ ἐν τοῖς φύλλοις, τῶν δὲ ἐν τοῖς ἄνθεσιν καὶ τοῖς κλωσίν, καὶ μᾶλλον αἱ ὀσμαὶ τῶν χυλῶν, ἐπεὶ καὶ ἐν τοῖς φλοιοῖς ἐνίων·

[1] N aP.
[2] U repeats αὐτοῖς.
[3] U : τῶν Schneider.

---

[1] Their aim is to serve human needs, and to do so they make the flavour depart from its nature.

[2] The treatment is reserved for the lost seventh book.

[3] The affections *CP* 6 9. 1–6 11. 2; the modes of production *CP* 6 3. 4–6 8. 8.

gence that applies it.[1] The products of intention and art, however, must be studied by themselves.[2] On the other hand, the character and modes of production of the natural flavours are to be studied in the light of what we have said.[3]

A Problem: Why Does not the Corresponding Part of Every Plant have the Good Flavour (or Fragrance)?[4]

11.3 At all events, now that these premises have been taken and conclusions drawn,[5] one might raise a problem: why good flavour and fragrance are not produced in the corresponding parts of all plants, but in some plants in the parts above,[6] in others in the parts below and roots[7]; again the good flavour and odour are not even in the corresponding upper part in all plants; in some they are in the pericarpia, in some in the leaves, in some in the flowers and twigs—a variation still greater in odours than in flavours, since odours actually occur in the bark of

[4] The discussion passes to plants in which the important flavour is not (as in trees) in the fruit or (as in cereals) in the "seed."

[5] About the natural savours and their production: *CP* 6 3. 3–6 11. 2.

[6] *Cf.* Theophrastus, *On Odours*, ii. 6 ". . . of odours some are in plants and their parts, such as twigs, leaves, bark, fruit and exudations . . ."

[7] *Cf. ibid.* vi. 27: "All perfumes are compounded either from flowers or from leaves or from the twig or the root or wood or fruit or exudations."

ἐπὶ[1] δὲ τῶν εὐόσμων ὅλως ἥκισθ' (ὡς εἰπεῖν) εὔοσμα τὰ ἄνθη (καθάπερ ἑρπύλλου σισυμ-

11.4 βρίου ἑλενίου). καίτοι κατὰ λόγον ἦν ὃ καὶ τοῖς ἄλλοις εὐοσμότατον, ἀόσμοις οὖσιν, τοῦτο καὶ ἐν τοῖς εὐόσμοις εὐωδέστατον εἶναι. θαυμαστὸν δὲ καὶ τὸ ἐνίων τὸ μὲν ἄνθος ἥδιστον ὄζειν, τῶν δ' ἄλλων μορίων ὅλως μηδέν, ὥσπερ ἐπὶ τῶν ἴων καὶ τῶν ῥόδων.

ἔχει δὲ καὶ τὸ πρῶτον λεχθὲν ἀπορίαν, ὅσων[2] ἐν ταῖς ῥίζαις ἡ εὐχυλία καὶ ἡ εὐοσμία τυγχάνει· ἄμφω μὲν γὰρ ταῦτα πέψει γίνεται, τὸ δὲ πλεί-στην[3] ἔχον καὶ ἀεὶ καινὴν τροφήν, ἀφ' οὗ τοῖς ἄλ-λοις ἡ διάδοσις, ἥκιστ' εὔλογον εἰς πέψιν ἥκειν ἢ εὐχυλίας ἢ εὐοσμίας, ὡς οὐδὲ τῶν ζῴων αἱ κοιλίαι.

11.5 περὶ δὴ τούτων καὶ τῶν τοιούτων, ἀρχῇ χρωμέ-νους τῇ πολλάκις εἰρημένῃ, διότι πέψει τινὶ ταῦτα γίνεται, τὸ μετὰ τοῦτο δεῖ λαβεῖν ἐπὶ τοῦ

§ 11.4: Pliny, *N. H.* 21. 37.

[1] U[r] aP : ἐπεὶ U[ar] N.
[2] u : ὅσον U.
[3] Schneider : πλείστον U.

some,[1] whereas if the entire plant is fragrant the flowers are commonly the least fragrant part (as in tufted thyme, bergamot mint and calamint). Yet one would have expected the part that is most fragrant in non-aromatic plants to be the most fragrant in aromatic plants as well. It is also odd that whereas the flower in some plants has a delightful odour, no other part of the plant has any odour at all, as in violet and rose.

11.4

Again the group mentioned first[2] poses a problem, the plants where good flavour and fragrance are in the root: they are both produced by concoction; yet it is highly unreasonable that a part which contains more food than any other, food that is constantly renewed, and which distributes their food to the rest, should manage to concoct either good flavour or fragrance, any more than the digestive tract does in animals.

The Roots: Why They Have Good Flavour and Odour When the Other Parts Have None

11.5

About these matters and the like we must begin with the principle that we have often mentioned,[3] that good flavour and aroma arise from a certain type of concoction. We must take our next step in

[1] As cinnamon and cassia: *HP* 9 5. 1, 3.
[2] *CP* 6 11. 3.
[3] *CP* 6 6. 2; 6 8. 4; 6 11. 4.

τελευταίου[1] λεχθέντος πρῶτον, ὅτι "αἱ ῥίζαι ὡς[2] κοιλίαι τοῖς φυτοῖς εἰσι πάντως·" εἰ γὰρ[3] καὶ τοῦτό τις θείη διὰ τὸ τὴν τροφὴν ἀλλοιοῦσθαί πως ἐν αὐταῖς, ἀλλ' ἐκεῖνό γε φανερόν, ὡς οὐκ ἔχουσιν οὐδὲν περίττωμα, δύναμιν δὲ ἔχουσιν εἰς τὸ πέττειν. τοιαύτας δ' οὔσας οὐδὲν κωλύει εὐχυλίαν καὶ εὐοσμίαν ἔχειν, ὅσαι κρᾶσιν εἰλήφασιν τοιαύτην· ἐπεὶ[4] καὶ ἐν ταῖς τῶν ζῴων κοιλίαις, περιττώματα ἐχούσαις, ὅμως ὕπεστιν ὑγρότης, εὐχυλία τις οὖσα καὶ πέψις, ὡς μάλιστα τῆς τοιαύτης ἀλλοιώσεως ἐνταῦθα γινομένης.

11.6 καὶ ἐπὶ τῶν δένδρων ἐστὶν ὅσα πίονα τυγχάνει,

[1] ego : τευπλειου U : τευτλίου u. [2] οὐχ ὡς Wimmer[c].
[3] ego (εἴ γε Moreliana) : εἴτε U. [4] u : ἐπι U.

---

[1] It has not been made before in the *HP* or *CP*. Aristotle says that the roots are in plants what the mouth is in animals: *cf. On the Soul*, ii. 1 (412 b 3–4): ". . . the roots are analogous to the mouth, since both draw in the food"; *On Longevity*, vi (467 b 2): ". . . the upper part of the plant and head is the root . . ."; *On Youth and Age*, i (468 a 9–11): ". . . roots are for plants as the so-called mouth for animals . . ."; *On the Parts of Animals*, iv. 10 (686 b 34–687 a 1): ". . . the roots have for plants the power of mouth and head." *Cf.* Plato, *Timaeus*, 90 A 1 – B 1.

[2] *CP* 6 11. 4, last paragraph.

[3] A formula chosen to avoid contradicting Aristotle, for whose views *cf. CP* 6 10. 3, note 2. To account for the

connexion with a statement first made[1] in what was last said,[2] "the root is definitely like a digestive tract to a plant"; even if one might defend this thesis because the food undergoes a certain alteration there,[3] the fact is obvious that the roots have no excrement[4]; they do, however, have a power conducive to concoction.[5] So even in the gut of animals, which contains waste matter, there is nevertheless present a fatty substance,[6] and so a type of good flavour and concoction, a fact that shows that this sort of alteration[7] definitely occurs there.

The Roots: (1) Trees[8]

The phenomenon is also found in trees when they 11.6

absence of excrement in plants, Aristotle had asserted that their food was taken from the ground already concocted. [4] *Cf.* Aristotle, cited in *CP* 6 10. 3, note 2.

[5] Aristotle mentions or implies the concoction of fruit (or pericarpion) at *Meteorologica*, iv. 3 (380 a 11–12); *On the Generation of Animals*, i. 1 (715 b 23–25).

[6] Literally "a fluidity." Fat and suet are counted as "fluid" by Aristotle: *On the Parts of Animals*, ii. 2 (647 b 11–14), ii. 7 (653 b 9–10). Theophrastus has the following passage in mind (*On the Parts of Animals*, iii. 14 [675 b 9–11] of the course of the gut): "The next portion of the gut extends in a straight line to the place of exit of the excrement, and in some animals this portion, the so-called rectum, is rich with fat, in others without fat."

[7] Concoction to good flavour.

[8] Apparently tree roots were not eaten and so Theophrastus must make do with torchwood.

καθάπερ ἡ πεύκη· πᾶσα γὰρ ἔνδᾳδος ταῖς ῥίζαις (ὥσπερ ἐλέχθη καὶ πρότερον). αἴτιον δ' ὅπερ ἐπὶ τῶν ζῴων, ὅτι τὸ διαθερμαινόμενον ἀεὶ καὶ πεττόμενον, καθαρώτατον ὄν, προσίζει, καὶ ἀθροιζόμενον καὶ πυκνωθέν, ἐποίησέν τινα πιότητα·[1] τὸ δέ, διιὸν[2] εἰς τὰ ἄνω μέρη, τροφὴ γίνεται τοῖς ὑπὲρ γῆς, οὐ διὰ τῆς πιότητος[3] ταύτης, ἀλλ' ἔχον τινὰς ἑτέρους πόρους, ἐπεὶ[4] πάντων γε δᾳδωθέντων ἀπόλλυται τὰ δένδρα (καθάπερ ἐλέχθη) συμπνιγόμενα, καὶ οὐδεμίαν ἔχοντα δίοδον τῷ

11.7 πνεύματι. τοῦτο δὲ συμβαίνει καὶ ἐπὶ τῶν ζῴων ὅσα διαπιαίνεται· ξυμφράττονται γὰρ οἱ πόροι διὰ τὴν πυκνότητα τῆς πιμελῆς, ὥστε μὴ διιέναι εἰς τέλος τὴν πνοήν. οἷς μὲν οὖν μὴ ἔνεστιν ὅλως λιπαρότης καὶ πιότης[5] (ἢ αὕτη[6] μὴ πολλή, μηδὲ σωματώδης), τούτοις οὐ γίνεται τοιαύτη πάχυν-

[1] Gaza (*pinguedinem*), Itali : ποιότητα U.
[2] ego (*transmissum Gaza* : ἰὸν Schneider) : ἴδιον U.
[3] Gaza (*pinguedinem*), Itali : ποιότητος U.
[4] u : ἐπι U.
[5] N aP : ποιότης U.
[6] Schneider : αὐτὴ U.

---

[1] *HP* 9 2. 3 ("... for every pine has torchwood in its roots"); *cf. HP* 9 2. 7; *CP* 5 11. 3.
[2] *CP* 5 11. 3.
[3] *Pneuma*; it leads to expansion (*i.e.* growth).

are fatty, as the pine; for every pine gets the torchwood in the roots (as we said before).[1] The reason is the same as for the animals: the portion of the intake that from time to time gets thoroughly warmed and concocted, this being the purest portion, settles, and as it accumulates and is thickened, produces a certain kind of fattiness. The rest, passing through to the upper parts, becomes food for the parts above ground. It does not make its way through this fattiness, but has certain other passages; indeed if all the passages turn to torchwood the trees die of "suffocation" (as we said),[2] affording no route for the breath.[3] (This also happens with animals that fatten out: the passages are blocked by the density of the fat and at last do not let the breath pass through.)[4] Now in trees with no oiliness or fattiness at all (or where it is not plentiful and has no body) no such thickening occurs; but

11.7

[4] Aristotle, *On the Parts of Animals*, ii. 5 (651 a 36–651 b 8), in discussing fat and suet, gives a different reason: "Now if these are of moderate amount in the parts of animals, they are beneficial; but if they are excessive in their amount, they destroy and do harm. For if the whole body should become fat and suet, it would perish. For what makes an animal is the part that is capable of sensation, and flesh and its analogue have sensation; but blood . . . has no sensation; hence fat and suet have also none, for they are concocted blood; so that if the whole body should become of this description, it would have no sensation."

σις· οἷς δὲ ἔνεστι, γίνεται κατισχυούσης τῆς θερμότητος.

11.8 παρόμοιον δὲ τούτῳ καὶ τὸ παρὰ[1] τὴν οὐλότητα τῶν ξύλων ἐστίν· οὐλότερα γὰρ ἀεὶ τὰ μὲν ἐν τοῖς στελέχεσι τῶν[2] ἄνω, τούτων δὲ αὐτῶν, τῶν ἄνω,[3] καθάπερ καὶ πυκνότατα, καὶ παχύτατα. συμβαίνει δὲ τοῦτο διὰ τὸ μᾶλλον ἀπολύειν[4] καὶ ξυνίστασθαι τὴν τροφήν, ὥσπερ[5] ὅλως καὶ ἡ εἰς βάθος αὔξησις· ἡ δ'[6] εἰς τὸ ἄνω διιοῦσα πρὸς τὴν βλάστην καὶ[7] μῆκός ἐστιν. ἐφισταμένης οὖν καὶ ὥσπερ εἰλουμένης ἐνταῦθα, καὶ ἡ πυκνότης καὶ ἡ οὐλότης γίνεται τῶν ξύλων· ἐκείνη δὲ ἀεὶ διίεται πρὸς τὸ πόρρω.

διὸ καὶ οὐκ ἔστιν ἐν τοῖς νέοις οὐλότης, ἅτε καὶ ἐπ' αὔξησιν ὡρμηκότων, ἀλλ' ὅταν στῇ τὰ τοῦ

§11.8: Pliny, *N. H.* 16. 231.

[1] U : περὶ u.
[2] Gaza (*quam* [*rami*]), Moreliana : τοῖς U.
[3] τῶν ἄνω U : *primae partes* Gaza : τὰ ἄνω Itali : τὰ κάτω τῶν ἄνω Schneider : τὰ κάτω Wimmer.
[4] U : *immoratur* Gaza : καταλύειν Scaliger.
[5] Wimmer (*qua de causa* Gaza : δι' ὅπερ Heinsius): ὅπερ U.
[6] ἡ δ' u : εἰ δ' U : δ' N : aP omit. [7] καὶ τὸ aP.

the thickening occurs in the trees that have this character when the heat prevails over the intake.[1]

The Parallel of Curly Grain in Wood

Similar to this thickening is the difference that goes with curly grain in the wood: for the wood in the trunk has always a curlier grain than in the upper parts, and in the upper parts themselves the parts of closest texture[2] are always the thickest. This curly grain occurs because the food breaks away here more than elsewhere and acquires consistency, just as it is in this way that lateral increment in general occurs, whereas the food that passes on upwards makes for foliage and increment in height. And so when the food stops advancing and is (as it were) packed into an eddying mass,[3] both the close texture and the curly grain of the wood are the result; whereas the rest of the food is constantly transmitted onward. 11.8

This is why there is no curl in the grain of young trees, since their impetus is toward increase in height, and occurs instead when they have stopped

[1] The heat is the agent of concoction.

[2] All wood with curly grain is of close texture: *cf. HP* 5 3. 3.

[3] The words *oulós* "curly" and *eiléō* "to pack, to swirl" are related by etymology.

μήκους, ὥσπερ ἐπὶ τῶν ζῴων.

11.9 ἡ αὐτὴ δὲ καὶ σύνεγγυς αἰτία καὶ διὰ τί δᾷδα καὶ πίτταν καὶ ῥητίνην ὅλως οὐκ ἔχει τὰ νέα· καταναλίσκεται γὰρ ἡ τροφὴ πᾶσα πρὸς τὴν αὔξησιν καὶ καρπογονίαν, ὕστερον γὰρ ἐκδᾳδοῦνται[1] καὶ ὅλως τὴν τοιαύτην ὑγρότητα τῆς καρπογονίας λαμβάνουσιν (ὥστε γε καὶ πλῆθος εἰπεῖν). τότε γὰρ οἷον περίττωμα γίνεται τοῦτο φυσικόν, ἀφῃρημένης μὲν τῆς εἰς τὸ μῆκος ὁρμῆς, ἰσχυόντων δὲ μᾶλλον τῶν ῥιζῶν καὶ τοῦ ὅλου δένδρου.

ταῦτα μὲν οὖν διὰ τὸ συγγενὲς τῆς αἰτίας ἔλαβε τὴν χώραν ταύτην.

11.10 ἡ δ' εὐχυμία καὶ ἡ γλυκύτης καὶ ἡ εὐοσμία ἐνίων ῥιζῶν ἄνευ τῶν ἄλλων μερῶν διὰ τὴν εἰρημένην αἰτίαν γίνεται.

[1] ἐκδαδοῦνται aP : ἐδοδοῦνται U : δαδοῦνται u : ἐδωδοῦνται N.

---

[1] That is, just as in animals growth in thickness occurs when they have stopped growing taller.

[2] *Cf. HP* 9 2. 8: "The pines do not bear and produce torchwood at the same time; for they bear from their early youth, but produce torchwood much later, when they are reaching a riper age."

[3] "The same" when formulated in general terms, "close"

growing taller (just as with animals).[1]

11.9 The reason why young trees have no torchwood[2] or pitch or resin at all is also the same as this and close to it[3]: the food is all expended on increase in growth and on producing fruit, since it is after their fruit-bearing that the trees get torchwood and acquire such fatty fluid in general (at least in any considerable amount), this being then produced as a sort of natural residue,[4] when the impetus to grow taller has disappeared and the roots and the whole tree are stronger.

These remarks[5] have been given a place here because of the close connexion of the causation.

11.10 Good flavour and sweetness and fragrance of certain roots without the other parts of the tree is due to the reason mentioned.[6]

when formulated more particularly.

[4] Aristotle's description of semen: *cf. On the Parts of Animals*, ii. 5 (651 b 8–17) [continuing the passage cited on *CP* 6 11. 7]: "For this reason moreover animals that are very fat age quickly: they have little blood, since the blood is expended on producing fat, and animals with little blood are already on the road to decay . . . Again fat animals are less fertile for the same reason: what should have passed from blood to generative fluid and semen is expended on producing fat and suet . . . , so that either no residue at all is produced or little."

[5] *CP* 6 11. 8–9 (why young trees lack torchwood and curly grain).

[6] *CP* 6 11. 5.

φανερὸν δὲ καὶ ἐπὶ τῶν ἐλαττόνων ἐστίν, οἷον ποιωδῶν καὶ λαχανωδῶν καὶ ἐνίων ὑληματικῶν, ὧν αἱ μὲν ῥίζαι γλυκεῖαι,[1] τὰ δὲ ὑπὲρ γῆς οὐχ ὅμοια. λέγω δ' οἷον ἀγρώστιδος κυπείρου τευτλίου σελίνου ἱπποσελίνου τῶν ἐν ταῖς λίμναις καὶ τῶν ἐν τοῖς ποταμοῖς τούτων φυομένων ἐδωδίμων· αἱ μὲν γὰρ ῥίζαι γλυκεῖαι[1] πολλῷ[2] καὶ ἐδώδιμοι, καὶ οἱ καυλοί, τὰ φύλλα δ' οὔ.

11.11 τὸ αἴτιον ἐπὶ πάντων τῶν τοιούτων ἐν δυοῖν· ἢ[3] γὰρ ὑγρότερα καὶ ὑδατωδέστερα (καθάπερ ἐπὶ τοῦ τευτλίου καὶ τῶν λιμναίων)· ἅμα γὰρ ὑδατώδη καὶ λεπτά, καὶ οὐκ ἔχει[4] πέψιν, μὴ ἔχοντα δ', οὐδὲ χυλὸν ἔνδηλον εἰς ἡδονήν· ἡ δὲ ῥίζα καὶ ὁ καυλὸς ἔχουσιν.

ἢ πάλιν διὰ ξηρότητα τῶν ἄνω, καθάπερ ἡ ἄγρωστις καὶ ἁπλῶς πάντα τὰ καλαμώδη. ξηρὰ γὰρ τὰ ἐπάνω, τὰ κάτω δ' ἔνυγρα· χυλὸς δὲ οὔτε ἐν ξηρότητι γίνεται, οὔτε ἐν ὑγρότητος πλήθει.[5] διὸ καὶ τῆς ἀγρώστιδος καὶ τοῦ σισυριγχίου καὶ

[1] u : -κύαι U.
[2] U (*cf. CP* 3 3. 3; Aristotle, *PA* iv. 10 [686 b 27]) : πολλῶν Schneider.
[3] Gaza (*aut*), Itali : ἡ U.
[4] οὐκ ἔχει Wimmer : οὐχι U. [5] u : πλήθο U.

The Roots: (2) *The Lesser Plants*
(a) *With Sweet or Pleasant Flavours*

This limitation is also seen in the lesser plants (such as herbaceous plants, vegetables and a few woody plants), where the roots are sweet, but the parts above ground have no similar sweetness; I mean for instance dog's tooth grass, galingale, beet, celery, alexanders and these familiar edible plants that grow in ponds and rivers. For the roots are sweet and edible, and also the stems (the roots with sweetness to spare), but not the leaves.

The Reason

11.11 In all such plants the reason lies in one or the other of two things:

(1) Either the leaves are too fluid and watery (as in beets and pond plants). For to be watery and thin is to lack concoction, and to lack concoction is to lack any flavour noticeable enough to be pleasant. The root and stem on the other hand have concoction.

(2) Or on the other hand the dryness of the upper parts is responsible, as in dog's tooth grass and in fact all reed-like plants: in these the parts above are dry, but the parts below have fluid, and flavour is compatible neither with dryness nor with abounding fluidity.[1] This is why in dog's tooth grass, bar-

[1] As in the first group.

τῶν ἄλλων τῶν τοιούτων αἱ μὲν ῥίζαι γλυκεῖαι, τὰ δ' ἄνω ξηρὰ καὶ οὐχ ἡδέα, καθάπερ ἄχυλα. ταὐτὸ δὲ καὶ ἐπὶ τῶν σελίνων καὶ ἱπποσελίνων· αἱ μὲν γὰρ ῥίζαι σαρκώδεις καὶ εὔστομοι, τὰ δὲ φύλλα ξηρότερα, καὶ ὥσπερ δριμύτερα. καὶ ἐπὶ τῶν ἄλλων ὁμοίως.

ἅπαντα δ' (ὡς εἰπεῖν) ἐν ταύταις ταῖς αἰτίαις

11.12 ἐστίν. διὸ καὶ ὅπου πλείων[1] εὐτροφία, καὶ τὰ ἄνω τῶν φύσει ξηρῶν ἐδώδιμα, καθάπερ ἐν Αἰγύπτῳ τοῦ καλάμου τοῦ ἐν τοῖς ἕλεσιν. ἔχει μὲν γάρ τινα γλυκύτητα καὶ ὁ ἄλλος ἐπὶ τῶν ἄκρων, ἀλλ' ἐπὶ βραχὺ πάντων·[2] ἐκεῖνος δὲ διὰ τὴν εὐτροφίαν ἁπαλός τ'[3] ἐπὶ πλεῖόν ἐστι καὶ γλυκύς. ἔχουσι δὲ καὶ αἱ ῥίζαι τὴν γλυκύτητα μέχρι οὗ ἂν ξηρανθῶσιν, ἀναξηρανθεῖσαι δ' οὐκέτι· τὸ δὲ ξηρὸν οὔτ' ἐδώδιμον οὔτ' ἔγχυλον (διὸ καὶ τῶν καυλῶν ἀκμή τις).

11.13 ἡ δ' ὀσμὴ σχεδὸν ἀνάπαλιν, ἐπί γε τούτων· χλωραὶ μὲν γὰρ [ἢ][4] οὐκ ὄζουσιν, <ἢ>[5] οὐχ

[1] u aP : πλείον U : πλεῖον N.
[2] U : πάντως Gaza (*admodum*), Schneider.
[3] Moreliana : γ' U. [4] Wimmer. [5] N aP.

---

[1] "Agreeable" (*eústomos*) is often used of pleasant tastes that are not sweet.

bary nut and the like the roots are sweet, but the upper parts are too dry and give no pleasure, having too little juice as it were to possess a taste. The same holds for celery and alexanders: the roots are fleshy and agreeable,[1] but the leaves too dry and (as it were) too pungent. Similarly with the rest.

All these plants (so to speak) come under one or the other of these two causes. This is why where the feeding is better even the upper parts of naturally dry plants are edible, as with the swamp reed in Egypt.[2] The rest of the reeds[3] to be sure have a certain sweetness at the tips, but in all for only a short distance; the swamp reed on the other hand is tender and sweet for a greater distance because it feeds well. Roots too have their sweetness until they get dry, but when dry they have it no longer, since what is dry is neither edible nor has it any flavour (which is why the stems too at a certain moment are at their peak). 11.12

The reverse (one might say) is true of odour, at least in these[4]: the roots when fresh have no scent 11.13

[2] The swamp reed is the papyrus: *cf. HP* 4 8. 4: ". . . the papyrus stalk itself has a great number of uses . . . But most of all it provides a most extensive supplementation of food. For the natives all chew it both raw and boiled and roasted; and they swallow the juice but spit out the quid . . ."

[3] Such as the sari: *cf. HP* 4 8. 5 (of the stalks of sari): "They chew this too and get rid of the quid . . ."

[4] The lesser plants.

ὁμοίως· ἀποξηρανθεῖσαι δ' ὄζουσιν, ὥσπερ καὶ ἡ τῆς ἴριδος, καὶ ἡ τῶν καλάμων δὲ καὶ σχοίνων[1] καὶ ἁπλῶς τῶν ἐνύγρων. (ἐπεὶ καὶ τὸ κύπειρον, καίπερ ὀσμῶδες ὂν καὶ ξηρὸν[2] ὄν,[3] ἧττον ὄζει πρόσφατον.) ἀκμὴ δέ τις καὶ τούτων, ὥστ' ἀπομαραινόμενα μᾶλλον ἀοσμότερα γίνεσθαι.

11.14 ἀλλὰ τοῦτο μὲν ὥσπερ σύμφωνον·

οἱ δὲ χυλοὶ καὶ ἐν τοῖς ἄλλοις ὁμοίως ἔχουσι τὰς δυνάμεις· ἐνίων γὰρ ἐν ταῖς ῥίζαις μάλιστά εἰσιν, καθάπερ τῶν δριμέων (οἷον σκόρδων κρομμύων ῥαφανίδων), ὡσαύτως δὲ καὶ τῶν φαρμάκων.[4] ἅπασαι δὲ καὶ αὗται σαρκώδεις· οὐ μὴν ἀλλὰ διαμένουσιν αἱ δυνάμεις τῶν φαρμακωδῶν ἀποξηραινομένων μέχρι τινός, εἶτ' ἀκμάζουσιν.[5] ἐπεὶ τό γ'[6] ὅλον ἰσχυρότεραι ξηρανθεῖσαι, διὰ τὸ ἀφῃρῆσθαι τὸ ὑδατῶδες· οὕτω γὰρ καὶ οἵ γ' ὀποὶ[7] πάν-

[1] u : σχοινίων U. [2] μὴ ξηρὸν Wimmer. [3] [ὂν] aP.
[4] U : *venenosis* Gaza (φαρμακωδῶν Itali).
[5] U : εἶτα παρακμάζουσιν Schneider.
[6] u : ἐπι το γ' U : ἐπεὶ τ' N : ἔπειθ' aP.
[7] Wimmer (*succi* Gaza : οἱ ὀποὶ Itali : οἵ τ' ὀποὶ Basle ed. 1541) : οἱ τόποι U.

---

[1] *Cf. CP* 6 14. 8 (with note 1). [2] *Cyperus rotundus.*
[3] That is, plants with flavours not sweet or agreeable.

(or not so good a one), but have scent on being dried, as the root of iris and so too of reeds and rushes[1] and aquatic plants in general. (Indeed *kypeiron*,[2] which has scent also when dry, has less of it when freshly cut.) Here too there is a moment when the plants are at their peak, after which, as they waste away more and more, they come to be more odourless.

11.14 But this wasting away after reaching a peak is in agreement (one might say) with what occurs with flavours.

The Roots: (2) *The Lesser Plants*
(b) *Powerful Flavours and Medicines*

In the other plants too[3] the flavours have their powers under the same conditions. Thus in some these flavours are mainly in the root, as in the pungent plants (such as garlic, onion, radish), and so too in the plants used as medicines. All these roots too[4] are fleshy; nevertheless in the medicinal plants the powers remain for a while as the drying of the root proceeds, and then reach their peak. In fact medicinal roots are as a rule more powerful after drying, since the watery part is then removed; for on the same grounds saps as well are all

[4] The sweet (or savoury) roots are also fleshy: for celery and alexanders *cf.* *CP* 6 11. 11.

τες χρήσιμοι καὶ σωματωθέντες, ἡ δὲ σωμάτωσις ἐκκρινομένου τοῦ ὑδατώδους. διὸ καὶ παρασκευάζουσιν[1] *αὐτοί,*[2] *τὰ μὲν ἐγχυλίζοντες καὶ ξηραίνοντες, τὰ δ' ἐντέμνοντες ὅπως ὁ ἥλιος καὶ ὁ ἀὴρ πήξῃ (καθάπερ τὰς ῥητίνας καὶ ὅσα ἄλλα τοιαῦτα, καὶ τὸν λιβανωτὸν καὶ τὴν σμύρναν, καὶ τὸν ὀπὸν τοῦ σιλφίου).*

11.15 *ἔστιν δὲ καὶ τῶν μὲν καὶ ἄνω καὶ ἐν ταῖς ῥίζαις ἡ ἐντομή, καθάπερ φαρμακωδῶν τέ τινων καὶ τοῦ σιλφίου (καὶ γὰρ ἡ ῥίζα καὶ ὁ καυλὸς ἐντέμνεται, καὶ ἑκατέρωθεν ὁ ὀπός)· τῶν δὲ μεμερισμένη,*[3] *τῶν μὲν ἐν ταῖς ῥίζαις, τῶν δὲ ἐν τοῖς καυλοῖς, ὡς ἂν ἔχωσιν ἑκάτερα*[4] *φύσεως· ἐὰν μὲν τὰς ῥίζας εὐχυλοτέρας, ταύτας, ἐὰν δὲ τὰ ἄνω, τοὺς καυλούς*

[1] u : *παρασκιαζουσιν* U.
[2] U : *αὐτοὺς* Scaliger (Gaza omits).
[3] Wimmer: *-ων* U.
[4] ego : *ἑκάτερα τῆς* Wimmer (*cf. CP* 6 12. 1 line 4) : *ἑκατέρας* U.

---

[1] *Cf. HP* 9 8. 3 (of saps with medicinal or other potencies): "In some plants there is not even any collection of the sap, but rather a kind of extraction of juice, as when the plant is chopped or ground and water is poured on; the result is then put in a filter and the liquid sediment is retained . . ."

[2] The liquid is "drier," that is, thicker, because it con-

adapted to our use when body has been added to them, and body is added as the watery part separates out (which is why producers take steps to bring the process about: with some plants they catch the flavour in water[1] and make the liquid drier[2]; with others they make incisions to permit hardening by sun and air, as with resins and the like, frankincense and myrrh, and the sap of silphium[3]).

11.15 In some the incision is made in both the upper part and the root, as with certain medicinal plants and silphium (for here both root and stalk are incised, and the sap comes from both)[4]; in others the incision is limited to the one part or the other, being in the root in some, in the stalk in others, depending on how the one part or the other stands with regard to its nature[5]: if the plant is juicier in the root, the root is tapped; if in the upper part, the

tains more solid particles. Evaporation or boiling sometimes follows.

[3] *Cf. HP* 9 2. 1 (raisins); 9 4. 1–10 (frankincense and myrrh); 9 1. 4, 7 (silphium).

[4] *Cf. HP* 9 1. 3: ". . . in some of these plants the exudation is both in the stalk and the root, for people take the sap in some by tapping both the stalk and the root, as they do with silphium"; *HP* 9 1. 7: ". . . in plants where both the stalk and the root are incised the stalk is incised first, as in silphium . . . So too do herbalists and those who gather medicinal saps, for these too take the sap first from the stalks."

[5] Explained below, *CP* 6 12. 1, end of first paragraph.

(ἐπεὶ ὧν γε ξηραὶ καὶ ξυλώδεις, οὐκ ὀπίζουσιν).

ὁμοίως δὲ καὶ τῶν αὐτομάτως ἐπιπηγνυμένων δακρύων, οἷον ἐπί τε τῇ σχίνῳ[1] καὶ ἀκάνθαις τισίν, καὶ ἐπὶ τῶν δένδρων διαδίδωσιν, ὥσπερ ἀμυγδαλῇ καὶ ἡ ἄρρην ἐλάτη καὶ τέρμινθος· τούτων δὲ καὶ τὰ μὲν καὶ εὔστομα καὶ εὐώδη, τὰ δ' ἄχυλα καὶ ἄοσμα, καθάπερ τὸ κόμμι τὸ τῆς ἀκάνθης τῆς ἐν Αἰγύπτῳ.

11.16 καλοῦσι δὲ τὰ μὲν ὀπούς, τὰ δὲ δάκρυα, κοινότερον δὲ ὀπός· διαφέρει δὲ ἴσως οὐδέν, ἐπεὶ τό γε κοινότατον ἀνωνύμως λεγόμενον, ἡ ὑγρότης ἡ οἰκεία καθ' ἕκαστον, πέψιν ἔχουσα. διὰ δὲ τὸ[2] σωματωδεστέρας εἶναι καὶ γλίσχρας, τὰς <δ'>[3]

[1] ego (τῆς σχίνου Heinsius) : τῆσπίνου U.
[2] τὸ τὰς μὲν Itali.
[3] aP.

---

[1] The shrub *Pistacia lentiscus.*

[2] "Thorn" is any thorny plant. Here the thorns are (1) "Indian thorn" (Balsamodendron Mukul), for which *cf. HP* 4 4. 12; 9 1. 2, and (2) "Pine-thistle" (Atractylis gummifera): *cf. HP* 9 1. 2 (of exudations): "They also form on the mastic (*schînos*) and on the so-called *ixínē* (*sc.* containing *ixía* or bird-lime) thistle, and from these mastic comes"; for the name *ixínē cf.* also *HP* 6 4. 9. The thistle is also called *ixía* (*HP* 9 1. 3) and white chameleon (*HP* 9 12. 1).

[3] *Cf. HP* 9 1. 2.

[4] *Cf. HP* 9 1. 3: "All these exudations are fragrant, and

stalk (since the root is not tapped in plants where it is dry and woody).

The conditions are similar too with the gummy exudations that solidify spontaneously on the surface, as on mastic[1] and certain thorns,[2] and they also come out on trees (as the almond, the male silver-fir and the terebinth).[3] Of these exudations moreover some have an agreeable taste and are fragrant, others lack flavour and odour, as the gum of the acacia in Egypt.[4]

11.16 Some of these products are called "saps,"[5] others "exudations,"[6] "sap" being the name more widely applied. But the name perhaps makes no difference,[7] since the most inclusive designation is no name at all, but the phrase "the fluid, proper to each different kind of plant, that has received concoction"; and because some of these fluids have more body and are viscous, whereas others are watery

indeed all (one may say) are fragrant that have a certain fattiness and oiliness; those that have none are odourless, as the gum of the acacia . . ."

[5] Those that do not harden of their own accord; they are discussed in *CP* 6 11. 14–15.

[6] Those that harden of their own accord; discussed in *CP* 6 11. 15 (last paragraph).

[7] At *HP* 9 1. 3 the product of silphium is treated with exudations, though the verb *opízō* ("to tap," literally "to get the sap") is used of obtaining it; at *HP* 9 1. 4 we are told that ". . . the so-called sap of silphium is an exudation"; and at *HP* 9 1. 7 the producers are said to speak of the "stalk-sap" and "root-sap" of silphium.

ὑδατώδεις κἀγλίσχρους,[1] τῶν μὲν γίνεται πῆξις, τῶν δ' οὔ· πρὸς ἐνίας δὲ καὶ παρεμβάλλουσίν τι τοῦ πῆξαι καὶ συλλέγειν.

12.1 ὅτι δὲ τὰ μὲν ἐν ταῖς ῥίζαις, τὰ δ' ἐν τοῖς καυλοῖς τὰς δυνάμεις ἔχει ταύτας, ἐκείνην χρὴ τὴν αἰτίαν ὑπολαβεῖν τὴν μικρῷ πρότερον λεχθεῖσαν· ὧν[2] ἡ φύσις ἑκατέρων[3] σύμμετρος εἰς τὴν δύναμιν, ἔνθα μὲν ὑγρὸν ἱκανόν, ἔνθα δὲ ἔλαττον ἔχουσα, καὶ ξηρὸν ὡσαύτως.

ἔτι δ' ὧν[4] ἡ οὐσία μᾶλλον ἐφ' ἑκάτερα ῥέπει· καὶ γὰρ ἐπὶ τῆς αὐξήσεως καὶ τοῦ μεγέθους τοῦτ' ἔστιν, τὰ μὲν ἐν τοῖς ἄνω, τὰ δ' ἐν ταῖς ῥίζαις ἔχειν[5] μᾶλλον. οἱ μὲν γὰρ σίκυοι καὶ κολοκύνται καὶ ἄλλ' ἄττα πλείω τὰ ἄνω μείζω, ῥίζαν δὲ μικρὰν ἔχουσιν· σκίλλα δὲ καὶ βολβὸς καὶ ἁπλῶς τὰ κεφαλόρριζα τὰ μὲν ἄνω λεπτὰ καὶ ἀσθενῆ, τὰς δὲ ῥίζας μεγάλας καὶ σαρκώδεις.

[1] ego : καὶ ἀγλίσχρους Gaza, Schneider : ἀγλίσχρους U.
[2] u (ὤν U) : ὧν γὰρ Gaza, Schneider : ὡς Wimmer.
[3] U : *parte utraque* Gaza : ἑκατέρα Schneider.
[4] u (ὤν U) : ὡς Wimmer. [5] U^ar : -ει U^r N aP.

---

[1] *Cf. HP* 9 1. 7: "The stalk-sap is too fluid, and this is why flour is sprinkled on it to make it harden."

and without viscosity, the former will harden, the others not. To obtain some of these fluids producers add a substance to harden them[1] and facilitate gathering.

Reason for Production in Root or Stalk

12.1 We must suppose the reason that some plants have these powers of production in the root, others in the stalk, to be the one mentioned a moment ago[2]: the powers belong to plants in which the nature of the one part or the other possesses the right quantity for the power, in the one part enough fluidity, in the other too little (and in the same way with the dry).

The powers depend again on which of the two directions is taken by the essence[3] of the plant. In fact the greater inclination in the one direction or the other occurs also in the matter of growth and size: some plants have growth and size to a greater extent in the upper part, some in the root. So cucumber and gourd and a good many others have the upper part large and the root small, whereas squill and purse-tassel and in a word bulbous plants have the upper part thin and weak but the root[4] large and fleshy.

[2] *CP* 6 11. 15.
[3] Called "nature" at *CP* 6 12. 2.
[4] The bulb: *cf. HP* 1 6. 8–10.

12.2 οἷς <δ'>[1] ἐνυπάρχει δριμύτης ἢ[2] καὶ ἄλλη τις τοιαύτη δύναμις, ἐν ταῖς ῥίζαις γίνεται μᾶλλον, ὥσπερ τοῖς κρομμύοις καὶ τοῖς σκόρδοις καὶ ταῖς σκίλλαις (τοῦτο δ' εὐλόγως, ἐνταῦθα τῆς φύσεως ὡρμηκυίας μᾶλλον, ἀκολουθοῦσιν γὰρ ταύτῃ καὶ αἱ δυνάμεις), ὃ[3] ἐπὶ τῶν φαρμάκων[4] ἐστὶ σχεδὸν τῶν πλείστων· αἱ γὰρ ῥίζαι φαρμακωδέστεραι καὶ μᾶλλον ἔχουσι τὴν δύναμιν.

ἡ δ' αἰτία δυοῖν[5] ἐκείνοιν νῦν τηρεῖται,[6] τῷ τε μὴ κάθυγρα, καὶ τῷ μὴ κατάξηρα γίνεσθαι· τὰ μὲν γὰρ οὐκ ἔχει πέψιν διὰ τὸ πλῆθος, τὰ δ' οἷον ὕλην εἰς τὴν πέψιν διὰ ξηρότητα. πέψις δὲ ἑκάστων ἐστὶν (ὥσπερ εἴρηται) πρὸς τὴν οἰκείαν φύσιν καὶ δύναμιν.

12.3 ὅτι μὲν οὖν ὥσπερ ἐναντίως τῶν μὲν ἄνω, τῶν δὲ κάτω συμβαίνει τοὺς χυλοὺς καὶ τὰς ὀσμὰς καὶ ἁπλῶς τὰς τοιαύτας δυνάμεις,[7] ἐκ τῶν εἰρημένων δεῖ[8] θεωρεῖν.

[1] Wimmer (*prorsus in quibus* Gaza : οἷς γὰρ Schneider) : οἷς U : οἷον N : ὅπου γὰρ aP. [2] u : εἰ U. [3] ὃ καὶ aP.
[4] U : φαρμακωδῶν Itali. [5] ἐν δυοῖν Schneider.
[6] νῦν τηρεῖται U : ἃ εἴρηται Schneider (*quae superius diximus* Gaza). [7] U : δ. εἶναι Schneider. [8] u : δὴ U.

---

[1] That is, strong taste.
[2] The bent of the plant's nature. [3] *CP* 6 12. 1.
[4] *CP* 6 4. 3–4; 6 6. 3. The sweet is not the only fully

12.2 Plants containing pungency or some other potency of the sort[1] tend to have it rather in the roots, as onion, garlic and squill. (It is reasonable that this should be so, since their nature is bent on greater increment here; for the potency takes the same direction as the nature.) This also holds of the majority (one might say) of medicinal plants, for their roots are more medicinal than the upper parts and have more of the potency of the plant.

The causation[2] is in this case preserved by those other two processes[3]: by not letting the part get too fluid, and by not letting it get too dry. For parts with excessive fluid get no concoction because there is too much fluid in them to concoct; whereas the others have no "matter" (as it were) to concoct, by reason of their dryness. In each plant (as we said[4]) concoction is in the direction of the proper nature and power of the plant.

12.3 That flavours, then, and odours (and in general such powers as these) turn out to be produced[5] in what one might call opposite fashion,[6] above in some, below in others, is to be understood in the light of the foregoing discussion.[7]

concocted flavour.

[5] I understand "are produced" (γίνεσθαι) from *CP* 6 12. 3 (Schneider supplies "are": εἶναι); Theophrastus leaves the verb to be understood.

[6] That is, at opposite ends of the plant.

[7] *CP* 6 11. 3–6 12. 2.

ὅτι δ' οὐκ ἐν τοῖς αὐτοῖς μέρεσιν τοῖς ἄνω πᾶσιν οὔθ' οἱ χυλοὶ τυγχάνουσιν οὔθ' αἱ ὀσμαί, σχεδὸν οὐ πόρρω τῶν εἰρημένων αἰτιῶν ἐστιν·[1] ὡς μὲν γὰρ ἁπλῶς εἰπεῖν ἔχει τὴν[2] ὁμοιότητα καὶ κατὰ τοὺς χυλοὺς καὶ <κατὰ>[3] τὰς ὀσμὰς ἕκαστον τῶν

12.4 μερῶν, διαφέρει δὲ τῷ[4] μᾶλλον καὶ ἧττον. ἐμφανὲς δὲ τοῦτο μάλιστα ἐπὶ τῶν ἀκρατεστέρων ταῖς ὀσμαῖς καὶ τοῖς χυλοῖς (οἷον ἐλάτης πεύκης κυπαρίττου πίτυος, ἔτι δὲ τῶν ἡμέρων συκῆς).

ἐν δὲ τοῖς[5] ὑδαρεστέροις οὐχ ὁμοίως, ἀλλ' ἐν τούτοις καὶ διαφέρουσιν ὥστε τὰ μὲν ἔγχυλα, τὰ δὲ ἄχυλα καὶ δύσχυλα, καὶ εὔοσμα, τὰ δ' ἄοσμα γίνονται.

καὶ διαφέροντα δὲ ταῖς ὀσμαῖς καὶ τοῖς χυλοῖς, ὥσπερ τῆς ἀμπέλου τό τε οἴναρον καὶ ὁ βότρυς,

[1] Wimmer : εἰσιν U.
[2] U : τιν' Schneider (*aliquid* [*similitudinis*] Gaza).
[3] aP.
[4] τῷ u P : τὸ U N a.
[5] Basle ed. of 1541 : ταῖς U N aP.

---

[1] *CP* 6 12. 1–2.
[2] And so the reasons explaining the presence of flavour

Upper Parts: Differences in Flavour and Odour
(1) *In Degree*

The reason that not all plants have either their flavour or their odour in the same upper part, is (one may say) not far removed from the reasons given[1]: broadly speaking, all upper parts of the same plant have a similar flavour and a similar odour and differ only in degree.[2] This difference in degree is most noticeable in trees with the more concentrated odours and flavours (as silver-fir, pine, cypress and Aleppo pine; further, among cultivated trees, the fig). 12.4

(2) *In Presence or Absence*

Difference in degree is not found to the same extent in trees with odours and flavours that are more watery; but here the upper parts also differ among themselves to the extent that in the same tree one part possesses flavour, the other possesses no flavour or a poor one, and the one part possesses a good odour, the other none.

(3) *In Presence of Different Ones*

Along with this the different upper parts also have distinct flavours and odours, as in the vine the

(and odour) in the upper parts, and not in the lower, will suffice.

καὶ ἁπλῶς δὲ πάντων καὶ τῶν φύλλων καὶ τῶν καρπῶν.

12.5 αἴτιον δ', ὅτι τὸ μὲν ἄπεπτον, τὸ δὲ πεπεμμένον.

τάχα δὲ μᾶλλον, ὅτι οὐδὲ ἐκ τῆς αὐτῆς ὕλης ἑκάτερον, ἀλλὰ τὸ μὲν ὥσπερ ἐκ καθαρᾶς τινος καὶ εἰλικρινοῦς, τὰ δ' ἐκ περιττωματικῆς· εἰσὶ δ' ἐκ βλοσυρωτέρας καὶ σωματωδεστέρας οἱ βλαστοὶ καὶ ἀκρεμόνες καὶ ὁ ὅλος ὄγκος, ὥστ', ἀνομοίων οὐσῶν, ἀνόμοιον καὶ τὸ τέλος, ἑκάστου πρὸς τὴν ἰδίαν φύσιν, ὥσπερ ἐν τοῖς ζῴοις. ἴσως δ' ἀνάπαλιν· ἐπεὶ[1] καὶ τὸ τέλος ἴδιον ἑκάστου,[2] καὶ ἡ

[1] u : ἐπὶ U.
[2] ἴδιον ἑκάστου ego : ἐκαστου ἴδιον U.

---

[1] That is, the evil taste is not due to failure to concoct, it was aimed at.

[2] *Cf.* Aristotle, *On the Generation of Animals*, ii. 6 (744 b 11–27): "Of the other parts each is produced from the food, the highest parts and those that partake in the supreme government of the animal being produced from the concocted, purest and first-grade food, the parts which are merely necessary and exist for the sake of the first, from the second-best food and scraps and left-overs. For nature, like a good householder, is not in the habit of throwing away anything from which some good can come, and in the management of households the best of the food that is pro-

leaf and cluster differ; and in general leaf and fruit differ in all.

Upper Parts: Reasons for the Differences

12.5 The reason is that the one part is unconcocted, the other concocted.

Or perhaps the reason is rather that the two are not even produced from the same matter: instead the one comes from a pure (as it were) and unadulterated matter, the leaves from left-overs, and the shoots and branches and general bulk of the tree are from matter that is still coarser and has still more body. Hence the different quality of the matter leads to a different quality of the final product, which in each is the development of the special nature of the part,[1] just as in animals.[2] But perhaps it is the other way round: since the final product is peculiar to each part, the matter too is different: in a

duced goes to the freemen, the second-best and what is left over from the best to the slaves, and the worst is also distributed to the beasts that are raised on the estate. So just as the intelligence outside (*i.e.*, as that of the manager is outside the estate) acts to promote increase, so the nature within the productions themselves composes flesh (*i.e.* the sense organ of touch) and the bodily part of the other senses from the purest food, but from the left-overs the bones, sinews and hair, and furthermore the nails, hoofs and all such parts as these (*i.e.* non-sentient). This is why these are the last formed, when a left-over of the nature is already available."

ὕλη διάφορος, ὡς δὲ ἁπλῶς εἰπεῖν, πάντων πρὸς τὴν ὑποκειμένην φύσιν.

12.6 ἐπεὶ δ' αὗται[1] διάφοροι καὶ κατὰ χυλοὺς καὶ ἀχυλίας,[2] καὶ ὀσμὰς καὶ ἀοσμίας, ἡ δ' ἀναφορὰ πρὸς τὴν ἡμετέραν αἴσθησιν, ἐν ταύτῃ παραλλαγὴ γίνεται τῶν μερῶν· οὐδὲν γὰρ κωλύει τὴν μὲν ἐν τοῖς φύλλοις κρᾶσιν ἐνίοτε σύμμετρον εἶναι τῇ γεύσει, τὴν δ' ἐν τοῖς καρποῖς ἀσύμμετρον καὶ σφοδροτέραν ἢ κατὰ στρυφνότητα καὶ αὐστηρίαν ἢ κατὰ πικρότητα καὶ ἄλλην τινὰ διάθεσιν.

ὅπερ φανερὸν καὶ ἐπὶ τῶν λαχάνων ἐστίν· τὰ μὲν γὰρ[3] σύμμετρα καὶ προσφιλῆ, τὰ δὲ σπέρματα δριμέα καὶ πικρότερα καὶ τὸ ὅλον ἰσχυρότερα 12.7 πρὸς τὴν αἴσθησιν. ὥστε σχεδὸν ἀνάπαλιν ἔχειν καὶ ἐπὶ τῶν δένδρων καὶ ἐπὶ τοῦ σίτου· τῶν μὲν γὰρ οἱ καρποὶ χρήσιμοι, τὰ δὲ φύλλα ἀχρεῖα· τῶν δὲ ταῦτα μόνον χρήσιμα, τὰ δὲ ἄλλ' ὡς[4] ἂν ἐπὶ φαρμάκου λόγον.

ὅπερ οὖν ἐπὶ τούτων, οὐδὲν κωλύει καὶ ἐπὶ τῆς φιλύρας καὶ ἐπ' ἄλλων δένδρων τινῶν, <ὧν>[5] ὁ μὲν[6] καρπὸς ἄβρωτος, τὸ δὲ φύλλον γλυκὺ καὶ

[1] N aP : αὐταὶ U.
[2] ego (ἀχυμίας Wimmer) : ἀχύλους U.
[3] γὰρ φύλλα Moreliana.
[4] U : ἄλλως N aP.
[5] ego. [6] μὲν γὰρ Schneider.

word, in all the parts the matter depends on the nature to be achieved.[1]

Since the nature of one part differs from the nature of another both in flavour and the lack of it and odour and odourlessness, and since the appeal is to our sense, our sense will make a difference between one upper part and another; for there is nothing to keep the temper of the qualities in the leaves from sometimes being exactly adjusted to the sense of taste, whereas the temper in the fruit is not adjusted and is too strong, either in astringency and dry-wine flavour or else in bitterness and some other disposition. 12.6

The distinction in adjustment is also evident in vegetables: thus the leaves are well adjusted and hence agreeable, but the seeds are pungent and too bitter (and in short too powerful for our sense). Consequently the case here is just about the reverse of what it is with both trees and cereals: in these it is the fruit that is useful to us, whereas the leaves are useless; but in vegetables the leaves alone are useful, whereas the rest practically rate as a drug. 12.7

There is then nothing to prevent what is true of vegetables from also being true of the linden and certain other trees with inedible fruit but with sweet and edible leaves[2]; since the fruit has the con-

[1] For a similar shift (this time away from the teleological explanation) *cf. CP* 2 10. 2 *ad fin.*

[2] *Cf. HP* 1 12. 4: "The most isolated case is that of the linden: its leaves are sweet, and many beasts eat them, but no animal can eat the fruit . . ."

βρωτόν· ἐκεῖνος[1] *μὲν γὰρ τὸ ἄκρατον ἔχει καὶ καθαρὸν τῆς φύσεως (εἴτε ξηρὸς ὢν ἄγαν καὶ ξυλώδης, εἴτ' αὐστηρὸς καὶ πικρός, ἢ κακώδης, ἢ καὶ ἄλλην τινὰ ἔχων δυσχέρειαν, ἄβρωτος γίνεται)· τὸ δὲ φύλλον, ὑγρότερον ὄν, καὶ ἅμα κεκραμένον,*[2] *ἔχει τινὰ συμμετρίαν.*

12.8 *ὃ καὶ ἐπὶ τοῦ σιλφίου καὶ ἄλλων δριμέων ἐστίν· ἡδὺ γὰρ τὸ φύλλον αὐτοῦ, καὶ ὁ καρπὸς ἁπαλὸς ὤν, διὰ τὴν ὑπάρχουσαν ὑγρότητα, σύμμετρος γὰρ ἡ δριμύτης γίνεται κραθεῖσα τῷ ὑδατώδει, καὶ ποιεῖ τινα χυλόν· ἀποξηραινομένου δὲ σφοδρότερον,*[3] *καὶ ἡ τοῦ σπέρματος δ' ἔτι μᾶλλον.*

[1] u : ἐκεῖνο U. [2] Scaliger : κρεμάμενον U.
[3] U : σφοδροτέρα Schneider.

---

[1] *Cf. HP* 6 3. 1: "Silphium has ... its leaf, which they call 'maspeton,' like that of celery; and a flat seed, resembling a leaf ... At the coming of spring the plant sends up this 'maspeton,' which purges the sheep, makes them very fat and makes the meat extremely good eating; next it sends up the stalk, which is eaten in every way, boiled or roasted, and this too is said to purge people ..." Theophrastus continues with another report at *HP* 6 3. 4–6: "Such then is the report of some. Others assert that the root gets to be a cubit long or a little longer, and has at its centre a head, which is the highest point of the root and just about above ground, and from this the plant grows; it is called the 'milk.' After this time the stalk grows, and from the stalk the 'magydaris' and the so-called 'leaf'; and to this belongs the seed ... Another contradiction with the other account is the assertion that the sheep do not

centrated and undiluted character of the nature of the tree (whether the inedibility comes from extreme dryness and woodiness in the fruit, or from its having the dry-wine or bitter flavour, or from its evil odour, or from some other offensiveness to sense), whereas the leaf, which is more watery and also tempered in its quality, possesses a certain adjustment.

12.8 This is also the case of silphium (and other pungent plants): its leaf is pleasant eating, and so too its fruit when it is tender, because of the fluid then present, since the pungency when tempered by the wateriness becomes adjusted and this results in a certain succulence. But when the plant dries out the leaf is too concentrated, and the pungency of the seed is still more so.[1]

get purged when they eat the leaf: they say that both in spring and in winter the sheep are let out to graze on the mountain and feed on this leaf and on another resembling southernwood. Both are regarded as heating in their effect, and do not bring about purgation, but cause drying out and assist concoction; and if (they say) a sheep that is sick or in a bad condition enters the silphium grounds it is quickly cured or dies, but for the most part tends rather to be saved. We must examine to see which of the accounts is true." In both accounts the sheep eat the leaf; in the second the leaf contains the seed. Both accounts also mention a medicinal effect: purging in the first, heating, with killing or curing for some, in the second. Theophrastus in our passage is apparently following the first account: the leaf is pleasant eating (for sheep), the stalk (that is, the fruiting-stalk, called "fruit" in our passage) for man.

12.9 ὅλως δ᾿ ἐν πολλοῖς τοῦτ᾿ ἔστιν, ὥστε χλωρὰ μὲν ὄντα, βρωτὰ γίνεσθαι καὶ ἔχειν τινὰ χυλὸν διὰ τὸ ἀναμεμῖχθαι τῷ ὑδατώδει, καταξηραινόμενα δὲ καὶ λυομένης τῆς κράσεως, ἄβρωτα· καὶ γὰρ οἱ τῶν ἀμπέλων βλαστοὶ τοιοῦτοι καὶ οἱ τῶν ἀκρεμόνων,[1] ἔτι δὲ ἄλλων τινῶν, ἀκανθωδῶν ὄντων, καὶ τῶν χεδροπῶν δ᾿ ἐνίων (οἷον ὤχρων κυάμων), καὶ ἁπλῶς ὅσα γλυκύτητά τινα ἔχει. σχεδὸν δ᾿ ἔτι[2] κοινὸν τοῦτο ἐπὶ πάντων ὧν στρυφνότεροι καὶ δριμύτεροι καὶ πικρότεροι καθ᾿ αὑτοὺς οἱ χυλοί· κεραννύμενοι γὰρ ὑπὸ τῆς φύσεως τῷ ὑδατώδει, βελτίους καὶ γλυκυτέρους ποιοῦσι τοὺς χυλούς, ὅπερ καὶ ἐπὶ τῶν μύρτων γίνεται καὶ τῶν ῥοῶν.

12.10 εἰ δέ τις[3] τῶν δένδρων (καὶ ὅλως τῶν ὑλημάτων[4]) τοιαύτη φύσις ἐστὶν ὥστε τοὺς μὲν καρποὺς ἀμίκτους ποιεῖν, ἔχοντας τὴν οἰκείαν δύναμιν καθαράν, τὰ δὲ φύλλα καὶ ἄλλο τι τῶν μορίων μικτά,[5] τούτων οὐδὲν κωλύει (καθάπερ εἴρηται)

[1] U : ἀσφαράγων Wimmer.
[2] U : ἔστι aP.
[3] Gaza (*qua*), Itali : τι U.
[4] aP : ἐνυλημάτων U N.
[5] Gaza (*creet per mixtionem*), Moreliana : μικρὰ U.

Upper Parts: A Part is Edible when Tender, Inedible when Dry

Indeed this is in general true of the upper parts of many plants: when the part is green it comes to be edible and have a certain flavour because of its admixture with the watery, but as it gets dry and the tempered character breaks down, it becomes inedible. Thus the shoots of the vine are of this description (even those of the branches); so too those of certain other plants which are thorny,[1] and again those of some legumes (as birds' pease and bean)—in a word of all plants that have a certain sweetness. And this extends further (one may say) to all parts with flavours that when unmixed are too astringent or pungent or bitter; for when these are mixed with the watery by the nature of the plant, they make the flavour of the part better and sweeter, which is what occurs with the myrtle-berry and pomegranate. And if among trees (and woody plants in general) any have such a nature as to produce fruit with no admixture of other qualities, fruit having its own power in the pure state, but to produce the leaves or another part with an admixture, there is nothing (as we said[2]) to prevent 12.9 12.10

[1] *Cf.* Aristotle, *History of Animals*, ix. 2 (610 a 5–6) [of the ass]: "... it eats thorny plants when they are tender"; *HP* 6 4. 2: "... the stalk comes up from the asparagus plant in spring and is edible; later it gets rough and thorny with the advancing season ..."

[2] *CP* 6 12. 7.

τὰ μὲν φύλλα βρωτά, τὸν δὲ καρπὸν ἄβρωτον εἶναι.

παραπλήσιον δέ τι τούτοις καὶ ἐπὶ τῶν ῥιζῶν τῶν ἐδωδίμων ἐστὶν ὧν τὰ μὲν φύλλα σχεδὸν ἄβρωτα διὰ τὸ ξηρὰ εἶναι καὶ ἀκανθώδη, οἱ δὲ καυλοὶ ἐδώδιμοι, αὐταὶ[1] *δὲ αἱ ῥίζαι καὶ ἡδεῖαι.* 12.11 *τῶν <δ'>*[2] *ἀνάπαλιν τὰ μὲν ἄλλα βρωτὰ*[3] *καὶ ἔχοντα τὴν*[4] *ἡδονήν, αἱ ῥίζαι δὲ ἄβρωτοι διὰ τὸ ξυλῶδες ἢ πικρὸν ἢ ὅλως δύσχυλον ἢ ἄχυλον.*

διαφορὰ δὲ οὐ μικρὰ καὶ διὰ τὰς φύσεις τῶν ζῴων· τὰ μὲν γὰρ ἡμῖν ἄχυλα τοῖς ἄλλοις ἔγχυλα φαίνεται διὰ τὴν ἰσχὺν καὶ τὴν κατεργασίαν, <ἄλλα δὲ>[5] *ἄλλοις κἀκείνων προσφιλῆ, καὶ κατὰ* 12.12 *τοὺς χυλοὺς καὶ κατὰ τὰς διαθέσεις. χαίρει γὰρ τὰ μέν, τοῖς ἁπαλοῖς, τὰ δέ, τοῖς ξηροῖς μᾶλλον.*

ἔνια δὲ καὶ ἀνάπαλιν· ἁπαλὰ μὲν ὄντα,

[1] u : *αὗται* U. [2] Schneider.
[3] u by expunction (*βρωτά τε* Itali) : *βρώματα* U.
[4] U : *τιν'* Schneider. [5] Wimmer (*alia tamen* Gaza).

---

[1] Smaller plants with useful roots were sometimes simply called "roots": *cf. HP* 9 8. 1, where we hear of the fruits and leaves of such "roots."

[2] *Cf.* Aristotle, *On the Parts of Animals*, iii. 14 (675 b 4–5): ". . . all horned animals (one may say) have large

the leaves of these trees from being edible but the fruit inedible.

Something similar is also found in the edible "roots"[1] where the leaves are practically inedible because they are dry and thorny, the stalks edible, and the roots themselves quite pleasant. In some plants, on the other hand, the situation is reversed: the other parts are edible and provide pleasant eating, but the roots are inedible, owing to their woody character or to their bitterness (or general poorness of flavour or absence of it).

12.11

No small difference is also due to another factor: the different natures of the animals that do the eating. Thus a part that is flavourless to us has flavour to them because they are strong and prepare their food more thoroughly[2]; again there are different preferences among the beasts too, depending on the flavour of the part and the disposition of the beast. For some prefer the tender parts, others the dry.

12.12

Upper Parts: A Part is Inedible when Tender, Edible when Dry

But in some parts the reverse is the case: when

intestines because of the working up of their food"; iii. 14 (674 a 28–30) [the camel has several stomachs] ". . . owing to the size of its body and the strength needed for the food, which is not easily concocted, but thorny and woody . . ."

ἄβρωτα, ξηραινόμενα δέ, ἐδώδιμα, διὰ τὸ συνεκπέττεσθαι καὶ ἐκκρίνεσθαι τὸ πικρὸν ὑπὸ τοῦ ἡλίου, καθάπερ τὸ σήσαμον καὶ τὸ ἐρύσιμον. τούτων γὰρ χλωρῶν ὄντων οὐδὲν ἅπτεσθαι δοκεῖ ζῷον διὰ τὴν πικρότητα καὶ δυσχυλίαν, ξηρανθέντων δὲ μᾶλλον· οἱ δὲ δὴ καρποί, καὶ ἡδεῖς.

ἀλλὰ δὴ τούτων μὲν τοιαῦταί τινες αἰτίαι.[1]

13.1 περὶ δὲ δὴ τῶν καρπῶν τῶν τροφίμων δῆλον ὅτι ῥᾴων[2] ὁ λόγος· οὗτοι γὰρ καθαροὶ καὶ εἰλικρινεῖς, οἷον ἐξηθημένοι[3] πως ὑπὸ τῆς φύσεως, ὥστε πλείοσι καὶ μᾶλλον ἁρμόττειν.

ἀλλ' ἐπὶ τῶν φαρμακωδῶν καὶ ὅλως <τῶν>[4] κατὰ τὰς δυνάμεις θεωρητέον· καὶ γὰρ ἐν τούτοις οὔθ' αἱ αὐταὶ δυνάμεις οὔτ' ἴσαι πάντων εὐθὺς ἀπὸ τῶν ῥιζῶν ἀρχομένοις· οὐδὲ δὴ τοῖς ἄνω πᾶσιν (οἷον φύλλοις καὶ κλωσὶ καὶ καρποῖς). καὶ τὸ μὲν τῷ μᾶλλον καὶ ἧττον διάφορον ἔχει τιν' εὐλο-

[1] *αἱ αἰτίαι* Schneider.
[2] U (*ῥάων* aP) : *ῥᾷον* u (*ῥᾶον* N).
[3] Gaza (*liquati purgatique*), Itali : *ἐξηνθημενοι* U.
[4] ego (*his . . . quae* Gaza).

[1] *Cf. HP* 8 7. 3: ". . . but nothing eats sesame when it is green or lupine either. Whether also nothing will eat

tender the part is inedible, but edible when dry, because the sun helps to bring about concoction and to eliminate the bitterness, as with sesame and hedge-mustard. For when these are green it is held that no animal will touch them by reason of their bitterness and evil flavour[1]; whereas animals are readier to eat the plant when it is dry, and the fruit is even pleasant.

Such then are the reasons for these variations in the other upper parts; as for the nutritious fruit[2] the explanation is easier: the fruits are pure and unadulterated, as if the nature of the plant had put them through a filter,[3] and thus agree with more animals and agree with them better. 13.1

Upper Parts: Medicinal Plants and the Like

We must however study the question in the case of medicinal plants and in general those that work by their potencies[4]: here too the potencies of all the parts are neither the same nor equal, beginning with the roots, nor yet again are they the same or equal in all the upper parts (as leaves, twigs and fruit). Now the difference in degree has a certain reasonableness; one is more likely to be surprised at

hedge-mustard or clary is a point for investigation; these too are bitter."

[2] *CP* 6 12. 12: ". . . and the fruit is even pleasant."

[3] *Cf. CP* 6 6. 5. [4] And not as food.

γίαν· τὸ δ' ὅλως ἔνια μηδὲ δύνασθαι ποιεῖν (ὥσπερ τὰ τῶν ῥιζῶν τὰ σπέρματα καὶ τοὺς καυλούς, ἢ πάλιν τὰ τῶν φύλλων τοὺς καρποὺς καὶ τὰς ῥίζας) μᾶλλον ἄν τις θαυμάσειεν.

13.2 *τὰς δ' αἰτίας καὶ τούτων ἐν τοῖς εἰρημένοις ἀνασκεπτέον. ἑκάστου γὰρ ἰδία κρᾶσις οὖσα καὶ φύσις, διαφόρους ποιεῖ καὶ τὰς δυνάμεις, ὥστε τὰ μὲν συντήκειν*[1] *καὶ διακρίνειν, τὰ δὲ μή, καὶ τὰ μὲν μᾶλλον, τὰ δὲ ἧττον· καὶ θερμαίνειν τε καὶ πέττειν, καὶ ψύχειν, καὶ ξηραίνειν, καὶ τἆλλα ὡσαύτως.*

τὸ[2] *δὲ δὴ μάλιστα τοὺς καρποὺς εἶναι τοιούτους οὐκ ἄλογον, εἴπερ μηδ' ἡ ὅλη φύσις ὁμοία ῥίζης τε καὶ καρποῦ καὶ τῶν ἄλλων μορίων, ἀλλὰ τὰ μὲν ὅλως ἄπεπτα, τὰ δὲ πεπεμμένα, καὶ ὅλως ἐκ πλειόνων σύνθετα, διὸ καὶ τοῖς χυλοῖς καὶ ταῖς*

13.3 *δύναμεσιν ἕτερα. τοῦτο γὰρ καὶ ἐπὶ τῶν ἀγρίων καὶ τῶν ἡμέρων ὁρῶμεν· ὧν αἱ ῥίζαι πικραὶ καὶ ὀπώδεις, τούτων τοὺς καρποὺς γλυκεῖς, ὡς ἂν ἐξ ἀπέπτου τινὸς πεττομένους ὄντας.*

ταὐτὸν δὲ καὶ ἐν ταῖς τῶν φαρμάκων δυνάμεσιν

[1] N aP : (*συν* illegible) *τικειν* U. [2] u : *τῶ* U.

---

[1] *CP* 6 12. 1, 5–6.

another difference: some parts quite lack the power to do what other parts do (thus the seeds and stalks lack the potency of the roots, or again the fruit and roots the potency of the leaves).

13.2 Here too the reasons must be sought in the preceding discussion[1]: each part has its special tempering and nature, and the tempering and nature make the potency differ as well, so that some parts bring about colliquescence[2] and separation,[3] others do not; and some do it to a greater, some to a lesser degree; and so with heating, concocting, chilling, drying and the rest.

That the fruit differs most of all from the rest in this way is not unreasonable, since the whole nature of root and fruit and the other parts is also dissimilar, some parts being quite unconcocted, others concocted; all moreover are composed of different constituents, a fact that makes them differ in flavour and in potency. 13.3 For we observe this distinction in both the wild and cultivated kind: plants with bitter roots that are full of fig-like sap have fruit that is sweet,[4] and this suggests that the fruit is concocted from something unconcocted.

We must suppose that the same variation occurs

[2] *Cf. CP* 1 22. 6; *cf.* also "melt" (*CP* 6 1. 3 with note 1).

[3] *Cf. CP* 6 1. 3 (1) "expand."

[4] *Cf. HP* 1 7. 2: "... the roots are bitter in some plants whose fruit is sweet ..."

ὑποληπτέον συμβαίνειν, ὥστ᾽ οὐκ ἄλογον εἰς ἔνια μὲν τὰς ῥίζας ἰσχύειν[1] μᾶλλον, εἰς ἔνια δὲ[2] τῶν λοιπῶν τι μερῶν.

ἐπεὶ[3] καὶ ῥίζαι ῥιζῶν ἐν τοῖς ὁμογενέσι, καὶ σπέρματα σπερμάτων, καὶ τἆλλα μέρη, πολὺ διαφέρουσιν ἰσχΰϊ διὰ τὸν ἀέρα ἑκάστης χώρας, ὥσπερ καὶ ὁ σῖτος καὶ οἱ ἄλλοι καρποὶ τῇ βαρύτητι καὶ κουφότητι διὰ τὰς τροφάς.

13.4 καὶ διὰ τοῦτο φάρμαχα φαρμάκων ἀμείνω κατὰ τοὺς τόπους, οὐ μακρὰν ἀπέχοντας· οἷον ὁ ἐκ τῆς Οἴτης ἑλλέβορος τοῦ ἐν τῷ Παρνασῷ (δοκεῖ γὰρ ἰσχυρότερος εἶναι οὗτος καὶ οὐχ ἁρμόττειν εἰς τὴν χρείαν). ἡ δὲ ἰσχὺς ἐκ παραπλησίας τινὸς αἰτίας γίνεται καὶ ἐπὶ τῶν καρπῶν· καὶ γὰρ ἐνταῦθα σκληρότητί τε τοῦ ἀέρος καὶ πλήθει τῆς τροφῆς βαρεῖς γίνονται διὰ τὸ πολὺ τὸ γεῶδες ἔχειν (ὥσπερ οἱ ἐν τῇ Βοιωτίᾳ), κἀκεῖ διὰ τὰς ὁμοίας αἰτίας.

13.5 ἄλλη δὲ πρὸς ἄλλην δύναμιν οἰκεία χώρα, καθάπερ ἐπὶ τῶν καρπῶν· ἔνιαι[4] γὰρ οὐδὲ

[1] aP : ἴσχυει U. [2] aP : μεν U.
[3] u : ἐπι U. [4] Heinsius : ἐνια U.

---

[1] Literally "fruits"; legumes are meant: *cf. CP* 4 12. 4.
[2] *Cf. CP* 4 9. 4.

in the potencies of medicinal plants as well, so that it is not unreasonable that the root is stronger in some directions, another part in others.

In fact in plants of the same kind one root differs greatly from another, and so with the seeds and other parts, owing to the weather of the different countries when the plant grows: so cereals and other seed-crops[1] differ in the matter of indigestibility and digestibility[2] by reason of the differences in their food.

13.4 For this reason the excellence of a drug varies with the region where it grows, though one region may be at no great distance from the other: so the hellebore of Mt Oeta is better than that of Parnassus,[3] which is considered too strong to be suitable for use. So too with the grains[4] strength arises from a similar cause: their heaviness comes from the harshness of the air and the abundance of their food, which lead to a large earthy component, as is true of the grains of Boeotia[5]; and with medicinal plants similar causes apply.

13.5 Different countries are suited to the production of different powers, as with the seed-crops[6]; so some

[3] *Cf. HP* 9 10. 4: "Best of all . . . is the hellebore of Mt Oeta. That of Parnassus and of Aetolia . . . is harsh and excessively strong."

[4] Mentioned at *CP* 6 13. 3 (with note 1).

[5] *Cf. CP* 4 9. 5.

[6] *Cf. CP* 3 21. 1–5.

ἐκπέττουσιν ὅλως τὰς φαρμακώδεις δυνάμεις· ἐπεὶ[1] πολλαχοῦ καὶ μέλας ἑλλέβορός ἐστι καὶ ἄλλαι τῶν ῥιζῶν, ἀλλ' ἀμβλεῖαι[2] κἀδύνατοί[3] τινες. ᾗ[4] καὶ δόξειεν ἂν ψυχροῦ τινος ἀέρος ἅμα καὶ εὔπνου δεῖσθαι, καὶ ἔτι τροφῆς συμμέτρου· φαίνεται δ' οὖν ἐν τοῖς ὄρεσι πλεῖστα φάρμακα γίνεσθαι, καὶ ἐν τοῖς ὑψηλοτάτοις καὶ μεγίστοις μάλιστα.

καὶ περὶ μὲν τούτων ἱκανῶς εἰρήσθω.

14.1 περὶ δὲ τῶν ὀδμῶν, ἐπειδὴ καὶ αὗται κατὰ μέρη γίνονται, τὰς αὐτὰς καὶ παραπλησίας αἰτίας ὑποληπτέον· ἡ γὰρ πέψις τοῖς μὲν ἐν τῷ ἄνθει μᾶλλον, τοῖς δὲ καὶ ἐν τῷ καρπῷ, τοῖς δὲ καὶ ἐν τοῖς φύλλοις καὶ ἐν τοῖς κλωσὶ γίνεται· τοῖς μὲν

§14.1: Pliny, *N. H.* 21. 37.

[1] u : ἐπὶ U.
[2] Gaza (*hebetes*), Schneider : ἀμελείαι U.
[3] ego (καὶ ἀδύνατοί Itali : *imbecillesque* Gaza) : καὶ δυνατοί U.
[4] aP : ἡ U (ἦ u N).

---

[1] *Cf. HP* 9 10. 3: "The black hellebore grows everywhere; thus it occurs both in Boeotia and Euboea and many other regions. The best is that of Mt Helicon, and this mountain is in general well supplied with drugs."

countries do not even bring medicinal powers to full concoction. Thus both black hellebore[1] and other medicinal roots are found in many places, but are of a dull and ineffectual sort. Hence it would appear that drugs require a type of air that is not only cold but in movement, and again the right amount of food and no more; at all events we see that most drugs are produced on mountains, and especially on the highest and greatest.[2]

For these matters let the present discussion suffice.

Odours: The Parallel with Flavours

14.1 As for odours, since these too arise in different parts,[3] we must suppose that the reasons are the same and similar as those that account for the distribution of flavours. So the concoction[4] in some plants tends more to occur in the flower, in some in the fruit as well, in some in the leaves too and in the

[2] *Cf. HP* 9 10. 3 for Mt Oeta and Mt Helicon, and *cf. HP* 9 15. 4: "Of districts in Greece the richest in drugs are Mt Pelion in Thessaly, Mt Telethrion in Euboea, and Parnassus; . . ."

[3] For the moment "upper parts"; the roots are dealt with later (*CP* 6 14. 3).

[4] Fragrance (like flavour) is due to concoction: *cf. CP* 6 6. 2; 6 11. 4; 6 11. 5 and Theophrastus, *On Odours*, i. 3: "Fragrant . . . are things concocted . . ."

ἐν τοῖς[1] ὑδαρέσιν (καὶ ὥσπερ ἐν τοῖς[2] ἀχύλοις) ἢ δριμέσιν, ἐν τοῖς ἄνθεσιν, οἷον ῥόδῳ κρόκῳ καὶ τοῖς τοιούτοις· ὅσα δὲ αὐτῶν μὴ εὔχυλα, καὶ ἐν ταῖς ὀσμαῖς ἐμφαίνει γέ[3] τινα βαρύτητα, καθάπερ τὸ κρίνον καὶ τὸ λείριον καὶ τὰ παραπλήσια τούτοις·

14.2 ἔνια δ' ὅλως οὐδ' εὐοσμίαν, ἀλλὰ βαρύτητα ἔχον[4] κατὰ τὴν ὄσφρησιν.

καὶ φανερὸν ὡς ἀπὸ τοῦ χυλοῦ πως γίνεται, καὶ οὐκ ἔστιν ἀπηρτημένον τὸ τῆς ὀσμῆς· ὅσα γὰρ ὑδαρῆ καὶ ἄχυλα, καὶ ἄοσμα ὡς ἐπίπαν. ἐν δυσὶν γὰρ τούτοιν (ὡς εἰπεῖν) ἡ ἀοσμία· τῷ τε τὸν χυλὸν ὑδαρῆ τιν'[5] ἔχειν φύσει, καὶ τῷ πολὺ τὸ ὑδατῶδες. τὸ μὲν γὰρ ὥσπερ ἄμικτον, τὸ δ' ὑπὸ τοῦ πλήθους ἀφανίζεται, διὸ τρόπον τινὰ καὶ τοῦθ' ὥσπερ ἄμικτον, ἐν μίξει δέ τινι καὶ τὸ τῆς ὀσμῆς.

[1] ἐν τοῖς U : οὖν Schneider.
[2] [ἐν τοῖς] Schneider.
[3] aP^{c} (Wimmer deletes) : τέ U N P^{ac} (?).
[4] N : εχων U : ἔχοντα u : ἔχει aP. [5] u : τινὲς U.

---

[1] *Cf. HP* 1 12. 4: "For in some plants the flowers are more fragrant than the leaves, whereas in others it is the other way round, and the leaves and twigs are the more fragrant, as in coronary plants; in others the fruit is fragrant; in others neither flower nor leaf is so; and in some it is the roots or some part of them"; *CP* 6 11. 3–4.

[2] Concoction begins with absence of flavour or with the

twigs.[1] In plants where the concoction occurs in the watery (and as it were "flavourless") parts or in the pungent parts,[2] the fragrance is in the flower (as in rose, saffron crocus and the like). On the other hand, when the flower has no agreeable flavour, it conveys a certain oppressive quality in its odour as well (as lily, autumn narcissus and the like); and a few impart no fragrance at all, but something oppressive to the sense of smell. 14.2

And it is evident that the smell is somehow produced from the flavour and is not detached from it, since everything watery and flavourless is also for the most part odourless. Lack of odour in fact depends on these two conditions (one may say): on having flavour of a watery[3] kind, and on having a large watery component. For the watery flavour is as it were unmixed, and the watered flavour is drowned out by the great amount of watery component, and so here too we have something that in a certain sense is "as it were unmixed"[4]; and odour (like flavour) resides in a type of mixture.[5] As for

presence of a "privative" or unconcocted flavour: *CP* 6 6. 6.

[3] That is, feeble, "unconcentrated."

[4] The word *ámiktos* ("unmixed") can mean either without any mixture at all or with a poor one: *cf.* Aristotle's remarks on "inaudible," "invisible," "footless," "stoneless" and "non-gustible" (*On the Soul*, ii. 10 [422 a 25–31]).

[5] *CP* 6 1. 1.

ὅσα δὲ ἐν τροφῇ πλείονι καὶ βλοσυρωτέρᾳ (καθάπερ τὰ λιμναῖα), ταῦτα καὶ τὰς ὀσμὰς ἔχει παραπλησίας, βαρείας τινὰς καὶ θολεράς.

14.3 ἐπεὶ[1] οὖν τῶν μὲν ἐν ταῖς ῥίζαις ἡ τοιαύτη μῖξις καὶ συμμετρία, τῶν δὲ ἐν τοῖς κλωσὶ καὶ τοῖς φύλλοις, τῶν δ' ἐν τοῖς καρποῖς, πλείστων[2] δ' ἐν τοῖς ἄνθεσιν, εὐλόγως εὔοσμα ταῦθ' ἑκάστων. καὶ διὰ τοῦτ' ἐν τοῖς δένδροις τὰ ἐλάττω τῶν ἀνθῶν εὔοσμα, καὶ ἧττον ἐν τοῖς ἄλλοις, ὅτι καὶ τὸ γεῶδες πλέον καὶ ἡ ὑγρότης· εἰ δὲ μή, ζητητέον αἰτίαν.

ἴσως μὲν οὖν καθόλου καὶ ἁπλῶς εἴρηται.

ὡς[3] δὲ καθ' ἕκαστον δὴ[4] σκοπεῖν τὰς εἰρημένας συμμετρίας, οὐκ ἂν δόξαιεν[5] ἀλόγως ἐν τῷ ἄνθει τοῖς πλείστοις· ἐνταῦθα γὰρ οἷον πρώτη καὶ

[1] u : ἐπι U.
[2] Moreliana : πλείστον U (πλεῖστον u).
[3] U : εἰ Schneider.
[4] ego : δεῖ U.
[5] a (δόξειεν Schneider) : δόξαι (no accent U) N P[c] : δόξ P[ac].

---

[1] If the matter is coarse (that is, too earthy) or in too great quantity there is little or no concoction.

[2] *Cf.* the remarks about the heavy-scented and unfra-

plants with food that is too abundant and coarse, like pond plants, these have odours that match their food, heavy and dank.[1]

Why Certain Parts are Fragrant: A Somewhat General Explanation

14.3 Consequently, since this mixture and nice adjustment of quantity is in some plants in the roots, in some in the twigs and leaves, in some in the fruit, in most in the flowers, it is reasonable that in each group these parts are fragrant. This is why in trees the smaller flowers are fragrant, the larger less so, because they have both too much earthy matter[2] and too much fluid. When this is not so, we must look for a special reason.

Perhaps, then, this explanation has been put too generally and with too little qualification.

The Parts Dealt with Severally: (1) The Flower

We take the parts one by one and examine the aforementioned[3] niceties of adjustment.[4] It would not appear unreasonable to find these niceties in the flower in most plants, since it is here that the first

grant flowers in plants with coarse food (*CP* 6 14. 2–3).

[3] *CP* 6 14. 2.

[4] Of wateriness (either as a quality of the flavour or as an added component) and of the amount of earthy matter.

ἐλαφροτάτη πέψις, ἀποξηραινομένου[1] θ' ἅμα καὶ μεταβάλλοντος.

14.4 ὅταν δὲ εἰς τόνδε[2] [τὸν καρπὸν][3] ἔλθῃ, πλείονος τῆς ὑγρότητος οὔσης, οὐκέτι διαμένει τὸ τῆς ὀσμῆς ἐφ' ὧν μὴ φύσει τοιοῦτος ὁ χυλός. ὡς δ' ἐπὶ τὸ πᾶν ἔν τε τῶν οἰνωδῶν τισι καρπῶν ἡ εὐοσμία (τοιοῦτον γὰρ καὶ τὸ μῆλον καὶ ὁ ἄπιος[4] καὶ τὸ μέσπιλον) καὶ τῶν εὐστόμων διὰ δριμύτητός τινος[5] (οἷον κεδρίδος τε καὶ τερμίνθου[6] καὶ τῶν τοιούτων) καὶ λιπαρῶν (ὥσπερ ὅ τε τῆς ἐλάτης καὶ τῆς πεύκης καὶ τῆς πίτυος καὶ τῆς δάφνης).

14.5 τῶν δὲ γλυκέων οὐδεὶς (ὡς εἰπεῖν) ἢ ἐπὶ μικρόν· ἐπεὶ[7] καὶ τὰ μῆλα τὰ γλυκέα [οὐδεῖς ὡς εἶπειν][8] πάντων ἥκιστα εὔοσμα, καὶ ὅσῳ ἂν γλυκύτερα, ἧττον· ἅμα δὲ καὶ ταῦτ' ἔχει πως οἰνώδη τὸν χυλόν.

αἴτιον δ', ὅτι παχύτερος καὶ γεωδέστερος ὁ γλυκὺς χυλός, ἄλλως τε καὶ ἐν τῷ περικαρπίῳ

[1] aP : -ομένη U N.
[2] [τόνδε] Gaza, Scaliger.
[3] ego.
[4] U (*cf. CP* 6 16. 4) : ἡ ἄπιος N : τὸ ἄπιον aP.
[5] U : δριμύτητα τινά aP.
[6] U : τερεβίνθου Schneider.
[7] u : επι U. [8] aP.

and lightest (as it were) of the concoctions takes place, the fluid being reduced by drying and at the same time changing.[1]

(2) *The Fruit*

14.4 But when the tree passes from the flower to this, the fluid is now greater in amount, and the odour does not remain in trees where the flavour-juice of the fruit is not naturally of the character described.[2]

Fragrance is found on the whole:

(1) in certain fruits with a vinous flavour (for apples, pears and medlars are of this sort),

(2) in some with an agreeable taste appearing through a certain pungency (as in juniper, terebinth and the like)

(3) and in some oily fruit (as that of silver fir, Aleppo pine and bay).

14.5 But no sweet fruit (one may say) is fragrant, or it is so only slightly. Thus the sweet apple is of all apples the least fragrant, and the sweeter it is, the less its fragrance. (Yet at the same time the sweet apple too has a certain vinous character in its taste.)

The reason for the absence of odour is that the sweet flavour is too thick and too earthy (especially

[1] The earthy part is changed and the watery part diminished.

[2] *CP* 6 14. 2: not too watery and not too earthy.

μεμιγμένος· ἡ δὲ ὀσμὴ λεπτοτέρου καὶ ξηροτέρου καὶ διαπνεομένων[1] *μᾶλλόν ἐστιν (ὑπὲρ ὧν καὶ ὕστερον διασαφητέον).*

14.6 *ἀλλ' ὡς ἐπὶ κεφαλαίων ἐν τούτοις τοῖς γένεσιν ἡ εὐοσμία.*

τῶν δὲ φύλλων[2] *καὶ κλωνῶν,*[3] *καὶ ὅλως τῶν δένδρων καὶ ὑλημάτων ἐν οἷς καὶ δριμύτης τις ὑπάρχει καὶ λιπαρότης, ὧνπερ ἀμέλει καὶ οἱ προειρημένοι καρποί, καὶ ἐπὶ τῶν οἰνωδῶν ἐνίων· ὁ γὰρ τῆς μυρρίνου*[4] *καρπὸς εὐώδης, εὔοσμος δὲ καὶ αὐτὴ καὶ εἴ*[5] *τις ἄλλη τοιαύτην ἔχει δριμύτητα, εἴτε λιπαροῦ τινος εἴτ' οἰνώδους εἴτε καὶ ἄλλου χυλοῦ κατὰ τοὺς καρπούς.*

14.7 *ὁμοίως δὲ καὶ ἐπὶ τῶν στεφανωμάτων (οἷον ἑρπύλλου σισυμβρίου καὶ ἐπὶ τῶν ἄλλων) καὶ ἐπὶ λαχάνων (οἷον πηγάνου σελίνου μίνθης) τό τε*

§14.7–8: *Cf.* Pliny, *N. H.* 21. 38.

[1] U N : *-μένου* aP.
[2] U : *φύλλα* G. R. Thompson.
[3] Gaza (*ramorum*), Itali : *κλῶνες* U.
[4] ego (*cf. CP* 1 13. 10; *τοῦ μυῤῥίνου* Scaliger : *τῆς μυῤῥίνης* Moreliana) : *τουμυρινου* U.
[5] u : *ἤ* U.

when mingled with the pericarpion); whereas odour belongs to a thinner and drier flavour and to bodies that evaporate more readily. These matters will be treated in greater detail later.[1]

14.6 Fragrance on the other hand is found in the classes mentioned[2] (to give the main headings).

(3) *Leaves and Twigs*

Fragrance is found in the leaves and twigs and in the whole plant

(1) in those plants in which a certain pungency and oiliness are present—the very ones in fact with the fragrant fruit just mentioned,[3]

(2) and in some trees with vinous-flavoured fruit. So the fruit of the myrtle is fragrant,[4] so too is the tree itself and any other tree with pungency of the sort (whatever the flavour in the fruit: vinous, oily or of another kind).[5]

14.7 Similarly in coronary plants too (as tufted thyme, bergamot mint and the rest) and in vegetables (as rue, celery and spearmint) a dry, pungent and

[1] *CP* 6 16. 5–8.

[2] *CP* 6 14. 4.

[3] *CP* 6 14. 4 (2).

[4] *Cf. HP* 1 12. 1 (of flavours): "... some are vinous, as those of the vine, the mulberry and the myrtle-berry ..."; *CP* 6 7. 4.

[5] The "agreeable taste appearing through a certain pungency" of *CP* 6 14. 4.

ξηρὸν καὶ δριμὺ καὶ στυπτικόν πως ἔνδηλον (οὐκ ἐπὶ[1] τοῖς καρποῖς, ἐν δὲ[2] τοῖς ἄλλοις ἐχόντων[3] τὴν ὑγρότητα τοιαύτην).

ἐν ἅπασιν δ' ἐνυπάρχει καὶ τούτοις καὶ τοῖς ἄλλοις τὸ κατὰ τὸν διορισμὸν τοῖς εὐώδεσιν, ὥστε τὸ ἔγχυλον ξηρὸν[4] ἀπομιγνύμενον ἐμφαίνειν τινὰ δύναμιν.

14.8 εὐοσμότερα δὲ καὶ ἡδίω, καὶ ὅλα <καὶ>[5] κατὰ μέρος, ἐν τοῖς εὔπνοις καὶ ξηροῖς τόποις, ἀφῃρημένου τοῦ ὑδατώδους, καὶ τοῦ καταλοίπου πεπεμμένου μᾶλλον.

ὡς γὰρ ἁπλῶς εἰπεῖν ἡ ξηρότης οἰκειοτέρα ταῖς ὀσμαῖς, καὶ μᾶλλον ἐνταῦθ' ἀποκλίνει πάντα. σημεῖον δέ, καὶ τὸ ἐν ταῖς θερμοτέραις χώραις πλείω γίνεσθαι, καὶ μᾶλλον, τὰ εὔοσμα (πεπεμμένα γὰρ δῆλον ὅτι μᾶλλον)· καὶ ἔνια ξηραινόμενα <μὲν>[6] ὄζει, χλωρὰ δ' οὐκ ὄζει (καθάπερ ὁ κάλαμος καὶ ὁ

[1] επι U : ἐν Schneider.
[2] ἐν δὲ ego (μόνον, ἀλλὰ καὶ ἐν Schneider) : οὐδ' ἐν U.
[3] μέρει τῶν ἐχόντων Schneider (V, p. 137).
[4] ego (*cf.* Aristotle, *On Sense*, v [443 a 7] ἡ ἐν ὑγρῷ τοῦ ἐγχύμου ξηροῦ φύσις ὀσμή) : ξηρῶ U.
[5] Gaza, Scaliger.
[6] ego.

astringent quality is somehow noticeable (in these plants however this quality of fluid occurs not in the fruit,[1] but in the other parts).

What conforms to the rule[2] for fragrant objects is present in all these[3] and in the rest[4]: that the intermixture of the flavoured dry[5] produces a certain sense impression.

Dryness Favourable to Fragrance

14.8 Plants, both in their entirety and in this or that part, have a stronger and more agreeable fragrance in places well-ventilated and dry, since the watery part has then been removed and the remainder better concocted.

Dryness in fact is (broadly speaking) better suited to odours, and all fragrant plants and parts tend to be drier. Proof is this: (1) a greater number of them is produced in the hotter countries and their fragrance there is greater, evidently because they are better concocted there; and (2) some have odour when dried, but none when fresh (as the reed and

[1] Literally "not at the fruits."

[2] *Cf.* ἀφορισμόν ("distinction") *CP* 6 1. 1 *ad fin.*

[3] The ones mentioned under heating (3) in *CP* 6 14. 6–7.

[4] The ones mentioned in *CP* 6 14. 3–5.

[5] *CP* 6 1. 1: "odour (*sc.* is) the intermixture in the transparent of the flavoured dry . . ."

σχοῖνος), τὰ δὲ καὶ μᾶλλον ὄζει ξηρανθέντα (καθάπερ ἡ ἶρις[1] καὶ ὁ μελίλωτος· οὗτος δὲ καὶ οἴνῳ ῥανθεὶς[2] εὐοσμότερος).

14.9 οὐ μὴν πάνθ᾽ οὕτως, ἀλλ᾽ ἔνια καὶ ἀνάπαλιν, ἃ δεῖ διαιρεῖν. ἁπλῆ δέ τις ἡ διαίρεσις· ὧν[3] μὲν γὰρ ἀσθενεῖς αἱ ὀσμαί (τοιαῦτα[4] δὲ ὡς ἐπίπαν τὰ ἄνθη μάλιστα), ταῦτα[5] μὲν εὐοσμότερα χλωρὰ καὶ πρόσφατα, χρονιζόμενα δὲ ἀμβλύνεται διὰ τὴν ἀποπνοήν· ὧν δ᾽ ἰσχυρότεραι (τοιαῦτα[4] δ᾽ ὅσα γεωδέστερα καὶ οὗ[6] γεωδεστέροις [καὶ οὐ][7] μέμικταί τις ὑδατώδης δύναμις), ταῦτ᾽, ἀποξηραινόμενα καὶ παλαιούμενα μέχρι τινός (ὧν καὶ τὰ

[1] ἡ ἶρις ego : ἥρις U : ἴρις u.
[2] Schneider : οἰνωρανθης U.
[3] Itali : τῶν U. [4] Moreliana : τοιαῦται U.
[5] Wimmer : τὰ U.
[6] u (οὐ U) : ἐν Schneider after Gaza.
[7] ego : οὖσι Palmerius : καὶ οἷς Schneider after Gaza.

---

[1] *Cf.* *CP* 6 11. 13 and *HP* 9 7. 1: "The reed and rush occur when you cross the Libanus . . . They have no odour when fresh but only after drying, and in appearance do not differ from other reeds and rushes . . ."

[2] *Cf.* Theophrastus, *On Odours*, vii. 34 (of dried aromatic herbs): "Of the flowers some, like rose, have their potencies from the time when they are fresh, but others have their potencies after drying, as saffron crocus

the rush[1]), others have increased odour when dried (as iris and melilot[2]; dried melilot when sprinkled with wine even increases in fragrance).

Dryness also Unfavourable to Fragrance: a Distinction

14.9 Nevertheless not all plants and parts gain in odour when dry, but of some the opposite is true. We must therefore distinguish the two groups. The distinction is simple: (1) the plants and parts with weak odour (and this on the whole is especially the case with flowers) are more fragrant when fresh and recent, but when kept longer become fainter in odour because of evaporation; (2) whereas the plants and parts with strong odour (and this is the case with the more earthy plants and parts, and those where a certain watery power is mixed with the greater earthiness[3]), have stronger odour when dried and kept for a certain time (and to this group

and melilot, since when fresh they are too watery."

[3] Compare the conditions for lack of fragrance (*CP* 6 14. 2): (1) the flavour is intrinsically watery; (2) watery fluid has been added to it; (3) it has an abundance of earthiness. In the case of the stronger odours that get stronger with time the two conditions are (1) the odour is intrinsically earthy and (2) watery fluid has been added to the intrinsically earthy odour. In both cases the decrease in the proportion of the watery quality or element leads to greater strength.

μῆλα τὰ Κυδώνια καὶ ὅσα τῶν στεφανωματικῶν δριμείας ἔχει τὰς ὀσμάς, οἷον τὸ ἀβρότονον μάλι-

14.10 στα, καὶ τὸ ἀμάρακον καὶ ὁ κρόκος)· ἅμα γὰρ ἀποπνεῖταί τε τὸ[1] ὑδατῶδες καὶ ἡ τῆς τροφῆς ἐπιρροὴ παύεται, καὶ ὥσπερ πέψιν ἐν ἑαυτοῖς ἔνιά γε λαμβάνει (διὸ καὶ ἀφαιρεθέντα ἀπὸ τῶν φυτῶν εὐωδέστερα, καθάπερ τὰ μῆλα καὶ ἄλλ᾽ ἄττα). συμβαίνει δὲ καὶ τῶν ποιωδῶν[2] ἔνια χλωρὰ μὲν ὄντα μὴ ὄζειν διὰ τὴν ὑγρότητα, ξηρανθέντα δέ (καθάπερ ἄλλα τε καὶ τὸ βούκερας). ἐπεὶ καὶ ὁ οἶνος τότε μάλιστα παρίσταται καὶ ὀσμὴν λαμβάνει, ὅταν ἀποκριθῇ τὸ ὑδατῶδες αὐτοῦ. τὰ μὲν οὖν τοιαῦτα πάντα παλαιούμενα εὐοσμότερα.

14.11 τὰ δ᾽ ἀσθενῆ ταχὺ διαπνεῖται, καθάπερ τὰ ἴα· καὶ τά γε λευκὰ πικρὰ καὶ κακώδη παλαιούμενα, καὶ οὐχ ὥσπερ τὰ ῥόδα διατηρεῖ τὴν εὐοσμίαν ἀποξηραινόμενα μέχρι οὗ ἂν ἐκλίπῃ (πλὴν τὰ μὲν χλωρὰ καὶ πόρρωθεν ὄζει, ταῦτα δ᾽ οὔ). τὸ δ᾽ αἴ-

§14.9–10 : Pliny, *N.H.* 21. 37–8.
§14.11: Pliny, *N. H.* 21. 36, 39.

[1] u : τῷ U.
[2] Gaza, Scaliger : ὁποδῶν U : ὁπωδῶν u : ὀπωρῶν N aP.

belong quinces and the coronary plants with pungent odours, such as southernwood especially, sweet marjoram and saffron crocus[1]); for at the same time (1) the wateriness evaporates and the influx of food stops, (2) and some parts at least acquire concoction of a sort within themselves[2] (which is why the part is more fragrant after removal from the plant,[3] as with the quinces[4] and certain others). Among herbaceous plants too it happens that some have no odour when fresh because of the fluid, but have odour when dried (as with fenugreek and others). Indeed wine too becomes fit to drink and acquires odour when its watery part has been separated off. All plants of this type, then, become more fragrant when kept. 14.10

14.11 On the other hand where the odour is weak it is soon lost by evaporation, as in stock and violet (stock indeed when kept too long turns bitter and rank in smell), not as in the rose, which as it dries retains its fragrance to the end (except the smell of the fresh flowers carries far,[5] whereas that of the

[1] *Cf.* Theophrastus, *On Odours*, vii. 34 (cited in note 2 on *CP* 6 14. 8).

[2] The evaporation and cutting off of the influx reduce the alien wateriness; the concoction after removal from the plant reduces the wateriness within the flavour itself.

[3] For the effect of such removal on flavour *cf. CP* 2 8. 2–3; 6 7. 4. [4] *Cf. CP* 6 14. 9.

[5] *Cf. CP* 6 17. 1 (first sentence).

τιον, ὅτι κατέσκληκέ πως, ἐλλελοιπότος τοῦ οἰκείου θερμοῦ, καὶ οὐ δίδωσιν ἀποπνοήν· ἐπεὶ πρός γε[1] *τὰς χρείας οἱ μυρεψοὶ καὶ ταῦτα ἀποξηραίνουσιν μέχρι τινός, ὅπως ἄκρατον καταλίπωσιν τὴν ὀσμήν. ὁ δὲ μελίλωτος καὶ εἰς πλείω χρόνον εὔοσμος διαμένει.*

14.12 *οὐ μὴν ἀλλ' ἔνιά γε τῶν εὐωδῶν, ἄγαν καταξηραινόμενα, χείρους ἴσχει τὰς ὀσμάς, τῷ δριμυτέρας εἶναι καὶ σκληροτέρας. ἔστι γάρ τις, ὥσπερ καὶ οἴνου καὶ ἁπλῶς χυλοῦ σκληρότης καὶ ἰσχύς, οὕτω καὶ ὀσμῆς, ὃ καὶ τῶν ἀγρίων ἔνια ὅμοια*[2] *δοκεῖ πρὸς τὰ ἥμερα πεπονθέναι, καθάπερ ἕρπυλλος*

[1] Moreliana : *τε* U.
[2] U : Gaza omits : *ὁμοίως* Heinsius : *ὁμογενῆ* Wimmer.

---

[1] *Cf.* the explanation of another difference in Theophrastus, *On Odours*, iii. 12–13: "The following point also presents a problem: why in flowers and coronary plants does the odour, though weaker, carry farther, whereas iris-root, nard and the other fragrant dry things are stronger close at hand? . . . The cause is that in the flowers the odour-producing part is at the surface (the flowers being open in texture and having no dimension of depth), whereas in roots and all three-dimensional things this

dried ones does not). The reason is that the dried flowers have somehow become hard with the failure of their native heat and allow no exhalation[1]; as a matter of fact perfumers also dry out stock[2] to a certain extent for their purposes, to get the odour in concentrated form. Melilot continues fragrant for a considerable time.[3]

Drying Worsens the Odour

14.12 Nevertheless when the dryness goes too far the odours of some fragrant plants become worse, getting too pungent and harsh (there being such a thing as "harshness" and strength in odour, just as there is in wine and flavour-juice in general). This is considered to apply to certain wild plants when compared to the cultivated types that they resemble (as tufted thyme, bergamot mint,[4] and among

part is deep inside, and the surface is dried and close in texture. Hence the flowers send their exhalations to a distance, whereas the rest require an opening as it were of the passages . . ."

[2] *Cf. ibid.* vi. 27: "From flowers are made for instance perfume of rose and perfume of stock . . ."

[3] *Cf. ibid.* vii. 34 (translated in note 2 on *CP* 6 14. 8).

[4] *Cf. CP* 6 16. 7; 6 20. 1 and *HP* 6 6. 3 (of coronary plants): "The woody ones are each single in their kind, as tufted thyme, bergamot mint and calamint, except for being divided into wild and cultivated and fragrant and less scented."

καὶ τὸ σισύμβριον,[1] *καὶ τῶν λαχανωδῶν μάλιστα τὸ πήγανον· σκληραὶ γὰρ αἱ ὀσμαὶ καὶ ἀγλυκεῖς, αἱ δὲ τῶν ἡμέρων ἔχουσί τινα*[2] *ἅμα τῷ ὑφειμένῳ γλυκύτητα [τινα]*[3] *καὶ εὐμένειαν. ἔστιν γάρ, ὥσπερ ἐν χυλοῖς, καὶ ἐν ὀσμαῖς γλυκύτης, σχεδὸν δὲ καὶ αἱ ἄλλαι προσηγορίαι τῶν εἰδῶν, ὡς οὐ πόρρω τῆς φύσεως ἑκατέρας οὔσης.*

15.1 *ἀλλὰ γὰρ ποῖαι μέν τινες ὀσμαὶ χρόνιαι, καὶ ποῖαι ξηραινομένων εὐοσμότεραι, καὶ τἆλλα τὰ τούτοις ὅμοια, διὰ τῶν εἰρημένων θεωρείσθω·*[4] *καὶ γὰρ ὅσα μὴ εἴρηται ῥᾴδιον ἐκ τούτων συνιδεῖν.*

ὅσα δὲ μὴ κατὰ μέρος εὔοσμα, ἀλλ' ὅλα τυγχάνει, περὶ τούτων ἀπορήσειεν ἄν τις ὃ καὶ πρότερον ἐλέχθη· διὰ τί τὸ ἄνθος οὐκ εὔοσμον αὐτῶν (ἢ οὐ κατὰ λόγον). ἔδει[5] *γὰρ* <ὃ>[6] *καὶ τοῖς ἄλλοις μὴ οὖσιν εὐόσμοις, τούτοις μάλιστα, διὰ τὸ προϋπάρχειν τὴν φύσιν.*

[1] U : *cf.* ἀβροτόνου *CP* 6 16. 7 (a passage that refers to this).
[2] ἔχουσι τινὰ U : ἔχουσιν Schneider.
[3] aP.
[4] aP (*cf. CP* 1 4. 6) : θεωρεῖσθαι U N.
[5] Schneider : ἐνδεῖ U.
[6] ego.

vegetables most notably rue): their odours are harsh and lack sweetness, whereas those of the cultivated types have with their lessened intensity a certain sweet and gentle character (there being in odours too, as in flavours, such a thing as sweetness, and the rest, one may say, of the flavour-names are also applied to odours,[1] which suggests that the two natures coupled under each name are not widely separated[2]).

15.1 But enough. Let the questions about what types of odour are enduring, what types become more fragrant when the plant or part is dried, and the like be studied in the light of the foregoing discussion,[3] which renders easily comprehensible such points as have not been mentioned here.

A Problem: The Rest of the Plant is Fragrant, the Flower Not

Where the whole plant is fragrant, and not just a part, one might raise the question that was mentioned earlier[4]: why is the flower not fragrant, or not so fragrant as one would expect? One would expect the part that is fragrant in plants which are not fragrant to be most fragrant here, because the fragrant nature is already present.

[1] *Cf. CP* 6 9. 2.
[2] *Cf. CP* 6 9. 1.
[3] *CP* 6 14. 9–12. [4] *CP* 6 11. 3–4.

15.2 αἴτιον δὲ φαίνεται διότι πέψις τις ἡ ἄνθησις, τὸ δὲ πεττόμενον ἐν μεταβολῇ τοῦ ὑπάρχοντος. ὅσα μὲν οὖν αὐτὰ μὴ ὀσμώδη, τούτων, πεττομένων τῶν χυλῶν, λαμβάνει τινὰ τἄνθος[1] εὐοσμίαν, ἐπείπερ ἡ πέψις ἐν μεταβολῇ· ὅσα δ' εὔοσμα, τούτων διὰ τὴν μεταβολὴν ἀναγκαῖον ἐξαλλάττειν τὴν ὀσμήν, ὀσμωδῶν δὲ ὄντων, ἧττον εὔοσμον ἔσται τὸ ἄνθος· αὕτη[2] γὰρ ἡ ἐξαλλαγὴ ἔοικεν ὥσπερ ἐξάνθησίς τις εἶναι τῆς προϋπαρχούσης ὀσμῆς καὶ δυνάμεως[3] (διὸ καὶ αὐτά[4] φασιν ὄζειν τότε· τοῦτο μὲν οὖν εἰ γίνεται σκεπτέον, φέρει γάρ τινα πίστιν).

15.3 φαίνεται δὲ παρόμοιον συμβαίνειν, ὥσπερ ἐπὶ τῶν χυλῶν τῶν τοιούτων, καὶ[5] τῶν ἐν τέλει καὶ πεπεμμένων πρὸς τοὺς ἀπέπτους ὅταν πυρῶνται καὶ ἡλιῶνται· μεταβολῆς γὰρ γινομένης, οἱ μὲν εἰς τὸ βέλτιον, οἱ δὲ εἰς τὸ χεῖρον μεταβάλλουσιν.

[1] U^ac : το ἄνθος U^c.
[2] u N : αὐτη U : αὐτὴ aP.
[3] U^cc : δύναμις U^ac.
[4] U : ἧττόν Wimmer.
[5] [καὶ] Wimmer.

---

[1] The expression (*exánthēsis*) is carefully chosen: it can mean "breaking out into flower," "passing away from the

The Solution

The reason, it appears, is this: flowering is a kind of concoction, and what is undergoing concoction is undergoing change from what it was originally. Now where the plant itself is not fragrant, the flower, as the flavour undergoes concoction, acquires a certain fragrance, since concoction involves change. Whereas where the plant *is* fragrant the odour necessarily loses its character because of the change, and since the plant itself is fragrant, the resulting flower will be less fragrant. For this loss appears to be a "flowering away"[1] (as it were) of the odour and power that were there before (which is why at that moment the flowers themselves, it is said, give out an odour too; whether this in fact occurs must be investigated, since it has a certain plausibility). 15.2

Something similar to what happens with these flavours[2] happens also (it appears) with flavours fully developed and concocted as compared with unconcocted flavours when both are cooked or set out in the sun: in the resulting change the unconcocted flavours change for the better, the others for the worse: the heat concocts the unconcocted flowers 15.3

flower" or "breaking out as an excrescence."

[2] *CP* 6 15. 2 (non-fragrant flavours that become fragrant in the flower on concoction, and fragrant flavours that then lose their fragrance).

τοὺς μὲν γὰρ πέττει, τοὺς δ' οἷον ἐξίστησιν, τὸ θερμόν· ἀνάγκη γάρ, ὅταν ἐν τῷ τέλει γίνηται,[1] τὴν μεταβολὴν εἶναι πρὸς τὸ χεῖρον, ὃ δὴ καὶ ἐνταῦθα φαίνεται συμβαῖνον, μαλακωτέρας τῆς ὀσμῆς γινομένης, οἷον γὰρ ἄνεσίς τις γίνεται τῆς ἀκράτου.

15.4 ἅμα δὲ καὶ διυγραίνεσθαι συμβαίνει, ξηρὸν ὂν φύσει, κατὰ τὴν ἄνθησιν, ὧν[2] γένεσις οὐκ ἄνευ τούτων, κἂν ταύτῃ γίνοιτό τις ἄνεσις.

εἰ δὲ μὴ ὑγρότης, ἀλλ' οἷον γλυκύτης ἐγγίνεται πρὸς τὴν τοῦ καρποῦ γένεσιν (ἅπαντα γὰρ ἐκ γλυκέος γεννᾶται· διὸ καὶ τὰ πολλὰ τῶν ἀνθῶν ἐστιν γλυκέα), καὶ τοῦτο κατὰ λόγον, ἐκλελυμένην τε καὶ θηλυτέραν εἶναι τὴν ὀσμήν, οἷον ἐπιγλυκαίνουσαν, ἀφῃρημένης τῆς ἀκράτου δριμύτητος.

τούτων μὲν οὖν τοιαύτην τινὰ αἰτίαν ὑποληπτέον.

[1] u aP : γίνεται U N.
[2] u (ὧν U) : εἴπερ οὖν Schneider.

---

[1] When a fragrant plant puts forth a scentless or less fragrant flower.
[2] The whole of the plant is fragrant, and such plants are naturally dry: *cf. CP* 6 14. 7–8.

but makes the concocted ones depart from their nature (as it were), since the change occurring at the stage of full development is necessarily for the worse. And this we see happens here[1] too: the fragrance becomes feebler, since there is a kind of slackening of the full intensity of the odour.

15.4 Moreover the plant, which is naturally dry,[2] becomes fluid at flowering time, when it is one that does not reproduce[3] without a flower,[4] and this too would lead to weakening of the odour.

(If it is not fluidity, but a sweetness[5] as it were of flavour that arises in the flower in order to produce the fruit—since everything is generated from the sweet,[6] which is why most flowers are sweet—here too the consequence is reasonable: the odour loses its vigour and lapses into a female character,[7] getting sweetish, so to speak, with the removal of the full pungency.)

We must suppose, then, that the causation of these matters is much as described.

[3] For fluidity as prerequisite to generation *cf.* *CP* 1 4. 6.

[4] Some fruit was held not to be preceded by a flower: *cf.* *HP* 3 3. 8 (of the fig), *HP* 1 13. 5 (of the female date palm); again, some flowers were held to produce no fruit (*HP* 1 3. 4).

[5] Sweetness involves thickening: *CP* 6 11. 1; 6 16. 2.

[6] *Cf.* *CP* 6 5. 6 and Aristotle, *On Sense*, iv (442 a 8): plants and animals are fed by the sweet.

[7] *Cf.* *CP* 1 16. 6; 3 1. 3; 4 5. 3; 4 5. 6.

16.1 *διὰ τί δ' οἱ ἄγριοι καρποὶ τῶν ἡμέρων εὐοσμότεροι τῶν ὁμογενῶν, οἷον μῆλά*[1] *τε καὶ ἀχράδες καὶ οὖα καὶ μέσπιλα καὶ τἆλλα, καὶ αὐτῶν δὲ τῶν μήλων ἔν τε τοῖς ἀγρίοις καὶ τοῖς ἡμέροις τὰ στρυφνότατα εὐοσμότατα*[2]*; καίτοι, διὰ πέψιν γινομένης τῆς ὀσμῆς, προσῆκεν εὐοσμεῖν τὰ μάλιστα πεπεμμένα.*

περὶ δὴ τούτων, ἔστι μὲν ἁπλῶς εἰπεῖν ὅτι τὰ ἄγρια ὡς σκληρότερα εὐοσμότερα, καθάπερ ἐν τοῖς ἄλλοις.

ἔστι[3] *δὲ καὶ οἰκειοτέρως εἴ τις δύναιτο διελεῖν τὰς πέψεις, τήν τε*[4] *τοῦ χυλοῦ καὶ τὴν τῆς ὀσμῆς, ἐν τίνι ἑκατέρα*[5] *γίνεται· φαίνεται γάρ τινα ἔχειν διαφοράν, εἴθ' ἑτέρων ὄντων, εἴτε τοῦ αὐτοῦ μετα-*
16.2 *βάλλοντος.*[6] *τοῦτο μὲν οὖν καθόλου διαίρεσιν ἔχει.*

§16.1: *Cf.* Pliny, *N. H.* 21. 35.

[1] Gaza (*mala*), Heinsius : *μηλέα* U.
[2] U[cc] : *εὔοσμα* U[ac]. [3] Schneider : *ἔτι* U.
[4] u : *δε* U. [5] u aP : *ἐκάτερα* (*ἑ-* N) U.
[6] N aP : *μεταβάλοντος* U.

---

[1] Wines, incense, perfumes and coronary plants (*cf. CP* 6 16. 6–7). [2] The answer is that fragrance is found rather in the flavours that have not yet been completely concocted (*CP* 6 16. 8).

Why are Wild Tree-Fruits More Fragrant?

16.1 Why are wild fruits more fragrant than the cultivated fruits of the same kind (as wild apples, pears, sorbs, medlars and the rest)? And why, to take the apples, are the most astringent, both among the wild and among the cultivated, the most fragrant? And yet, since odour is produced by concoction, the best concocted should be the fragrant ones.

Answers

About these matters one can answer in simple fashion that the wild ones are more fragrant as being harsher in taste (as in the other things[1]).

But one can also give the answer in a more specific fashion if one can draw a distinction in that in which each of the two concoction occurs,[2] that of the flavour and that of the odour, since there appears to be some difference in it (whether the two concoctions occur in two distinct things or in the same thing in the course of change[3]). 16.2 Now this more specific way of answering involves the making of a general distinction.[4]

[3] The two are later spoken of as distinct: *CP* 6 16. 7 (second paragraph); 6 16. 8 (first sentence).

[4] Between all good flavour and all good odour, and not merely between the comparative excellence of odour and flavour in wild and cultivated tree-fruit.

πρὸς δὲ τὸ νῦν, ἱκανὸν τοσοῦτον (ὅπερ καὶ φανερόν)· ὅτι ἡ μέν γλυκύτης, καὶ ὅλως ἡ εὐστομία,[1] *παχύνσει γίνεται τῶν χυλῶν (διὸ καὶ γηράσκοντες οἱ καρποί, καὶ ἡλιούμενοι, πάντες γλυκύτεροι· παχύνονται γὰρ ἀφαιρουμένου τοῦ ὑδατώδους)· τὸ*[2] *δὲ τῶν ὀσμῶν ἐν προτέρᾳ τινὶ καὶ ἀτελεστέρᾳ πέψει. σημεῖον δέ, καὶ ὅτι αὐτὰ τὰ εὐώδη (καθάπερ ἄπιοι καὶ μῆλα*[3] *καὶ τὰ ἄλλα) μᾶλλον εὔοσμα γίνεται μὴ τελέως ἐκπεπανθέντα,*

16.3 *καίτοι τότε τοῦ χυλοῦ μάλιστα πέψις. καὶ ἔοικεν ἐν ταῖς προτέραις μεταβολαῖς τῶν χυλῶν (πλείους γάρ) εὐοσμία*[4] *γίνεσθαι*[5] *καθάπερ ἅμα πνευματική τις οὖσα καὶ οὔπω τοῦ χυλοῦ τὴν οἰκείαν ἔχοντος φύσιν· ὅταν* <δ’>[6] *εἰς ταύτην ἐπέλθῃ παχυνθεὶς καὶ πεφθείς, τὰ μὲν τῆς τροφῆς*[7] *ἐλάττω, τὴν δὲ γλυκύτητα τὴν οἰκείαν (καὶ ὅλως τὴν τῇ αἰσθήσει πρόσφορον) λαμβάνει. τὰ μὲν οὖν*

[1] Wimmer (*saporis suavitudinem* Gaza: *εὐχυλία* Scaliger): *εὐοσμία* U.
[2] ego (*ratio* Gaza : “deest nomen *γένεσις* vel *σύστασις*” Schneider): *ἡ* U. [3] Gaza (*mala*), Heinsius : *μηλεαι* U.
[4] U : *-αι* u. [5] Schneider : *γίνεται* U : *γίνονται* u.
[6] Schneider. [7] U : *ὀσμῆς* Gaza, Itali.

---

[1] “Agreeable taste” renders *eustomía*, which can be sweet or not. It takes the place of *euchylía* (“excellence of

For the question of the moment, however, it is enough to say (what is evident) that sweetness (and in general all agreeable taste[1]) arises from thickening of the flavour-juices (which is why all fruit gets sweeter both with age and with exposure to the sun, as it gets thicker as wateriness is removed); whereas odours depend on an earlier and less complete concoction. Proof of this is that the fragrant fruits themselves (as pears, apples and the rest) get more fragrant before they are fully ripe, although it is when they are ripe that the concoction of their flavour is most complete. And it would seem that fragrance arises in the earlier changes (there being more than one)[2] of the flavours since it has *pneuma*[3] and at the same time the flavour is not yet in possession of its own nature; but when the flavour has reached this nature through being thickened and concocted the portion of food in it is smaller[4] and on the other hand it acquires the sweetness that is proper to itself (and in fact suited to the sense of taste). Now the wild fruits get as far as this

16.3

flavour"), which suggests juiciness, when lusciousness is out of the question.

[2] *Cf. CP* 6 6. 7 (of tree-fruit, of which Theophrastus is here speaking).

[3] The *pneuma* accounts for expansion and volatility. At *CP* 2 9. 6 it accounts for the distension (and dropping) of the unripe fig.

[4] The food that has not yet been fully concocted (and so contains expansive and volatile *pneuma*).

ἄγρια, μέχρι τῆς προτέρας προϊόντα, καὶ τὴν εὐωδίαν ἔχει κατὰ λόγον· τὰ δὲ ἥμερα, τελεούμενα τῇ πέψει, καὶ εὐτροφοῦντα διὰ τὴν κατεργασίαν, μεταβάλλει τοὺς χυλοὺς εἰς γλυκύτητα καὶ ἀοσμίαν.[1]

16.4 ἅμα δὲ καὶ τὴν ὑγρότητα καὶ τὴν ξηρότητα, καὶ τὴν πολυτροφίαν καὶ[2] ὀλιγοτροφίαν, καὶ τὴν εὔπνοιαν εἰκός τι συμβάλλεσθαι πρὸς τὰς ὀσμάς, εἴπερ ὅλως καὶ τὰ ξηρά, καὶ ἐν ξηροῖς καὶ ὀλιγοτροφίᾳ,[3] καὶ ἐν εὔπνοις, εὐοσμότερα· μιγνυμένη γὰρ ἣ[4] πλείων ὑγρότης ἀμβλύνει τὴν ὀσμήν (διὸ καὶ ἐν τοῖς παλισκίοις καὶ ἐφύδροις οὐκ εὔοσμα). καὶ τἆλλα δὴ[5] τῇ παχύνσει (καθάπερ εἴρηται)

16.5 λαμβάνει τὴν μεταβολήν· ὡς δ' ἁπλῶς εἰπεῖν καὶ ξηρότερα τὰ ἄγρια, καὶ ὀλιγοτροφώτερα, κἂν εὐπνοίᾳ[6] μᾶλλον καὶ ἐν ἡλιώσει, καὶ ἔτι λεπτοχυλότερα, καὶ οὐχ ὁμοίως τὸν χυλὸν ἀναμεμιγμένον ἔχοντα τῇ σαρκὶ τοῖς ἡμέροις· ἅπαντα δὴ ταῦτα συμβάλλεται πρὸς εὐοσμίαν.

[1] Gaza (*odoris hebetudinem*), Scaliger (εὐστομίαν Wimmer) : εὐοσμιαν U.

[2] καὶ τὴν Schneider.

[3] U : ὀλιγότροφα Moreliana (after Gaza).

[4] ἣ U : ἡ N aP.

[5] τἆλλα δὴ ego : μάλα δὴ U N P : μάλιστα δὲ a.

[6] ego (καὶ ἐν εὐπνοίᾳ u) : καὶ ανεύπνοια U.

prior concoction, and so have the fragrance we expect; but the cultivated fruits, which are completed in concoction and well-fed by cultivation,[1] bring their flavours to the further stage of sweetness and absence of odour.

16.4 It is moreover likely that such matters as fluidity and dryness, much or little food and good ventilation[2] play a part in producing fragrance: fruit after all that is dry, grows in dry places, and gets little food and is exposed to the breeze is more fragrant, since an addition or excessive heightening of fluidity[3] dulls the odour (which is why fruit growing in the shade or on land with surface water has no fragrance). Again other fruit (as we said[4]) develops its flavour by thickening it; 16.5 whereas wild fruit (speaking broadly) is (1) drier, takes less food and is more exposed to breeze and sun; it (2) also has thinner flavour-juice and this does not mix with the flesh of the fruit to the same degree as in the cultivated kinds. All these circumstances contribute to its fragrance.

[1] Presumably the shorter supply of food in wild fruits prevents their full concoction (*cf. CP* 1 16. 3–6).

[2] For the preference of wild trees for mountain country *cf. HP* 3 2. 4.

[3] *Cf.* the two ways of being watery: *CP* 6 14. 2.

[4] *CP* 6 16. 2–3.

ἰδεῖν δὲ τοῦτ᾽ ἔστιν καὶ ἐπὶ τῶν οἴνων· οἱ μὲν γὰρ γλυκεῖς ὅλως ἄοσμοι,[1] καὶ οἱ μαλακοί· τῶν δὲ ἄλλων οἱ λεπτοὶ μᾶλλον εὐώδεις, καὶ θᾶττον

16.6 τῶν πιόνων παριστάμενοι καὶ παχέων. ἔτι δὲ ἅμα τῇ εὐοσμίᾳ συμβαίνει καθάπερ χωρισμόν τινα γίνεσθαι τοῦ ὑγροῦ καὶ τοῦ γεώδους, ὑφισταμένης τῆς τρυγός, καὶ πρὸς τούτοις ἀποπνοὴν τοῦ ὑδατώδους, ὥστε διὰ πάντων εἶναι τὴν εὐωδίαν, λεπτότητός τε, καὶ χωρισμοῦ, καὶ διαπνοῆς (ἐπεὶ καὶ αὐτοῦ τοῦ κεραμίου τὸ πρὸς τῇ τρυγὶ ἧττον εὔοσμον).

ἡ αὐτὴ δ᾽ αἰτία καὶ τοῦ θᾶττον παρίστασθαι τοὺς ἠθητικοὺς τῶν οἴνων· λεπτότεροι γάρ, καὶ ὥσπερ εὐθὺς ἄμικτοι. συνεπιμαρτυρεῖ δὲ καὶ τὸ[2] ἐκ τῶν λεπτοτέρων καὶ εὐείλων καὶ εὐπνόων εὐωδεστέρους γίνεσθαι, καὶ ἐκ τῶν πρεσβυτέρων ἢ νεωτέρων.

ἔτι δ᾽ ὅσοι[3] πολύοσμοι καὶ ἰσχυροὶ ταῖς ὀσμαῖς, οὐκ ἔχουσιν ἐν τῇ γεύσει τὸ μαλακόν, ὡς οὐκ οὔ-

§16.5: Pliny, *N. H.* 15. 110 and 14. 80.
§16.6: *Cf.* Pliny, *N. H.* 21. 35; 16. 117; 15. 110.

[1] Gaza (*caret odore*), Scaliger : εὔοσμοι U.

We can see this also in wines: sweet wines are quite odourless, and so are mild wines; whereas of the rest the thin wines tend to be more fragrant and are sooner matured than the ones that are full in body and thick. The production of fragrance is moreover attended by a separation[1] (so to speak) of the fluid part from the earthy as the lees settle, and in addition by an evaporation of the watery part, so that the fragrance comes from all these factors—thinness, separation and evaporation (indeed in the jugful of wine itself the part next to the lees is less fragrant). 16.6

The same reason accounts for the more rapid maturing of strained wines as well: they are thinner and unmixed (as it were) from the start. Further corroboration is this: wines from vineyards with thinner soil and in sunny and airy places are more fragrant, and those from older vines are more fragrant than those from younger.[2]

Again, wines with a rich and powerful aroma are not mild when tasted, which shows that agreeable

[1] Used of the removal of the juice from the pulp (*CP* 6 7. 3; for the verb *cf.* *CP* 6 19. 2); hence "so to speak."

[2] The older vine feeds less and is drier (*CP* 2 11. 20; 6 17. 4).

---

[2] Schneider : ὁ U.

[3] Wimmer (οἵ Schneider) : ὥς U.

σης ἅμα τῆς τε κατὰ τὸν χυλὸν εὐστομίας,[1] καὶ

16.7 πολυοσμίας καὶ ἰσχύος. ὅπερ καὶ ἐπὶ τῶν μύρων καὶ ἐπὶ τῶν θυμιαμάτων καὶ ἐπὶ τῶν στεφανωμάτων καὶ ἐπὶ τῶν ἄλλων τῶν εὐωδῶν[2] συμβαίνει· πάντα γὰρ πικρὰ καὶ δύσχυλα τὰ τοιαῦτα, καθάπερ καὶ αἱ ἀμυγδάλαι. καὶ οἱ ὀρεινοὶ δὲ τῶν οἴνων ὀσμὴν μὲν ἔχουσιν ἔνιοι, σκληροὶ δὲ καὶ οὐκ εὔχυλοι.[3]

ἐξ ἁπάντων οὖν τούτων δῆλον ὡς ἕτερον τὸ τὴν εὐοσμίαν ποιοῦν.

ἐὰν οὖν συμμετρίαν τινὰ λάβῃ τῆς κράσεως, ἥδιστον τὸ ἐξ ἀμφοῖν (ἄλλως τε καὶ ὧν κατὰ τὴν γεῦσιν ἡ ἀπόλαυσις)· ἐπεὶ καὶ ὧν[4] ἐν αὑταῖς ἀγρίων τινῶν ἔφαμεν λυπεῖν τὸ δριμὺ καὶ ἄκρατον, ὥσπερ ἑρπύλλου καὶ ἀβροτόνου[5] καὶ πηγάνου, μιχθείσης[6] δὲ ὑγρότητος συμφύτου, μᾶλλον

§16.7: *Cf.* Pliny, *N. H.* 21. 35.

[1] Wimmer : εὐοσμίας U.
[2] ego (εὐόσμων Schneider) : ευοσμοδῶν U.
[3] u : εὐχειλοι U. [4] U : τῶν Wimmer.
[5] aP : -ρω- N : ..ρο|τονου U) (*cf.* σισύμβριον *CP* 6 14. 12).
[6] N aP : -ας U.

[1] The bitter almonds from which the oil, perfume and unguent were made: *cf.* Theophrastus, *On Odours*, iv. 15–16. [2] For things smelt by tasting them *cf.* *CP* 6 9. 3.
[3] *CP* 6 14. 12.
[4] "Bergamot mint" at *CP* 6 14. 12.

flavour does not coexist with abundant and powerful odour. This holds also of perfumes, incense, coronary plants and other aromatic substances: all such things are bitter and of poor flavour (as are the almonds[1] too). Again mountain wines have in some cases a bouquet, but are harsh and not agreeable in flavour.

16.7

All this, then, shows that what produces fragrance is distinct from what produces good flavour.

If then the juice that produces the odour-flavour acquires a certain tempering in its taste-flavour, the combination of odour and taste is most pleasurable (especially when the odour is enjoyed by tasting[2]). Indeed in certain wild varieties, where we said[3] the aroma was so pungent and untempered as to become disagreeable (as in tufted thyme, southernwood[4] and rue), the odours, after fluid native to the plant[5] has been mingled with the odour-flavour,

[5] Obtained by cultivating the plant. For other ways of mixing *cf.* Theophrastus, *On Odours*, iii. 9: "Some mixtures (*sc.* in perfumes and unguents) are for the sake of the odour alone, and are aimed at the sense of smell; others aim at 'seasoning' (as it were) the sense of taste, as when persons pour perfume in wine or put spices into it, since the two sensations lie closer together and influence one another, which is why good odours are sought out for the objects of taste themselves"; xiv. 67: "Perfume is also considered to add to the pleasantness of wine . . . It is not unreasonable that since the sensations lie close together there should be some partnership between them when their object is the same."

εὐκράτους[1] *γίνεσθαι καὶ ἡδίους. μίγνυνται* <*δὲ*>[2] *καὶ*[1] *εὐχυλοτέρων γιγνομένων· ἐὰν δὲ ὑπεραίρῃ θάτερον, ἐπιζητοῦμεν τὸ ἐλλεῖπον (ἅμα γάρ πως ἡ ἀπόλαυσις, ἐν τοῖς τοιούτοις τῶν γευστῶν, ὀσμῶν καὶ χυλοῦ).*

16.8 *χρὴ δὲ πειρᾶσθαι καθόλου διαιρεῖν ὥσπερ εἴπομεν,* <*εἰ*>[3] *ἐν τῷ χυλῷ μὲν ἡ εὐστομία καὶ ἡ γλυκύτης, ἀπὸ τοῦ χυλοῦ δὲ ἡ ὀσμή, ποῖος ἑκάτερος καὶ πῶς ἔχων. ἐκεῖνο δ' οὖν φανερὸν ἐκ τῶν εἰρημένων, ὅπερ ἐξ ἀρχῆς ἠπορήθη· διὰ τί οἱ ἄγριοι*[4] *τῶν καρπῶν εὐοσμότεροι.*

φαίνεται δὲ κἀκεῖνο συμβαίνειν, ὥσπερ[5] *ἐν τοῖς ἀτελέσιν καὶ τοῖς στερητικοῖς χυλοῖς μᾶλλον εἶναι*

§16.8: Galen, *De Simpl.* iv. 22 (vol. xi, pp. 698. 4–699. 4 Kühn); Theophrastus, *On Odours*, ii. 5.

[1–1] *εὐκράτους . . . καὶ* U : *εὔοσμον γίνεται* Wimmer.
[2] ego : *γὰρ* aP.
[3] Gaza, Schneider.
[4] u aP (*ὄρειοι* Schneider) : *ἀέρειοι* U (*ἀέριοι* N).
[5] U : *ὥστε* Schneider.

---

[1] An amplification of *CP* 6 14. 12: "those (*sc.* the odours) of the cultivated type have with their lessened intensity a certain sweetness."

became (we said[1]) better tempered and so more pleasant. Similarly mixture occurs with improvement in taste-flavour[2]; but if the one partner prevails, we miss the other, since in gustibles of this type[3] enjoyment of odours and of taste-flavour is somehow simultaneous.[4]

The Generic Distinction Between the Taste-Flavour and the Odour-Flavour

16.8 But we must endeavour to draw the generic distinction that we spoke of[5]: if good taste and sweetness are *in* the flavour and odour *from*[6] the flavour, what is the character and condition of the flavour in each case? At all events the answer to the question with which we began,[7] why wild fruit is the more fragrant, is not evident from the discussion.[8]

It appears that the other answer is yielded as well: fragrance lies rather (as it were) in the unperfected and privative flavours, inasmuch as fragrant

[2] As the same plant changes from producing odour to producing flavour.

[3] Fragrant gustibles.

[4] *Cf.* Aristotle, *On Sense,* v (443 b 31–444 a 3): "Those who now mix such powers (*i.e.* odours that do not remind us of food) into their potions are forcing pleasure by habit until from two sensations the pleasurableness becomes as if it came from one" (*cf.* also 443 b 16–21).

[5] *CP* 6 16. 1–2. [6] *Cf. CP* 6 14. 2.

[7] *CP* 6 6. 1. [8] *CP* 6 16. 1–8.

τὴν εὐοσμίαν, εἴ γε τὰ μὲν πικρά, τὰ δὲ δριμέα, τὰ δ' ἀσθενῆ, τὰ δὲ στρυφνά, τὰ δὲ ἄλλην ἔχοντα δυσχέρειαν. οὐκ ἀλόγως δ' ἴσως· ἀποπνοὴ[1] *γάρ τις μᾶλλον ἀπὸ τούτων, ὥστ' ἐὰν μὲν εὔκρατα λάβῃ, σύμμετρον εἶναι πρὸς τὴν ὄσφρησιν, ἐὰν δὲ ὑπερβάλλοντα, δυσχερῆ καὶ βαρεῖαν ἤδη.*[2] *καὶ οὐκ ἂν δόξειεν ἐν στερήσει <τὸ>*[3] *τοιοῦτον εἶναι γένος, ἀλλ' ἑτέραν τινὰ φύσιν ἔχειν, εἴ γε ἔνια ποιητικὰ τῶν ἄκρων.*

ἡ[4] *μὲν οὖν εὐοσμία*[5] *διὰ τοῦτο ἐν τοῖς ἀγλυκέσι*[6] *καὶ δυσχύλοις μᾶλλον (ὡς [δ']*[7] *εἰπεῖν).*

τῶν <δ'>[7] *ὀσμῶν ἡ μὲν ἰσχὺς ἑκάστων δῆλον ὅτι κατὰ τὴν ὑποκειμένην φύσιν.*

17.1 *οὐ μὴν ἀλλ' ἔνιά γε τῶν ἀνθέων ἐξ ἀποστάσεως*

§17.1: [Aristotle], *Problems*, xii. 2 (906 a 30–33); xii. 4 (906 b 35–907 a 4); xii. 9 (907 a 24–27).

[1] *ἀπο|πνοὴ* U : *ἡ πνοὴ* N aP.
[2] u : *εἴδη* U.
[3] ego.
[4] u : *εἰ* U.
[5] *εὐοσμία* Itali (*odor* Gaza) : *αοσμια* U.
[6] Moreliana (*minus dulcibus* Gaza, *μὴ γλυκέσι* Itali) : *γλυκεσι* U.

things are some of them bitter, some pungent, some faint, some astringent, some with another unpleasant taste. This is not perhaps unreasonable, since there is more exhalation[1] of a sort from these unperfected flavours. Hence if the exhalation carries off particles that are well-tempered, the odour is properly adjusted to the sense of smell; whereas if it takes particles that run to excess, the odour becomes disagreeable and oppressive. And it would not appear that odours of this type are a mere matter of privation, but possess a nature of their own, inasmuch as the effects are in some cases extreme.[2]

Fragrance, then, for this reason lies more in non-sweet and ill-flavoured substances (so to speak).

The Strength of Odours

The strength of the odour in different plants evidently depends on the nature to be achieved.

The Strength Varies with (1) *Distance*

Some flowers are nevertheless more fragrant at a 17.1

[1] *Cf.* Theophrastus, *On Odours*, i. 3: ". . . odour depends on exhalation . . ."

[2] *Cf. CP* 6 5. 5.

---

[7] Schneider.

ἢ πλησίον ἐλθοῦσιν εὐοσμότερα (καθάπερ καὶ τὰ ἴα δοκεῖ). τὸ δ' αἴτιον, ὅτι πρὸς μὲν τὰ πόρρω καθαρὰ φέρεται καὶ ἀμιγὴς ἡ ὀσμή· πλησίον δ' ὄντων, ὅτι συναπορρεῖ τι[1] καὶ ἀπὸ τῶν ἄλλων μορίων, ἅπερ οὐ δικνεῖται πρὸς τὰ πόρρω διὰ τὸ γεωδέστερα καὶ παχύτερα εἶναι. καθόλου γὰρ ταῖς μὲν ἀσθενέσι τῶν ὀσμῶν ἡ ἀμιξία, τῶν δ' ἰσχυροτέρων ἐνίαις αἱ μίξεις αἱ οἰκεῖαι χρησιμώτεραι (καθάπερ τοῖς χυλοῖς). οἷον καὶ τῆς σμύρνης[2] <ἥδιον> δοκεῖ[3] τῆς ἀμίκτου[4] θυμιᾶσθαι καὶ[5] καταβραχεῖσα[6] μελικράτῳ ἢ γλυκεῖ· μαλακω-

17.2

§17.2: *Cf.* Theophrastus, *On Odours*, x. 44; xiv. 67; Pliny, *N. H.* 21. 36.

[1] U : τινα Gaza, Heinsius.
[2] ego (*myrra* or *mirra* [nom.] the MSS of Gaza : τῇ σμύρνῃ Wimmer) : τηῖ μυρρίνηι U.
[3] ego : δοκεῖ κρείττων Schneider.
[4] Schneider (γὰρ ἄμεινον Wimmer) : τῶ ἀμίκτωι U.
[5] U N : aP omit.
[6] N aP : -εχεισα U (-εχεῖσα u).

---

[1] *Cf.* Theophrastus, *On Odours*, iii. 12–13 (cited in note 1 on *CP* 6 14. 11) and [Aristotle], *Problems*, xii. 2: "Why are the odours of incense and of flowers less fragrant close at hand? Is it because particles of earth are carried along with the odour, and these drop out first because of their

distance than upon closer approach,[1] as violets are said to be. The reason is this: the odour is pure and unmixed when it carries to distant objects; but when an object is near by, some portion of other parts of the plant than the flower is carried along, particles too earthy and thick to reach the greater distance. For in general weak odours are preferable without admixture, whereas the appropriate admixtures are preferable for some of the stronger ones, as with flavours. So it is held that myrrh when well-soaked in hydromel or sweet wine burns with a more agreeable odour than when unmixed,

17.2

weight, so that the odour becomes pure at a greater distance?"; xii. 4: "Why do flowers and incense tend to have a more agreeable smell further off, but close at hand the smell of the flowers is too grassy, that of the incense too smoky? Is the answer this? Odour is a kind of heat, and fragrant objects are hot. What is hot is of light weight, so that for this reason as the fragrant parts proceed to a greater distance the odour becomes less mixed with the accompanying odours of the leaves and of the smoke (which is a watery vapour), whereas when the flowers or incense are close at hand, the objects mixed with them join in the smell of the things with which they are mixed"; xii. 9: "Why are the odours of both incense and flowers less fragrant close at hand? Is the answer this? When they are close at hand the earthiness is also carried along, and so by being blended makes the power weaker, whereas when carried to a distance the odour drops."

τέρα γὰρ ἡ ὀσμὴ κεραννυμένη καὶ γλυκυτέρα γίνεται. ταὐτὸ δὲ τοῦτο συμβαίνει καὶ ἐπὶ τῶν χυλῶν· ἔνιοι[1] γὰρ δέονται μίξεως πρὸς εὐστομίαν.

ὡς δὲ ἁπλῶς εἰπεῖν, ἕωθεν ὀσμαὶ καὶ[2] πλεῖσται καὶ ἀκρατέστεραι, προσιούσης[3] δὲ τῆς μεσημβρίας ἧττον,[4] μεσημβρίας δ' ἥκιστα, διὰ τὸ ἀναξηραίνειν τὸν ἥλιον.

17.3 ὡς δ' ἐν ταῖς[5] τῆς ἡμέρας ὥραις, ὁμοίως ἐν ταῖς τοῦ ἐνιαυτοῦ κατὰ λόγον (πλὴν εἴ τινων

§17.3: Pliny, *N. H.* 21. 39, 36.

[1] N aP : ἔνιο (?) U : ἔνια u.

[2] ὀσμαὶ καὶ ego (μὲν αἱ ὀσμαὶ Schneider after Scaliger) : καὶ ὀσμαὶ U.

[3] Wimmer : προϊοῦσης (from -ού-?) U.

[4] aP (ἧττον and a blank of 5 letters N) : ητ...|U.

[5] ἐν ταῖς U (legible on the opposite page, where the ink has come off) : N omits in a blank of 5–6 letters : καὶ ἐν ταῖς aP.

---

[1] *Cf.* Theophrastus, *On Odours*, x. 44: "Of perfumes the Egyptian, myrrh-oil and others with strong odours are pleasanter when mixed with fragrant wine, since their oppressiveness is thus removed. Indeed myrrh itself is more fragrant for burning after soaking in sweet wine, as was said earlier [*i.e.* at *CP* 6 17. 2]"; *ibid.* xiv. 67: "Also mingling with wine makes certain perfumes as well as

for the odour when tempered becomes gentler and sweeter.[1] The same is true of flavours: some require mixture to be agreeable to the taste.

(2) *With the Time of Day, Year and Life*

Generally in the morning odours are most numerous and more intense; with the approach of noon[2] they become less so; at noon they are least of all so because of the drying effect of the sun.[3]

17.3 As fragrance varies with the time of day, so too it varies with the time of year[4] (except where a plant

certain incense, such as myrrh, more fragrant."

[2] So Xenophon (*Cynegeticus*, iv. 9. 11) advises taking the hounds hunting in summer up to noontime, in winter throughout the day, in autumn at any time except noon, and in spring at any time before nightfall.

[3] *Cf.* Theophrastus, *On Odours*, ix. 40: "Perfumes are spoilt in a warm season, in a warm location, and when they are set out in the sun; this is why perfumers look for houses with an upper storey that do not face the sun but are as far as possible in the shade, for sunlight and heat remove the odour and do more than cold to denature the perfumes entirely; whereas cold and frost, even if they make the perfume less fragrant by contracting it, do not however remove its power completely."

[4] *Cf.* Xenophon, *Cynegeticus*, v. 5: "Spring, which is properly blended in its seasonal qualities, makes the trail (*sc.* of the hare) clear . . . It is light and indistinct in summer, since the earth is permeated with heat and does away with the warmth of the track, which is thin . . . In autumn the trail is unmixed with other odours . . ."

ἀκμαὶ[1] καὶ πεπάνσεις κατ᾽ ἄλλην ὥραν, ἢ εἴ τινων αἱ κράσεις τοῦ ἀέρος σύμμετροι πρὸς τὰς ὀσμάς, ὥσπερ[2] καὶ καθ᾽ ἡμέραν[3] ἐστὶν ἐπὶ τῆς ἑσπερίδος καλουμένης· αὕτη[4] γὰρ τῆς νυκτὸς ὄζει μᾶλλον).

κατὰ δὲ τὰς ἡλικίας,[5] οὐκ ἐν ταῖς ἄκραις, ἀλλ᾽ ἐν ταῖς ἀκμαῖς εὐοσμότατα, μικρόν τε[6] παρεγκλί-

17.4 νοντα[7] τὴν ἀκμήν (οὐ γὰρ ἴσως ἡ αὐτὴ χυλοῦ καὶ ὀσμῆς πέψις). τὸ μὲν γὰρ νέον, ἅτε πλείω[8] τροφὴν ἐπισπώμενον, οὐ πέττει, τὸ δὲ γεγηρακὸς ἐξασθενεῖ δι᾽ ἔνδειαν θερμότητος. οὐ μὴν ἀλλὰ ταῦτά γε εὐοσμότερα τῶν νέων· ἐλάττων[9] γὰρ ἡ ὑγρότης, ὥστ᾽ ἐπικρατεῖ μᾶλλον, ὥσπερ καὶ τὰ ἐν ταῖς χώραις ταῖς λεπτογείοις. ὡς δὲ ἁπλῶς εἰπεῖν, τὰ ἐν ἀκμῇ καὶ πρὸς εὐχυλίαν ἄριστα καὶ πρὸς εὐοσμίαν[10] (ὧν ἑκατέροις ἐνταῦθα τὸ τέλος).

17.5 ἐν δὲ τοῖς ψύχεσι καὶ πάγοις ἀμβλύτεροι καὶ οἱ

§17.5: *Cf.* Aristotle, *On Sense*, v (443 b 12–16); [Aristotle], *Problems*, xii. 6 (907 a 8–12); Xenophon, *Cynegeticus*, v. 1–2.

[1] U : ὀσμαὶ N aP.
[2] N aP : ὅσπερ U.
[3] U : ἑσπέραν Wimmer.
[4] u : αὐτη U.

reaches its prime and ripens at a different season, or where the air has the right tempering for the odour, as occurs daily in the so-called "evening flower,"[1] this being more fragrant at night).

As for the time of life, plants are most fragrant not in youth or old age, but at their prime, and when they are slightly past it (since the concoctions of flavour and of odour are perhaps not the same). Thus the young plant attracts more food than it can concoct, whereas the plant that has grown old is too weak to concoct, since it has not enough heat. Nevertheless the older are more fragrant than the young, since they have a smaller amount of fluid and so can master it better, just as plants that grow in countries with thin soil.[2] But speaking broadly, plants at their prime are not only the best producers of good flavour but also of good odour (among plants whose goal it is to produce the one or the other).

17.4

(3) *With Cold*

In cold spells and frosts both flavours and odours

17.5

[1] Night-scented stock (*hesperis*).
[2] *Cf. CP* 1 18. 1–2; 2 4. 3.

---

[5] U^c : ἡλιακὰς U^ac.
[6] M : γε U N aP.
[7] πα|ρεγκλίνοντα U : παγκλίνοντα N aP.
[8] U : ἡλίω N aP.
[9] u : ελαττον visible in U.
[10] N aP : εὐκοσμίαν U.

χυλοὶ καὶ αἱ ὀσμαὶ διὰ τὴν πῆξιν, πεπηγότα γὰρ οὐ διαδίδωσιν. ἅμα δὲ καὶ ἀφανίζεταί πως ὑπὸ τοῦ ψύχους, κατακρατοῦντος μᾶλλον, κἀκεῖνα καὶ ἡ αἴσθησις, ἀμφότερα γὰρ ἐξίστησιν, ἐκεῖνα δὲ μᾶλλον (ἐπεὶ καὶ τῶν καρπῶν ἐξαιρεῖται τὴν γλυκύτητα, διατμίζον τῇ πήξει καὶ ἐξαεροῦν [1]).

ὡσαύτως δὲ καὶ ἡ τῶν καυμάτων ὑπερβολὴ τοὺς χυλοὺς λυμαίνεται· [2] τὰ μὲν γὰρ κατακάει, τὰ δὲ οὐ πέττει, τὰ δ' ὥσπερ σήπει καὶ διυγραίνει (καθάπερ καὶ ἐπὶ τῶν συκῶν ἐλέχθη).

τὰς δ' ὀσμὰς ἴσως ἐνίων μᾶλλον διατηρεῖ· 17.6 τάχα δὲ καὶ ποιεῖ τάς γε ἐν τοῖς ξηροῖς γινομένας,

§17.6: *Cf.* [Aristotle], *Problems*, xii. 3 (906 b 12–16).

[1] Schneider : ἐξαῖρον U.
[2] aP (λιμαίνεται u : λημαίνεται N) : λαμβάνεται U.

---

[1] *Cf.* Theophrastus, *On Odours*, ix. 40, cited on *CP* 6 17. 2, note 2.

[2] *Cf.* Xenophon, *Cynegeticus*, v. 1–2: "In the winter no odour comes from it (*i.e.* from the trail of the hare) early in the morning, when there is hoar-frost or freezing; for the hoar-frost by its own strength holds the warmth back and keeps it in itself, and the freezing does so with a coat of frost. And the hounds have their nostrils numbed by the cold and cannot smell the trail when it is in this

are dulled because of the congealing, for when plants are congealed[1] there is no transmission. Again, not only the odours and flavours but also our sensations[2] are in a way obliterated by the cold when it prevails too greatly: it denatures both, but does so more to the flavours and odours. In fact it even removes the sweetness from fruits, turning it to air and vapour by congealing them.[3]

(4) *With Heat; Distinctions*

So too excessive hot weather spoils flavours: such weather sometimes burns them, sometimes fails to concoct them, sometimes causes decomposition (as it were) and dilutes with watery fluid (as we said[4] of the fig-trees).

But it perhaps tends rather to preserve the odours of some; and it may be that it even produces 17.6

state . . ."

[3] *Cf. CP* 2 3. 8: "Cold air . . . removes the fluid in some cases"; Aristotle, *On the Generation of Animals*, v. 3 (783 a 15–17): "For coldness hardens because it dries by congealing; for as the heat is pressed out the fluid evaporates along with it . . ."; *cf.* also Aristotle, *On the Parts of Animals*, ii. 4 (650 a 8–9). *Cf.* in general Aristotle, *On Sense*, v (443 b 12–16): "It is clear then that what savour is in water, odour is in air and water. And for this reason cold and congealing dulls savours and extinguishes odours; for the chilling and congealing extinguish the heat which imparts the movement [*i.e.* it is the efficient cause] and prepares them." [4] *CP* 2 3. 8.

ἄκρατοι γὰρ αὗται·[1] πολλὰς δὲ καὶ φθείρει.

τὰς δ' ἐν τοῖς καρποῖς[2] ἐγχυλοτέρας καὶ μεμιγμένας τινὰς εἶναι <δεῖ>[3] (μὴ πεπαινομένων δὲ καλῶς οὐ γίνεται[4]), τὸ δ' ὅλον[5] ἐν συμμετρίᾳ τινὶ τὴν μῖξιν ὑπάρχειν καὶ ἐμφαίνεσθαί πως τὴν τοῦ ξηροῦ φύσιν ἐν ταῖς ὀσμαῖς, τοῦτο γὰρ τὸ τὰς ὀσμὰς[6] ποιοῦν ἢ πάντων ἢ τινων, ὅπερ φανερὸν εὐθὺ καὶ ἐπὶ τῆς γῆς ἐστὶν ἐν τοῖς τοιούτοις ὑετοῖς· διακεκαυμένης γὰρ ἐν τῷ θέρει, τὸ θερμὸν πέττον[7] τὸ ὕδωρ ποιεῖ τὴν εὐωδίαν. 17.7 ταῦτα δὲ ποιεῖ καὶ ἐν ἄλλοις. καὶ γὰρ τὸ περὶ τὴν ἶριν λεγόμενον, ὡς ὅπου ἂν κατάσχῃ, ποιεῖ τὰ δένδρα καὶ τὸν τόπον εὐώδη, τοιοῦτόν ἐστιν· ποιεῖ γὰρ οὐ πάντως, ἀλλ' ἐὰν ὕλη τις ᾖ νεόκαυτος, οὐδ' ἴσως

§17.7: Pliny, *N. H.* 12. 110; 21. 39; Plutarch, *Quaest. Conv.* iv. 2, 2 664 E–F; [Aristotle], *Problems*, xii. 3 (906 a 36–b 27).

[1] N aP : αὐταί U.

[2] καρποῖς <τῷ Moreliana : διὰ τὸ Heinsius (*quia ... sunt* Gaza)>.

[3] ego.

[4] U : γίνεσθαι Schneider (taking it to be the reading of U).

[5] ὅλον χρὴ Heinsius after Gaza.

[6] N aP : ὁμᾶς U : ὁμὰς u.

[7] ego (πως καὶ Heinsius) : πῶς U.

the odours, at least those that occur in the dry parts, since these odours are concentrated; but there are many it also destroys.

The odours in the fruits on the other hand must contain more succulence and be of a mixed sort (but if the fruit is not being properly ripened there is no fragrance), and this mixture must in general have a certain balance so that the special character of dryness may be noticeably present in the odour: for it is this discernible presence of the dry that gives their odours to all fragrant things or to some. This can also be seen with no further ado in the case of the earth when rain of this sort[1] falls; for when the earth has been baked dry in summer the heat in it concocts the rainwater and produces the fragrance.[2] The heat also produces these results elsewhere: thus the story about the rainbow, that wherever it rests it makes the trees and the locality fragrant, is an instance. For the rainbow does not do this under any conditions, but only where there is recently

17.7

[1] A rain that leaves dryness discernible.

[2] *Cf.* [Aristotle], *Problems*, xii. 3: "The cause of the fragrance (*sc.* where the rainbow touches after a forest has been burnt) is the same as the cause of the fragrance of the earth: when the earth is full of heat and has been baked through, whatever place is at first rained on is fragrant. For the things that in a certain way are affected by fire, among those possessing little moisture, become fragrant, since the heat concocts the moisture."

καθ' αὑτήν, ἀλλὰ τρόπον τινὰ κατὰ συμβεβηκός, ἐφύει γὰρ ὅπου ἂν ἐφιστῇ·[1] *πεπυρωμένης γὰρ τῆς ὕλης, ἡ κατάμιξις ποιεῖ τινα ἀτμίδα καὶ εὐωδίαν. οὐδὲ γὰρ οὐδὲ γίνεται πλῆθος ὕδατος, ἀλλ' ὡς ἐπὶ τὸ πολὺ ψακάς,*[2] *ὥστε συμμετρίαν εἶναι πρὸς τὴν θερμότητα καὶ ξηρότητα. δεῖ δὲ καὶ αὐτὴν τὴν ὕλην ποιάν τινα προϋπάρχειν, οὐ γὰρ ἐν πάσῃ καὶ πάντως.*

17.8 *ὅλως δὲ καὶ ἐν ἄλλοις ἥ τε πύρωσις καὶ ἡ κατάμιξις τῶν πεπυρωμένων ποιεῖ τινας εὐωδίας, καὶ τὸ ὅλον ὀσμάς, ἐὰν ἔχῃ τὸ σύμμετρον· ἐπεὶ καὶ τὰ θυμιάματα ταύτας πυρώσει τὰς*[3] *εὐωδίας, μαλακῇ*

[1] U (the verb is treated as meteorological, like ἐφύει) : ἐπιστῇ Coray.

[2] U : ψεκάς u.

[3] ταύτας πυρώσει τὰς ego (*ex hac eadem ratione imposita igni redolent* Gaza : ταῦτα πυρωθέντα ἀνίει Wimmer) : ταῦτα πυρώδεις ταῦτας U.

---

[1] *Cf.* [Aristotle], *Problems*, xii. 3: "It is said that the trees become fragrant on which the rainbow rests . . . We must assign the cause as incidental to the rainbow . . . That the result does not occur with all trees, or always occur, is evident . . .; and when it occurs, it does not occur in every kind of wood . . . And we must assign the cause as only incidental to the rainbow . . . The result does not

burnt wood, nor does the rainbow do so perhaps directly, but in a way only incidentally, since there is rain wherever the rainbow comes to rest. Thus the mixture of the rainwater when the wood has been exposed to fire produces a certain vapour and fragrance; in fact there is no great amount of rainwater, but usually only a drizzle, so that the quantity is of the right amount for the heat and dryness in the wood. And the wood itself must first be of a particular character, since the result does not occur in every kind of wood and under every condition.[1]

17.8 And in general elsewhere too exposure to firing and intermixture of fired substances produces certain fragrances and odours in general, if the firing and intermixture is not excessive. Indeed incense produces its familiar fragrance by firing, but the

occur in any state of the forest, but the herdsmen say that the fragrance becomes noticeable in a forest that has been burnt, after the rain that ensues on the rainbow . . . The cause of the fragrance is the same as the cause of the fragrance of the earth: . . . For the fluid must neither be too great (since a great amount would not be concocted) nor too small (for then there is no vapour). This occurs with a wood that has recently been burnt . . . The belief that the fragrance occurs in trees on which the rainbow rests is due to the fact that nothing can happen without rainwater; for when the wood has been wet and has concocted the wetness by means of the heat contained within the wood, the wood releases the vapour that arises within itself . . ."

δὲ καὶ οὐ κατακαιούσῃ, τὸ γὰρ σύμμετρον οὕτω πρὸς ὀσμήν·[1] ἐκείνως δὲ φθορά.

17.9 πάρεγγυς δὲ καὶ ταὐτό πως τούτῳ καὶ <τὸ>[2] ἐπὶ μὲν τῶν δένδρων (καὶ ὅλως τῶν ὑλημάτων) εὔοσμα πολλά, ζῷον δ' οὐδέν (εἰ μὴ τὴν πάρδαλίν φασι, καθάπερ ἐλέχθη, τοῖς θηρίοις). τὰ μὲν γὰρ θερμὰ καὶ ξηρὰ τὴν φύσιν, ὥστ' εὐπεπτοτέρα[3] καὶ καθαρωτέρα τις ἡ αὐτῶν ἀποπνοή· τὰ δ' ἐν ὑγρότητι πλείονι[4] καὶ πηλωδεστέρᾳ,[5] διὸ καὶ ἡ ἀποπνοὴ τοιαύτη· καὶ ὅλως ἡ τροφὴ τῶν μὲν ἁπλῆ καὶ ἀπερίττωτος, τῶν δὲ ποικίλη καὶ περιτ-

§17.9: *Cf.* [Aristotle], *Problems*, xiii. 4 (907 b 35–908 a 19); Pliny, *N. H.* 21. 39.

[1] Itali (*odoribus* Gaza) : ὁρμὴν U.
[2] ego (<διὰ τί> Moreliana).
[3] ego (*cf.* ὥστε εὐπεπτοτέρα [Aristotle], *Problems* 908 a 12 : ὡς πεπτότερα or ὡς λεπτότερα Scaliger : *itaque* . . . *tenuior* Gaza : ὥστε λεπτοτέρα Moreliana) : ὥσγυ|ποπτότερα U : ὡς γυπόπτερα N : ὥς γ' ὑπόπτερα aP.
[4] Heinsius : πλεῖον U.
[5] ego (πηλωδέστερα Moreliana) : μηλοδέστερα U.

---

[1] *Cf.* Theophrastus, *On Odours*, iii. 13 (preceded by the passage cited in note 1 on *CP* 6 14. 11): ". . . Frankincense and myrrh, which have a nature still more closely packed,

firing must be a gentle one[1] and not a conflagration, since it is then of the right degree for producing odour, whereas otherwise it simply destroys.

A Related Point: Why no Animal is Fragrant

17.9 Closely connected with this and the same in a way is another point: that whereas in trees, and in woody plants in general, many are fragrant, there is no animal that is fragrant (except for the report that the panther, as was said,[2] is fragrant to other animals); for the trees are in their nature hot and dry, so that the exhalation from them is better concocted and purer, but the animals have fluid that is more abundant and muddier, so that their exhalation is of this character too. And in general the food of the trees is simple and yields no excrement,[3] but that of the animals varied and productive of it.[4]

require a gentle firing, which by gradually heating them will produce the smoke . . ."

[2] *CP* 6 5. 2.

[3] *Cf. CP* 6 10. 3 (with note 2); 6 11. 5.

[4] *Cf.* [Aristotle], *Problems*, xiii. 4: "Why is it that whereas no animal is fragrant except the panther (and this is fragrant even to the beasts, for it is said that they like to sniff at it), and when they decay they are actually foul-smelling, many plants on the other hand, even when they decay and wither become still more fragrant? Is it because the reason for the evil smell is some inconcoction of waste food? . . . But plants contain no excrement."

17.10 τωματική. συμβάλλεται δέ τι καὶ τοῦτο πρὸς εὐοσμίαν, ὅταν ἡ φύσις ἅμα προϋπάρχῃ ποιά τις, ἀρχὴ γὰρ αὕτη[1] καὶ πρῶτον· ἐπεὶ οὐδὲ τὰ δένδρα πάντα εὐώδη, διὰ τὸ μὴ τὰς ὁμοίας ἔχειν κράσεις.

τάχα δὲ καὶ ἡμῖν οὐκ ἔνδηλον ἐκ τῶν ζῴων τὸ εὔοσμον διὰ τὸ χειρίστην εἶναι τὴν ὄσφρησιν· ἐπεὶ τά γ' ἄλλα καὶ πόρρωθεν αἰσθάνεται καὶ τὰ σύνεγγυς ἀκριβέστερον. ᾗ[2] καὶ τὰ μὲν ἴσως αὐτοῖς εὐώδη, τὰ δὲ καὶ ἀηδῆ πάντα[3] φαίνεται, καθάπερ καὶ ἡμῖν ἑτέρων ἕτερα μᾶλλον, ἔνια δὲ καὶ ὅλως φεύγομεν.

17.11 ἀλλ' ἐκεῖνο ἄτοπον (ὃ καὶ πρότερον εἴπομεν), εἰ τὸ ἡμῖν κακῶδες καὶ ἄοσμον ἐκείνοις εὔοσμον γίνεται. τάχα δ' οὐ ἄτοπον· ὁρῶμεν δ' οὖν τοῦτο καὶ ἐφ' ἑτέροις συμβαῖνον ἐν[4] αὐταῖς[5] εὐθὺ ταῖς τροφαῖς· ὧν μάλιστ' ἄν τις αἰτιάσαιτο τὰς κράσεις, ἀνωμαλεῖς γε[6] οὔσας. ἐπεὶ τά γε σχήματα Δημοκρίτου (καθάπερ ἐλέχθη), τεταγμένας ἔχοντα

[1] u : αὐτὴ U.
[2] Gaza (*unde*), Moreliana : ἡ U.
[3] U : *penitus* Gaza : πάντως Itali : πάντῃ (or delete) Scaliger. [4] οἷον ἐν Schneider.
[5] Gaza (*ipsis*), Scaliger : ἑαυταῖς U. [6] aP : τε U N.

---

[1] *Cf.* the qualification at the end of *CP* 6 17. 7.
[2] *Cf.* *CP* 6 5. 2 (with note 2).

And it also contributes to fragrance when the nature too has a certain character to begin with,[1] since the nature is fundamental and primary. Thus not all tree are fragrant either, because they do not all have similar temperings. 17.10

Perhaps the fragrance from animals is not detected by us since man has the worst sense of smell,[2] the rest not only perceiving odours from afar but distinguishing more accurately between those close at hand. This superior sense of smell would perhaps make all other animals appear to them as in some cases fragrant, in others actually malodorous (just as for man too there are differences, and some animals are more malodorous than others, and some we avoid entirely).

What *is* odd, however, is the following point (also mentioned earlier[3]), that what is malodorous to us and what is odourless should turn out to be fragrant to them. But there may be no real oddity here: we observe after all the same occurrence in other things[4] in the very food of man and beast (to go no further).[5] Here one would mainly give as reason the variation of temperings[6] in man and beast. One would not appeal to the figures of Democritus[7]: here (as was said[8]) since the shapes do not vary, the 17.11

[3] *CP* 6 5. 2 (the panther has no fragrance to man); *cf. CP* 6 17. 9. [4] In the flavours.
[5] *Cf. CP* 6 4. 7. [6] *Cf. CP* 6 5. 4 *ad fin.*
[7] *Cf. CP* 6 5. 4. [8] *CP* 6 2. 1–2.

τὰς μορφάς, τεταγμένα καὶ τὰ πάθη καὶ τὰς αἰσθήσεις ἐχρῆν[1] ποιεῖν.

ἀλλὰ ταῦτα μὲν οὕτω λεκτέον.

17.12 ὃ δὲ καὶ μικρῷ πρότερον ἐλέχθη, καὶ νῦν πάλιν εἴπωμεν· ὅτι καὶ ἡ εὐοσμία καὶ ἡ εὐχυλία ζητεῖ τινα καὶ χώρας καὶ ἀέρος ποιότητα[2] καὶ τροφῆς· δεῖ γὰρ μήτε ὑπερβάλλειν, μήτ' ἐλλείπειν, μήτε ἀλλοτρίαν εἶναι τοῖς ὑποκειμένοις τὴν τροφήν, ὡς τὸ μὲν πλῆθος κωλύει τὴν πέψιν, τὸ δ' ἐνδεὲς οἷον ὕλην οὐκ ἔχει, τὸ δὲ μὴ οἰκεῖον οὐ ποιεῖ τὸ τῆς φύσεως· τοῦτο[3] δὲ μάλιστα ἐν τῷ ποιόν τι ποιεῖν[4] τὸ ἔδαφος εἶναι, καὶ τὸν ἀέρα, καὶ τὴν θερμότητα τοῦ ἡλίου· ταῦτα γάρ ἐστιν τὰ τὰς τροφὰς καὶ τὰς πέψεις οἵας δεῖ[5] ποιοῦντα. πρὸς ἑτέρους δὲ καρποὺς καὶ χυλούς, ἑτέρα καὶ διάφορος.

17.13 ὅτι δὲ ἀληθὲς τὸ λεγόμενον ἐκ πολλῶν φανερόν· ἄλλα[6] γὰρ ἐν ἄλλῃ χώρᾳ καὶ εὔοσμα καὶ

[1] καὶ τὰς αἰσθήσεις ἐχρῆν Diels : και τ a blank of 5–6 letters χρην (a low dot between χ and ρ) U : καὶ τ a blank of 3 letters χρῆν N : καίτοι γε οὐκ ἐχρῆν aP. [2] aP : -ας U N.
[3] u : τοῦτω U. [4] [ποιεῖν] Heinsius.
[5] U : οἷασδὶ u : οἵας δὴ N a (οἱασδὴ P).
[6] u aP : ἀλλὰ U N.

---

[1] *CP* 6 12. 1 first paragraph; 6 12. 2 last paragraph.

effects and sensations produced should not vary either.

This then is how we must treat these matters.

The Right Air and Soil are Requisite for Fragrance

17.12 What we said a short while ago[1] let us now say again: fragrance and good flavour both demand that the country and air, and hence the food, should have a certain character: the food must be neither excessive nor deficient nor ill-suited to the ends in view, for too much food prevents concoction, too little supplies no matter (as it were),[2] and food of the wrong sort does not produce what belongs to the plant's nature. This production rests mainly on this: that the soil, the air and the heat of the sun lend the plant a certain character, these being the things that produce the proper sort of food and give it the proper sort of concoction; and for different fruits and flavours distinct and different kinds of food and concoction are required.[3]

17.13 The truth of what we say here is evident from many considerations: in different countries the same plants vary not only in fragrance and good

[2] A virtual citation of *CP* 6 12. 2: "... have no 'matter' (as it were) to concoct..."

[3] Odour comes from the flavour.

εὔχυλα καὶ κάρπιμα καὶ ἄκαρπα (καὶ ἔτι[1] *βλαστητικὰ καὶ ἀβλαστῆ· καὶ γὰρ ἐνταῦθ'*[2] *ὑπὸ τὴν αὐτὴν αἰτίαν ὑποπίπτει*[3]*), περὶ ὧν οὐδὲν ἕτερον* <*ἄν*>[4] *τις αἰτιάσαιτο παρὰ τὴν ἐκ τοῦ ἀέρος κρᾶσιν καὶ τὴν ἐκ τοῦ ἐδάφους τροφήν. καὶ γὰρ τὰ κωλύοντα καὶ αὔξοντα ταῦτα, καὶ τὰ τὴν πέψιν ποιοῦντα καὶ εὐχυλίαν. ἐπεὶ δὲ καὶ αἱ ὀσμαὶ καὶ οἱ χυλοὶ πολυειδεῖς, διὰ τοῦτο οὐ μία κρᾶσις, οὐδὲ μία τροφή, πᾶσι πρόσφορος, ἀλλ' ἑκάστοις ἡ πρὸς τὴν ἰδίαν φύσιν.*

18.1 *ὅθεν καὶ ταύτης τῆς ἀπορίας λύσις· διὰ τί ποτε, μᾶλλον ὄντων εὐόσμων τῶν θερμῶν τόπων, οὐχ ἅπαντα ἐν ἅπασι τοῖς θερμοῖς, τὰ εὔοσμα δ'*

[1] *καὶ ἔτι* ego : *και τ* and a blank of 4 letters U : *καὶ τ* and a blank of 2–3 letters N : *καὶ τὰ* (with an index of corruption) M : *καὶ* aP.
[2] U : *ἂν ταῦθ'* Wimmer.
[3] ego (*πίπτοι* Wimmer) : *περιπίπτει* U. [4] Wimmer.

---

[1] *Cf. CP* 2 3. 2–4, 6–8; 2 4. 1–5.
[2] That is, the plants are otherwise the same.
[3] For the fragrance of whole countries *cf.* [Aristotle] *Problems*, xii. 3 (906 b 16–21): ". . . of the whole earth the parts facing the sun are more fragrant than those facing the north; and of these parts that face south the ones facing east are more fragrant, because the region of Syria and Arabia is more earthy, whereas Libya is sandy and

flavour but also in bearing or failure to bear (and again the sprouting or failure to sprout, for here too the plants come under the same cause).[1] For these differences one would give no other reason than the tempering that comes from the air and the food that comes from the soil.[2] For it is these that hinder and promote the growth of the plant and produce concoction and good flavour. But there are many varieties of both odour and flavour, and for this reason no single tempering and no single food is good for all, but for each that tempering and food is good which is conducive to the distinctive nature of the plant.

A Problem Solved: Why All Plants are not More Fragrant in Hot Countries

18.1 We thus obtain the solution of the following problem: Why is it, when hot countries produce more fragrances,[3] that not all plants in all hot countries

without moisture. For the moisture must neither be too abundant (since this would not get concocted), nor must there be none (since then no vapour occurs)"; xiii. 4 (908 a 11–16): ". . . Or do plants have excrement? But because plants are in their nature hot and dry, as a result their moisture is better concocted and not muddy (*sc.* as with animals). This is shown by the fact that the part of the earth that is in hot regions is fragrant, Syria and Arabia, and is shown by the fragrant substances that come from there, because they are dry and hot, and what is so will not decompose."

ἔτι μᾶλλον; διὰ τί ποτε τὰ μὲν παρὰ τοῖς ἄλλοις ἄοσμα, παρὰ τούτοις εὔοσμα (καθάπερ κάλαμος καὶ σχοῖνος ἐν Συρίᾳ)· τὰ δὲ παρὰ τοῖς ἄλλοις εὔοσμα, παρ' ἐκείνοις οὐδὲν μᾶλλον (ὥς δέ τινές φασι, καὶ ἧττον); καίτοι κατὰ λόγον καὶ ταῦτα μᾶλλον. ὑπὸ γὰρ τὴν λεχθεῖσαν αἰτίαν ἅπαντα ταῦτ' ἐστίν· ἡ γὰρ ἀνωμαλία τοῦ ἀέρος καὶ τῆς τροφῆς, ἑτέρων ὄντων τῶν εἰδῶν, ἄλλη πρὸς

18.2 ἄλλους ἁρμόττει. διὰ τοῦτο γὰρ καὶ αὐτῆς τῆς Συρίας βραχύς τις τόπος καὶ τοῦ καλάμου καὶ τοῦ σχοίνου (καὶ πάλιν τῶν[1] τὸ ὀποβάλσαμον, καὶ τῶν ἄλλων τῶν τὰς εὐοσμίας ποιούντων [τε][2])· <ἃ>[3] πρὸς <τὰ>[3] μὲν τὴν οἰκείαν,[4] πρὸς δὲ τὰ παρὰ τοῖς ἄλλοις εὐώδη παραπλησίαν ἢ χείρω κρᾶσιν ἔχει (καθάπερ συμβαίνει καὶ ἐπὶ τῶν καρπῶν)· οὐ γὰρ ἴσης πάντα δεῖται θερμότητος, οὐδ' ὁμοίας, ἀλλὰ τὰ μὲν πλείονος, τὰ δὲ ἐλάττονος καὶ μαλακωτέρας (ὥσπερ καὶ ἐπὶ τῶν ἑψομένων, καὶ ὅλως τῶν κατὰ τὰς τέχνας γινομένων).

[1] ὁ τῶν Schneider. [2] aP. [3] ego.
[4] U : τὰ οἰκεῖα (or τὴν οἰκείαν φύσιν) ἁρμόττει Schneider: τὴν οἰκείαν <. . .> Wimmer.

---

[1] *Cf. CP* 6 14. 8 (with note 1).
[2] *CP* 6 17. 12–13.

are fragrant, and that the fragrant ones are not more fragrant still? Why is it that whereas plants that are elsewhere odourless are fragrant in hot countries (as the reed and rush in Syria),[1] nevertheless plants that are fragrant elsewhere are no more fragrant in hot countries, but, as some say, are in fact less so? And yet it would be reasonable that these plants too should become more fragrant in such countries. For all the plants here come under the cause that we mentioned[2]: the types of odours being different, the variation in the air and food makes some regions suitable, others not. For this is why even in Syria the fragrant reed and rush are found in a fairly small district[3] (so too with the plants that produce balsam of Mecca[4] and the rest that produce the well-known fragrances); and these districts have a tempering that is suited for these products, but is no better, or is worse, for producing the fragrances found elsewhere (as also happens with the fruit); for perhaps not all products need the same amount or kind of heat, but some need a greater, some a lesser and more gentle heat (as in boiling and in general in the preparation of the products of the various arts).

18.2

[3] *Cf. CP* 6 18. 1; 6 14. 8 (with note 1).

[4] *Cf. HP* 9 6. 1: "Balsam of Mecca grows in the valley of Syria. It is said that there are only two plantations, the one of twenty plethra (*sc.* of 2000 feet or 12 acres), the other much smaller."

18.3 ἐνιαχοῦ δὲ καὶ τοῖς καθ' ἕκαστα εὔδηλον· ἐν Αἰγύπτῳ γὰρ χείριστα τὰ ἄνθη καὶ τὰ στεφανώμαθ' (ὡς εἰπεῖν) ὅτι ὁ ἀὴρ ὁμιχλώδης καὶ δροσοβόλος. ἐν δὲ ταῖς τοιαύταις χώραις οὐ γίνεται τὸ ὅλον εὐωδία διὰ τὸ μὴ γίνεσθαι πέψιν, ἀλλὰ μᾶλλον ἐν ταῖς καταξήροις, ἐν ταύτῃ γὰρ ἐκπέττουσιν. ἐπεὶ τὰ περὶ Κυρήνην[1] διὰ ταῦτα εὔοσμα, τά τ'[2] ἄλλα καὶ μάλιστα τὸ ῥόδον καὶ ὁ κρόκος· ἡ γὰρ χώρα λεπτὴ καὶ ξηρὰ καὶ οὐκ ἄγαν θερμή, καθαρῷ δὲ τῷ ἀέρι καὶ ἀνύδρῳ. πρὸς εὐωδίαν δὲ αἱ τοιαῦται τροφαὶ συμμετροῦνται·[3] τὸ δὲ ῥόδον καὶ ὁ κρόκος ὀλιγότροφα, διὸ καὶ τῶν ἄλλων εὐοσμότερα.

18.4 θαυμασιώτερον δ' ἐν Αἰγύπτῳ τὸ περὶ τὰς μυρρίνας, ὅτι τῶν ἄλλων ὄντων ἀόσμων, ὑπερβάλλου-

§18.3: Pliny, *N. H.* 21. 36.
§18.4: *Cf.* Athenaeus xv. 18 (676 E). *Cf.* also *HP* 6 8. 5.

[1] aP : κύρνην U N.
[2] Moreliana : τὰ δ' U.
[3] aP : συμμετρᾶται U N.

---

[1] *Cf. HP* 6 6. 5 (of roses): "The most fragrant are those of Cyrene, which is why the perfume is also the most delightful. In a word the odours both of violet and of other

In some hot countries specific difference in tempering is evident. So in Egypt flowers and coronary plants are (one may say) at their worst because the air is misty and sheds much dew; and in such countries no fragrance is produced at all because there is no concoction. Fragrance occurs rather in arid countries, for here concoction is complete. Indeed in Cyrene the plants are fragrant for this reason, above all the rose and saffron crocus.[1] For the country has thin soil, is dry, is not extremely hot, and has air that is clear and free from water. Such food as this is of the right quantity for producing fragrance; and the rose and saffron crocus are small feeders, which is why they surpass the rest in fragrance. 18.3

The Exceptional Case of the Myrtle in Egypt

But more astonishing is the case of the myrtle in Egypt: the other plants that are fragrant elsewhere are in Egypt without odour,[2] whereas the Egyptian 18.4

flowers are there most undiluted, and notably that of the saffron crocus, for this flower is held to be most extraordinary there."

[2] *Cf. CP* 6 18. 3 and *HP* 6 8. 5: "What contributes most greatly to fragrance of rose and violet and other flowers is the region and the air appropriate to each; thus in Egypt all the other flowers and dried herbs are without odour, whereas the myrtle is astonishingly fragrant."

σιν αὗται τῇ εὐοσμίᾳ· καὶ γὰρ ἡ ξηρότης καὶ ἡ θερμότης οὐχ ἧττον, ἀλλὰ μᾶλλον ἔν τισιν ὑπάρχει τῶν λοιπῶν, καὶ τὰ ἄλλα δὲ σχεδὸν τὰ κατὰ τὴν αἴσθησιν. ἔστιν μὲν δὴ καί τι γένος ἴσως ὃ μεταφυτευόμενον εἰς ἑτέρας χώρας (οἷον Κύπρον, Ῥόδον, Κνίδον) ἐμφαίνει τι τῆς δυνάμεως· οὐθὲν μέντοι τῶν αὐτῶν,[1] *ἀλλὰ καὶ πλατυφυλλότερα*[2] *γίνεται (λεπτὴ γὰρ ἐκείνη) καὶ τὴν εὐωδίαν οὐκ*

18.5 *ἐγγὺς πρὸς ἐκείνην. οὐ μὴν ἀλλὰ τό γε πλεῖστον τῇ*[3] *χώρᾳ τὸ αἴτιον, καὶ τὴν ἰδιότητα ποιεῖ*[4] *πρὸς τἆλλα τὴν θαυμαζομένην.*

τὸ γὰρ αὖ μικρόκαρπόν τε[5] *εἶναι, μήτε λευκὴν ἀλλὰ μέλαιναν, οὐχ ὑπεναντίον πρὸς τὴν εὐωδίαν· ἄμφω γὰρ ξηρότητος*[6] *(αἰτιῶνται δὲ ὡς μικρόμυρτον διὰ τὸ μὴ γίνεσθαι τὸ οὐράνιον ὕδωρ, ὡς ἐπιζητοῦντα τὸν καρπόν,*[7] *τὰς δὲ δρόσους οὐ βοηθεῖν).*

[1] U : *ἄλλων* Schneider.
[2] U : *-οτέρα* Moreliana.
[3] *ἐν τῇ* Schneider.
[4] Gaza, Moreliana : *ποιειν* U.
[5] ego (*fructu exiguo* Gaza : *μικρόν τε τὸν καρπὸν* Itali : *μήτε μεγαλόκαρπον* Heinsius : *μικρόκαρπον* Wimmer) : *μήτε καρπὸν* U.
[6] Moreliana (*ἡ ξηρότης* Schneider): *ξηρότης* U[c] (*ξ* from *ζ*).
[7] U : *ἐπιζητοῦντος τοῦ καρποῦ* Scaliger.

myrtle surpasses all other myrtles in fragrance[1]; in fact both dryness and heat[2] are found not less, but more in some of the other myrtles, and so too with practically[3] all the other sensible qualities. (Now there is moreover a certain variety perhaps which when transplanted to other countries, as Cyprus, Rhodes and Cnidus, gives a hint of this power. These myrtles however fall far short of the Egyptian, but their leaves become broader—the Egyptian myrtle having a narrow leaf—and the plants come nowhere near it in fragrance.) It is mainly[4] the country nevertheless that is responsible, and that produces the astonishing peculiarity of the myrtle in comparison to the rest of the plants in Egypt.

18.5

That the myrtle has a small fruit on the other hand and is not light but dark in colour[5] is not incompatible with its being fragrant, for both size and colour come from dryness (authorities account for the smallness of the berry by the absence of rain, alleging that the fruit requires rain, which the dews do not compensate for).

[1] *Cf. CP* 2 13. 4; *HP* 6 8. 5 (cited in the preceding note).
[2] Mentioned in *CP* 6 18. 3 as productive of fragrance.
[3] That is, excluding fragrance.
[4] Some of the responsibility lies with the plant: *cf. CP* 6 18. 8–9 *init.*
[5] It is incidentally the fruit that is dark: *cf. CP* 6 18. 8.

18.6 τοῦτο μὲν οὖν, <καὶ>[1] εἴ τι τοιοῦτον ἕτερον ᾧ[2] συμβαίνει δυνάμεις ἰδίας ἔχειν παρὰ τἆλλα κατὰ τόπους ἐνίους,[3] ἰδιωτέρων δεῖται λόγων· ὁ γὰρ κοινός,[4] ὅτι πολλὰς ἀνωμαλίας αἱ χῶραι φέρουσιν, ἀληθὴς μέν, προσαπαιτεῖ δὲ καὶ τὰς οἰκείας δυνάμεις καὶ διαφοράς.

ἔοικεν <δ'>[5] οὖν ὁμοίῳ[6] καὶ τὸ ἐν Κιλικίᾳ[7] περὶ τὰς ῥόας συμβαῖνον·[8] ὡς γὰρ ἐκεῖ περὶ τὸν χυλόν, ἐνταῦθα περὶ τὴν ὀσμὴν ἡ δύναμις καὶ ἡ 18.7 ἰδιότης. ἐκεῖ μὲν οὖν αἰτιάσαιτ' ἄν τις τὴν τοῦ ποταμοῦ γειτνίασίν τε καὶ φύσιν· ἡ γὰρ ῥόα φίλυδρον, καὶ μεταβάλλειν φασὶν ἐξ ὀξέος εἰς γλυκὺν ἐὰν ἔχῃ πολυυδρίαν, ὥστ' εἴ τι διαρρέων ὁ ποταμὸς ἅμα τῷ πλήθει καὶ τῇ ποιότητι ποιεῖ τοιαύτην μεταβολήν, οὐδὲν ἄτοπον. ἐνταῦθα δὲ πειρᾶσθαί γέ τινα δεῖ ζητεῖν ἰδίαν αἰτίαν ἥτις οἰκεία πρὸς τὴν εὐοσμίαν.[9] ἔοικεν δὲ μάλιστα τοιαύτη τις εἶναι, συντιθεμένων εἰς ταὐτὸ[10] πάντων, οἷον τοῦ τε κοινοῦ καὶ τοῦ ἰδίου γένους τῆς μυρρίνης, καὶ τοῦ ἐδάφους, καὶ τοῦ ἀέρος.

[1] Gaza a. [2] Heinsius : ὁ U (ὃν U^{ar}?).
[3] ἐ|ν̣ι̣ους U. [4] κ̣ο̣ινὸς U.
[5] ego. [6] ego : ὁμοίως U.
[7] u : λικια U : λυκία N aP. [8] U : συμβαίνειν Schneider.
[9] U^{c} : οἰκεῖαν U^{ac}. [10] Schneider (*unum* Gaza) : ταῦτα U.

18.6 Now this plant (and any other like it, that happens to have in some region or other a special power in comparison with the rest) requires a more special explanation. For the general explanation—that the countries are responsible for many irregularities—while true, needs to be supplemented by a consideration of the special powers and distinctions of the plant.

Thus what happens to the pomegranate in Cilicia[1] is a similar case: as in Cilicia the special power is in the flavour, so here it is in the odour. 18.7 Now one would account for the flavour of the Cilician pomegranate by the nearness and nature of the river: the pomegranate is a tree that likes water and is said[2] to turn from acid to sweet when well-watered. So there is nothing odd that the river as it flows through the orchards produces some such change by virtue not only of the great quantity, but also of the special quality, of its water. With the Egyptian myrtle we must endeavour to find some causation, peculiar to the tree, that conduces to fragrance; and it appears that the causation best meets these requirements when all factors are combined, to wit: the general class of myrtle, this particular variety of it, the soil and the air.

[1] *Cf. CP* 1 9. 2 (with note *a*); 2 13. 4 (with note *a*).
[2] *Cf. CP* 2 14. 2.

18.8 αὐτό τε <γὰρ>[1] τὸ φυτὸν ὅλως ξηρόν, καὶ τὸ γένος τοῦτο μᾶλλον τῶν ἄλλων· δηλοῖ[2] δὲ ἡ στενοφυλλία, καὶ ἡ μικροκαρπία, καὶ ἡ χρόα τοῦ καρποῦ, πάντα γὰρ ταῦτα ξηρότητος. τὸ δὲ ξηρὸν ὀλίγον ὑγρὸν ἔχει, τὸ δ' ὀλίγον εὔπεπτον· ἡ εὐωδία δ' ἐν πέψει[3] τῇ πρὸς ὄσφρησιν, αὕτη δὲ ἀγλυκὴς καὶ οὐκ εὔχυλος.

18.9 ἀπὸ μὲν οὖν τοῦ φυτοῦ ταῦτα προϋπάρχοντα.

ἀπὸ δὲ τοῦ ἐδάφους καὶ τοῦ ἀέρος, ὅτι τὸ μὲν ξηρόν, ὁ δ' ἀὴρ μαλακός· ἄμφω δὲ ταῦτα εἰς πέψιν, ὃ[4] μὲν τῷ συνέψειν, ὃ[5] δὲ τῷ <μὴ>[6] καθυγραίνειν. ἐπεὶ καὶ ἐνταῦθα αἵ ἄγριαί τε τῶν ἡμέρων εὐοσμότεραι καὶ ἐν προσείλοις τῶν ἐν παλισκίοις καὶ μάλισθ' αἱ πρὸς μεσημβρίαν· ἅπασαι δ' αὗται τροφὴν ἐλάττω καὶ τὸ πέττον[7]

[1] ego (*etenim ipsa* Gaza : Αὐτὸ γὰρ Schneider : αὐτό γε Wimmer) : αὐτό τε U N P : αὐτὸ δὲ a.
[2] Moreliana (*constat* Gaza [G^{ed} omits]) : δῆλον U.
[3] ευωδια δ' ἐν πεψει U^{c} : εὐπεψια U^{ac}.
[4] ego (τὸ Schneider) : ὁ U. [5] ego : ο U.
[6] ego.
[7] Schneider : πέττω U : πέττειν u.

---

[1] For the small size and dark colour of the fruit *cf. CP* 6 18. 5.
[2] The heat is the agent, and its action is here called

The Plant in General and This Variety of It

18.8 The myrtle itself is a dry plant, and the Egyptian myrtle is drier than the rest (as is shown by the narrow leaf and the smallness and colour of the fruit, for all three characters come from dryness[1]). A dry plant has little fluid, and that little is easily concocted; fragrance lies in concoction that is suited to our sense of smell; and this concoction lacks sweetness and has no succulence.

18.9 From the plant, then, the preliminary conditions for fragrance are these.

The Soil and the Air

From the soil and the air we have these conditions: the soil is dry and the air mild. Both conditions promote concoction: the mild air because it helps with the cooking,[2] the dry soil because it does not soak the plant.[3] (Indeed in Greece too wild myrtles are more fragrant than the cultivated,[4] and myrtles that grow in sunny places more fragrant than those growing in the shade, and most of all those with a southern exposure. All these classes have less food and more of the concocting agent

"cooking." The heat of the sun (the air is "mild") helps the native heat of the tree to do the work.

[3] The dry soil does not overfeed or "soak" the tree with fluid. [4] The wild trees get less food (moisture).

πλέον ἔχουσιν, διὸ καὶ οὐδ' εὐαξεῖς εἰς μῆκος, ἀλλὰ θαμνωδέστεραι (καὶ γὰρ πάχος λαμβάνουσιν, καὶ σχίζονται μᾶλλον διὰ τὸ πολλὰς λαμβάνειν ἀρχάς, πέττοντος τοῦ θερμοῦ καὶ μερίζοντος πανταχῇ)· τῷ[1] δὲ ἐν τοῖς παλισκίοις εἰς ἓν μόνον ἡ ὁρμή, διὸ καὶ μήκη μὲν αἱ ῥάβδοι μᾶλλον λαμβάνουσιν, πάχη δ' οὔ, καὶ ἐκ τούτων αἱ βακτηρίαι (καθάπερ ἄλλοθί τε καὶ ἐν Σκιάθῳ[2]).

18.10 κοινὸν δὲ τοῦτο ἴσως ἐπὶ πάντων τῶν εὐωδῶν[3] ἐστιν· λοιπὸν δ' οὖν εἰπεῖν διὰ τί ταῦτα οὐκ εὐώδη.

τούτων δ' αἰτιατέον δρόσον τὴν πρὸς ὅλον·[4] αὕτη[5] δὴ[6] πολλὴ πίπτουσα καθυγραίνει[7] τὰ μὲν ἄνθη (καὶ ὑγρότερα τὴν φύσιν ὄντα) μᾶλλον, τὰ δὲ ἄμικτα καὶ ξηρὰ ἔλαττον[8] καὶ ἀσθενέστερον,[9]

[1] ego (τῶν Basle ed. of 1541) : τὸ U.
[2] σκιάθω u (σκιαθω U) : σκίσθω N : κισσῷ aP.
[3] ego : δένδρων U.
[4] ego (*cf.* πρὸς ὀλίγον ; προσέλειον Schneider[c]) : προσολον U : προσυλον u (πρόσυλον N) : πρόσηλον a : πρόσειλον P.
[5] Schneider (αὔτὴ u) : αὐτὴ U N aP.
[6] U N : δὲ aP.
[7] Gaza (*humefacit*), Moreliana : καθυγραίνεται U.
[8] Gaza (*levius*), Moreliana : ἐλάττω U.
[9] Heinsius : καὶ ἀσθενεστερα U : Gaza omits.

than the rest; and this is why they do not grow tall, but are more shrub-like in their habit. Thus they not only spread out but branch out more[1] because they acquire many starting-points of growth, the heat concocting and distributing the food throughout the plant. Whereas the tree that grows in the shade has an impulse of growth that takes but one directon; this is why the branches acquire more length but no spread, and it is from these that walking-sticks are made, as at Sciathus and elsewhere.)

18.10 But this explanation applies perhaps to all fragrant plants. In any case it remains to be explained why the other plants, fragrant elsewhere, are not fragrant in Egypt.

The Scentlessness of the Rest

The reason that we must give for the lack of fragrance in these other Egyptian plants is the constant dew. The dew falls in great abundance, and whereas this soaks the flowers, already more fluid in their nature, to a greater degree, it wets what resists mixture and is dry to a smaller extent and less effectively, which is why, given the same

[1] *Cf. HP* 1 3. 3: "The myrtle if not pruned turns into a shrub."

διὸ καὶ κρατεῖ τι[1] μᾶλλον ἀπὸ[2] τῆς αὐτῆς ὑγρότητος ὅδε[3] [μύρρινος]·[4] σημεῖον δέ, καὶ τὸ εἰρημένον ὑπὲρ αὐτοῦ, μικρόκαρπος γὰρ εἶναι δοκεῖ διὰ τὸ μὴ λαμβάνειν τὸ κατὰ κεφαλὴν[5] ὕδωρ ἱκανόν.

ὥστε τοῦ μὲν μυρρίνου διὰ τούτων πειρατέον ἀναζητεῖν τὰς αἰτίας.

18.11 τῶν δ' ἐν τοῖς ψυχροῖς εὐόδμων τὴν θερμότητα τὴν ἐν τῇ γῇ νομιστέον εἶναι· πανταχοῦ γὰρ τὸ θερμὸν τὸ πέττον, ἀλλ' ὁτὲ μὲν εὐθὺς προσπῖπτον, ὁτὲ δὲ κατακλειόμενον (ὥσπερ καὶ ἐν τῇ τῆς ὀπώρας πεπάνσει, παραπλήσιον γὰρ τὸ συμβαῖνον· ἀντιπεριισταμένη γὰρ ἡ θερμότης εἰς τὴν γῆν καὶ συνελαυνομένη πέττει).

δεῖ δὲ καὶ τὴν γῆν μήτε πηλώδη μήτε πίειραν εἶναι μήτε γλίσχραν (καὶ <γὰρ>[6] διὰ τὴν ὑγρότητα καὶ τὴν γλισχρότητα οὐχ ὁμοίως ἐργάσεται τὸ θερμόν), ἀλλὰ [μὴ][7] τοιαύτην ὥστε μήτε τὸ ἐκ

[1] U : τὰ μὲν κρατεῖται Schneider : κρατεῖται Wimmer.
[2] U : ὑπὸ Schneider. [3] ego : ὁ δε U.
[4] ego : μύρρινος κρατεῖ Schneider : μύρρινος <. . .> Wimmer. [5] U N (*cf. HP* 4 10. 7) : κεφαλῆς aP.
[6] Gaza, Moreliana.
[7] Gaza, Heinsius (μὴν Itali : δὴ Moreliana).

amount of fluid, this Egyptian myrtle masters it somewhat better. Proof of this is the explanation given about it[1]: its berry is held to be small because the tree does not get enough rain.

These, then, are the approaches to use in the endeavour to find the causes that apply to the myrtle.

Fragrance and Cold Countries

18.11 On the other hand when plants are fragrant in cold countries[2] we must consider the cause to be the heat in the earth, heat being everywhere the agent of concoction; heat however sometimes concocts by direct contact,[3] and sometimes by being shut in (just as in the ripening of tree fruit,[4] for what occurs here is similar: the heat concocts when it is displaced and driven into the ground).

The ground moreover must not be muddy or fat or viscous, since both fluidity and viscosity will keep the heat from operating to the same extent; instead the ground must be such as not to admit the cold

[1] *CP* 6 18. 5.

[2] Iris root is the only aromatic fragrance in cold countries: *cf.* *CP* 6 18. 12 note 1.

[3] With the sunlight and air.

[4] That is, of tree fruit that ripens after the onset of cold weather: *CP* 2 8. 1.

τοῦ ἀέρος δέχεσθαι ψυχρόν, τό τε[1] *ἐν αὐτῇ*[2] *θερμὸν ἀποστέγειν καὶ τηρεῖν.*

18.12 *διὰ τοῦτο γὰρ καὶ ἐν τῇ Ἰλλυρίδι βελτίων ἡ ἶρις ἢ ἐν Μακεδονίᾳ. ἐν δὲ τῇ Θρᾴκῃ καὶ ταῖς ἔτι ψυχροτέραις ἅμα καὶ ἀπεπτοτέραις ὅλως ἄοσμος (καὶ γὰρ ἡ γῆ πίειρα, καὶ ἡ ὑγρότης πολλή, καὶ ὁ χειμὼν ἔκτοπος), ἐν δὲ ταῖς ἀλεειναῖς* <*καὶ*>[3] *λεπταῖς διὰ τὴν μαλακότητα τοῦ ἀέρος ἄοδμος (οὐκ ἴση γὰρ ἡ ἀντιπερίστασις)· ἐξ ἀμφοῖν δὲ ζητητέον (ὥσπερ καὶ ἐν τοῖς ὑπὲρ γῆς εὐόδμοις), καὶ ἐκ τοῦ ἐδάφους καὶ ἐκ τοῦ ἀέρος, τὰς δυνάμεις τὰς εἰς τὴν πέψιν συνεργούσας.*

ὑπὲρ μὲν οὖν τούτων ἱκανῶς εἰρήσθω.

19.1 *τὰς δ' ὀδμὰς ἤδη [δε]*[4] *τὰς ἀπ' ἀλλήλων οἱ καρποὶ μὲν οὐχ ἕλκουσιν, οὐδ' ὅλως οὐδὲν τῶν ἐν*

§19.1: *Cf. Geoponica*, xi. 18. 1; Michael Psellus, *De Omnifaria Doctrina*, clxxxviii.

[1] *τό τε* aP : *τοῦτο* U N.
[2] Gaza, Moreliana : *αὐτῆι* U.
[3] Gaza, Itali. [4] Schneider.

---

[1] *Cf. HP* 4 5. 2: "But there are none of the aromatic fragrances in these countries (*i.e.* in northern countries and Greece), except the iris in Illyria and on the Adriatic coast, for here it is fine and far superior to iris elsewhere

from the air and also to seal off and preserve the internal heat in the earth.

18.12 For this is why iris root is more fragrant in Illyria[1] than in Macedonia. In Thrace, and in countries still colder and still less effective in concocting, the iris is quite odourless, since the soil is fat, the fluid abundant, and the cold extreme; on the other hand in warm countries with thin soil the root is odourless because of the mildness of the air, since the heat is not displaced to the same extent[2]; and one must look to both sources (as with plants whose aromatic parts are above ground),[3] to soil and air together, for the forces that contribute[4] to concoction.

Let this suffice for the discussion of these matters.

Attraction of Odours

19.1 Passing to odours absorbed by one thing from another, no fruit and indeed no other part of a plant

. . ."; *ibid.* 9 7. 3–4: "From Europe itself comes nothing (*sc.* among plants used for producing scent) except iris. This is best in Illyria, not by the coast but in the interior, where the country tends more to face north."

[2] For this theory of compensatory displacement of heat by cold *cf. CP* 1 12. 3 with note *a*.

[3] *Cf. CP* 6 17. 13; 6 18. 1.

[4] With the native heat of the plant.

τοῖς φυτοῖς μορίων, ὥστε[1] γε καὶ ποιεῖν τι δῆλον· ὥστε καὶ παραφυτευόμεν γ' ἔνια τῶν δριμέων (οἷον σκόρδα καὶ κρόμμυα) τοῖς στεφανώμασιν· φασὶν γὰρ ὠφελεῖν εἰς εὐωδίαν. τοῦτο[2] δ' εἴπερ ἀληθές, δυοῖν θάτερον αἰτιάσαιτ' ἄν τις· <ἢ>[3] ὡς τὸ κακῶδες ἀφαιρουμένων (ἕκαστον[4] γὰρ δὴ τρέφεται τῷ οἰκείῳ καὶ ἕλκει τὸ συγγενές· ἀφαιρεθέντος δὲ τοῦ ἀλλοτρίου, καθαρώτερον καὶ εὐωδέστερον τὸ λοιπόν)· ἢ ὡς καταξηραινόντων τὸν τόπον διὰ τὴν θερμότητα καὶ πολυτροφίαν (ἐν δὲ τοῖς ξηροῖς ἅπαντα εὐοδμότερα).

τοῦτο μὲν οὖν ὡς ἂν καθ' ὑπόθεσιν εἰρήσθω.

19.2 οἱ δὲ χυλοὶ χωρισθέντες ἕλκουσιν, καὶ μάλιστα τό τ' ἔλαιον καὶ ὁ οἶνος. ὃ δὴ καὶ ἄλογον φαίνεται· τί δή ποτε τὸ μὲν ὕδωρ, καὶ λεπτομερέστε-

§ 19.2: *Cf.* Aristotle, *On Dreams*, ii (460 a 28–32).

[1] U : ὥς Schneider.
[2] U : τούτου Schneider.
[3] aP. [4] u : -ος U N aP.

---

[1] That is, the pungent plants remove the pungent food from the bed.

[2] *Cf. CP* 2 18. 1; 3 10. 3; 3 15. 4.

[3] For oil *cf.* Theophrastus, *On Odours*, iv. 14: "In fact it (*i.e.* oil) is not at all naturally receptive of odour because of its close texture and fattiness, and among oils the fattiest is the least receptive, like oil of almonds, whereas

attracts them (at least to any noticeable extent), so that we even grow certain pungent plants (as garlic and onion) next to coronaries, since they are said to improve the fragrance of the coronaries. If the statement is true, one of two reasons could be given. One, that the pungent plants remove bad odour, since every plant feeds on the food appropriate to it and attracts what is akin to itself,[1] and with the removal of the alien component the remainder is purer and more fragrant. The other reason is that the pungent plants dry out the bed[2] with their heat and great consumption of food, and all plants are more fragrant in dry ground.

Our explanation is to be taken on the assumption that the report is true.

19.2

On the other hand flavour-juices after separation from the fruit, especially oil and wine,[3] do attract odours. This appears unreasonable: why does water, not only thinner than these[4] but also

sesame oil and olive oil are the most receptive." For wine *cf. ibid.* iii. 11: "For wine, as was said before [at *CP* 2 18. 4], has a great tendency to attract odours." For wine and oil *cf.* Aristotle, *On Dreams*, ii: "For oil, once it is prepared, quickly acquires the odour of things near it, and wines are affected in the same way, for they not only absorb the odours of what is put into them or mixed with them, but also of things placed near, or growing near, the wine vessels."

[4] *Cf.* Aristotle, *On Sense*, iv (441 a 23–24): ". . . water is the thinnest of all things fluid in this sense (*i.e.* of being liquid), thinner than oil itself."

ρον τούτων,[1] καὶ ἄοδμον καὶ ἄχυλον (καὶ ὅλως οὖν[2] ἀειδές[3]), οὐ δέχεται, τὰ δὲ καὶ ἔγχυλα καὶ ὀδμώδη καὶ παχύτερα <δέχεται;>[4] δέχεται δὲ ὁ[5] οἶνος καὶ μὴ ἐμβαλλομένων,[6] ἀλλὰ πλησίον ὄντων ἕλκει τῷ[7] ἔχειν τινὰ θερμότητα ἐν ἑαυτῷ (τοῦτο γὰρ τὸ ἕλκον)· τὸ δ’ ὕδωρ ἥκιστα ἔχει, φύσει γὰρ ψυχρόν.

19.3 τὸ δὲ δὴ[8] δεξόμενόν[9] τε ἅμα καὶ διατηρῆσον <δεῖ> μὴ[10] οὕτω λεπτὸν ὡς στεγνόν τι[11] καὶ φυλακτικὸν [τι][12] εἶναι· τὸ μὲν <γὰρ>[13] διίησιν,[14] ὥσπερ ἠθμός,[15] οὐκ ἔχον ᾧ στέγει·[16] τὸ δὲ σωματωδέστερον[17] καὶ πυκνότερον εἴς τε τὸ δέξασθαι καὶ τηρεῖν εὐφυές, πρὸς ἄμφω συμμετρίαν ἔχον, ἅπερ ἀμφότερα τῷ οἴνῳ καὶ τῷ ἐλαίῳ συμβέβηκεν. καὶ οἱ μυρεψοὶ δὲ τὰς ὀσμὰς εἰς τὸ ἔλαιον τίθενται· καὶ γὰρ ἄλλως ἁρμόττον πρὸς τὴν χρείαν, καὶ ἅμα δύναται μάλιστα θησαυ-

[1] ego : λεπτομερες ον τοῦτων U : λεπτομερὲς ὂν τοῦτο N aP.
[2] U : ὂν Schneider.
[3] ego : διειδες U. [4] ego.
[5] δε ὁ U : ὁ δὲ Heinsius (beginning the sentence here).
[6] u (?) N aP : ἐμβαλλομενον U.
[7] τῶ u : τὸ U.
[8] τὸ δε δὴ U : δεῖ δὲ τὸ Wimmer.
[9] Gaza, Moreliana : δεξάμενον U.
[10] Schneider (*non . . . debet* Gaza).
[11] ego (λεπτὸν ὡς τὸ ὕδωρ, ἀλλὰ στεγνὸν Schneider, after Gaza) : στεγνὸν ὡς λεπτόν τι U.

odourless and flavourless (and so quite without form[1]), refuse to receive odours, whereas other substances that not only possess flavour and odour but are thicker attract them? But wine, even when a substance is not put into it but merely stands near, attracts[2] the odour because wine has a certain internal heat, heat being what attracts; but least of all does water possess heat, since it is naturally cold. To receive and also to retain an odour a substance must be not so much thin as impermeable and so retentive; for the thin substance lets the odours through like a sieve since it has nothing to keep them from leaking out. But the substance with more body and closer texture is well adapted both to receive and to retain, since it has these characters in the requisite amount for both results, and both characters are found in wine and oil. Again, perfumers put their odours in oil: besides its convenience for this use, it is also best able to store the odours

19.3

[1] And so negative or privative. *Cf.* the argument, from their strength, that the "privative" flavours and odours are not truly privative or negative (*CP* 6 5. 5; 6 18. 8).

[2] *Cf. CP* 2 18. 4.

---

[12] Gaza, Basle ed. of 1541.
[13] aP.
[14] Schneider : δίησιν U.
[15] Moreliana : ἴθμὸς U.
[16] ᾧ στέγει aP : ὡσγεγει U : ὡς γε γεῖ N.
[17] aP : σωσματωδεστερον U (-δέ- N).

ρίζειν διὰ τὴν ἀμεταβλησίαν. τὸ δ' ὕδωρ εὐθὺ διαπνεῖ, καὶ ὥσπερ ἐκπλύνει, καὶ διίστησιν. ἡ δὲ λεπτότης (ὥσπερ εἴπομεν) οὐ χρήσιμος,[1] ἐπεὶ οὐδ' ὁ ἀὴρ δύναται κατέχειν, ἀλλὰ διαπέμψαι[2] μόνον.

19.4 τῶν δὲ ξηρῶν, μάλιστα μὲν ὅσα μανὰ καὶ ἄοσμα καὶ ἄχυλα (καθάπερ ἔρια καὶ τὰ ἱμάτια καὶ εἴ τι ἄλλο τοιοῦτον)· οὐ μὴν ἀλλὰ καὶ ὅσα χυλοὺς ἔχει καὶ ὀσμὰς (ὥσπερ καὶ τὸ μῆλον· καὶ γὰρ τοῦτο δέχεται καὶ ἕλκει τὰς ἐκ τῶν χυλῶν ὀσμάς), ἔγχυλα[3] γὰρ αὐτὰ[4] μᾶλλον. χρὴ γὰρ (ὡς ἁπλῶς εἰπεῖν) μήτε κατάξηρον εἶναι τὸ δεξόμενον[5] (ὥσπερ τέφραν ἢ ἄμμον), μήτε κάθυγρον· τοῦ[6] μὲν γὰρ διιὸν οὐχ ἅπτεται, ἐν τῷ δὲ διαχεῖται καὶ ἐκκλύζεται.

19.5 διὰ τοῦτο γὰρ καὶ τὰ ἴχνη τῶν λαγῶν εὐσημότερα ψακάσαντος[7] μαλακῶς ὑπ'

§19.5: Xenophon, *Cynegeticus*, v. 3–4.

[1] οὐ χρήσιμος U : ἀχρήσιμος N aP.
[2] U : διαπέψαι u.
[3] ego (ἔγχυλον Schneider) : ἐγχυλεῖ U N a : ἐγχυλοῖ P.
[4] U : αὐτὸ Schneider.
[5] Gaza, Moreliana : δεξάμενον U.
[6] ego : τὸ U.
[7] ego : ψακασθέντα U (ψεκ- u).

because of its resistance to change.[1] Water on the other hand at once exhales the odours and (as it were) washes them out and lets them come apart; and its thinness (as we said[2]) is not serviceable; indeed air[3] too lacks the power to retain odour but can only transmit it.

19.4 Of solids[4] it is chiefly those that are open-textured, odourless and flavourless (such as wool, clothing and the like) that absorb odours. Nevertheless solids that have flavour and odour do so too (as the apple, for it too receives and attracts the odours from flavour-juices), since these solids have more the character of being flavour-juices themselves.[5] In a word what is to receive an odour must be neither quite dry (like ashes or sand), nor yet soaking wet, since the odour passes through the former without fastening upon it, whereas in the latter the odour is dispersed and washed out. 19.5 This moreover is why the tracks of hares are more easily made out when a light drizzle has fallen just before the hunt: the

[1] *Cf.* Theophrastus, *On Odours*, iv. 14: "The composition and the whole preparation of perfumes is directed to storage (as it were) of the odours, which is why they are put into the oil, since oil is longest lasting and also most convenient for use."

[2] In the first sentence of this section.

[3] Which is still thinner than water.

[4] Literally "dry (things)."

[5] And so resemble the separated flavour-juices of *CP* 6 10. 2.

αὐτὴν τὴν κυνηγίαν. ἐμμένει γὰρ μᾶλλον διὰ τὸ σφοδρῶς ἀπερείδεσθαι καὶ εἰς βάθος, καὶ οὐχ, ὥσπερ ξηρᾶς οὔσης, ἐπιπολῆς, οὐδ' αὖ πηλώδους, ὥσπερ ἐφ' ὕδατι καὶ νοτίοις,[1] ἐναντία γὰρ καὶ τὰ πνεύματα καὶ τὰ ὕδατα, καὶ ἀπόλλυσιν ἄμφω τὰς ὀσμάς. ἀλλὰ δεῖ,[2] καθάπερ <τὰ>[3] ἀπομάγματα τῶν δακτυλίων, μέσην τινὰ κρᾶσιν ἔχειν.

καὶ περὶ μὲν τούτων ἅλις.

20.1 ἐπεὶ[4] δὲ τὰ μὲν ἥμερα, τὰ δ' ἄγρια τῶν εὐόσμων, οὐκ ἀκολουθεῖ κατὰ τὸ γένος ὅλως τὸ εὐοσμότερον, <ἀλλ' ἔνθα μὲν τὸ ἥμερον>[5] (καθάπερ τὸ ῥόδον), ἔνθα δὲ τὸ ἄγριον, ὥσπερ τὸ ἴον τὸ μέλαν καὶ ὁ κρόκος (ὁ δ' ἕρπυλλος καὶ τὸ ἑλένιον δριμύτερα, καθάπερ καὶ τῶν λαχανωδῶν τὸ πήγανον).

αἴτιον δέ, ὡς μὲν εἰπεῖν καθόλου, τὸ καὶ πρότερον λεχθέν, ὅτι τὴν συμμετρίαν ἑκάτερα λαμβάνει

[1] ego (καὶ νότωι u) : καίνοντος U. [2] aP : δὴ U N.
[3] Schneider (taking it to be in U). [4] u : ἐπὶ U.
[5] ὅλως . . . ἥμερον ego (so Itali, omitting ἀλλ' : *praestantia odoris . . . : nam et urbanum praecellit* Gaza : ὅλως ἡ εὐοσμία, ἀλλ' ἔνθα μὲν τὸ ἥμερον εὐοσμότερον Schneider : ἀλλ' ἔνθα μὲν εὔοσμον τὸ ἥμερον Wimmer) : ὅλως τὸ εὐοσμότερον U.

track lasts better then because the imprint is firmly made and deep and not shallow as when the ground is dry. Again they do not remain when the ground is muddy, as when there is rain and the winds are southerly, for both winds and rain are unfavourable, and both destroy the scent. But the track must have a certain intermediate tempering, like the material that takes the impression of a signet ring.

But enough of these matters.

Fragrance is Sometimes in the Wild Plant, Sometimes in the Cultivated

20.1 Since among fragrant plants some are of the cultivated variety, some of the wild, greater fragrance is not simply a matter of belonging to the cultivated variety or to the wild, but in some plants it is found in the cultivated variety (as in rose[1]), in others in the wild, as in violet and saffron crocus (the wild variety however of tufted thyme and calamint is too pungent, just as wild rue is among vegetables).

The Cause, Put Generally

The cause, to put it generally, is the one mentioned before[2]: the cultivated plant or the wild acquires through the one or the other character the

[1] *Cf. HP* 6 8. 6 (many mountain flowers, roses, stocks and the rest, yield an inferior odour).

[2] *CP* 6 17. 12–14.

δι' ἑκατέρου τῆς ὑγρότητος καὶ ξηρότητος, ἐξ ὧν αἱ ὀσμαί.

20.2 ὡς δὲ καθ' ἕκαστα, φανερὸν ἐπισκοποῦσιν.

τὸ μὲν γὰρ ἴον τόδε[1] [μέλας][2] καὶ ὁ κρόκος οὔτε πολλῆς δεῖται τροφῆς, ἐξ αὐτῶν[3] θ' ἱκανή,[4] μεγαλόρριζα[5] γάρ, ὥστ' ἐν τοῖς ἡμέροις ἡ πλείων[6] ἀπεπτοτέρα[7] (διὰ τοῦτο γὰρ καὶ τὸ[8] μὲν τέφρᾳ[9] περιβάλλουσι, τὸν[10] δὲ πατοῦσιν[11]).

τὸ δὲ ῥόδον καὶ ὁ ἕρπυλλος καὶ ὅσα ὅμοια τούτοις, ἄγρια μὲν ὄντα, ξηρότερα τοῦ συμμέτρου γίνεται· διὸ τὸ μὲν ῥόδον ἀσθενές, οἷον τὸ[12]

[1] N : τὸ δε U : τό γε u : τὸ aP.
[2] ego : μέλας U : μέλαν u.
[3] N (-ων u^{ac}) : αυτων U : αὐτῶν (-ων u^{c}) aP.
[4] ego (ἱκανὴν ἔχει Gaza, Schneider) : ἱκανην U.
[5] ego : καιφαλόριζα U.
[6] u : πλεῖον U.
[7] a : ἀπεπτότερα U N : ἀπεπτότεα γὰρ P.
[8] ego (τοῖς Heinsius [*aliis* Gaza]): τὸν U: τῶ u N P: τῶν a.
[9] τέφρα U : τέφραν u.
[10] U (τοῖς Heinsius [*aliis* Gaza]) : τῶ u : τὰ N aP.
[11] U (δ' ἐπιπάττουσι Heinsius [*respergant* Gaza]) : δὲ πάττουσιν u.
[12] U : καὶ οἷον Schneider : οἷόν τι Wimmer.

right amount of fluidity and dryness, and from these odour arises.

The Cause Applied to Particular Cases

How the cause is to be put particularly is evident when we consider the cases. 20.2

Thus violet and saffron crocus (1) require no great amount of food and (2) provide enough from their own store,[1] since their roots are large[2]; consequently the additional food in the cultivated plants is less well concocted (this in fact is why growers put ashes around the violet and let the crocus get trampled[3]).

Rose on the other hand and tufted thyme and the like are in the wild state too dry. Hence the rose,

[1] That is, they need no help from agriculture.

[2] *Cf. HP* 6 6. 7: "The black *ion* (violet) differs from the white (stock) both in the other respects and in the plant itself, having . . . a great deal of root"; *HP* 6 6. 10: "The saffron crocus has a great deal of root and that fleshly . . ."; *HP* 7 9. 4 (saffron crocus has a fleshy root that is oblong and acorn-shaped); *HP* 1 6. 7 (saffron crocus has a fleshy root).

[3] *Cf. HP* 6 6. 10 (of saffron crocus): "It . . . is in general tenacious of life. It likes moreover to be trampled and grows finer when the root is worn down by the tread of feet. This is why the plant is best by the roadside and in well trampled ground." Presumably it is not the stalk or flower that is trampled, but the soil over the roots.

ἄοδμον, ἄνικμον ὄν (οὐδὲ γὰρ τὸ ἴον τὸ λευκὸν ἐν ταῖς ἄγαν ξηραῖς καὶ λεπταῖς εὔοσμον, οὐδ' ὅπου θερμὸς σφόδρα καὶ ἔμπυρος ὁ ἀήρ· ἀναξηραίνει γάρ).

20.3 ὁ δ' ἕρπυλλος καὶ τὸ ἑλένιον καὶ τἆλλα τὰ τοιαῦτα διὰ τὴν ξηρότητα δριμείας ἄγαν[1] ἴσχει καὶ σκληρὰς τὰς ὀδμάς, ἡμερούμενα δὲ μαλακωτέρας.

ἡ δὲ συμμετρία διότι καὶ πέψιν ποιεῖ καὶ εὐοσμίαν φανερόν (ἐπεὶ καὶ τῶν εὐωδῶν <αἱ>[2] ὀσμαὶ πρὸς τῷ[3] διὰ[4] φύσιν[5] καὶ τὴν τοῦ ἀέρος εὐκρασίαν ἀπαιτοῦσιν ὥστ' ἀναμίγνυσθαι, μήτε κωλύεσθαι μηδ' ὑφ' ἑνός).

20.4 ἔοικε γὰρ τοιοῦτόν τι συμβαίνειν καὶ περὶ τὰ ἴχνη τῶν λαγῶν (ὧν[6] καὶ ἀρτίως ἐμνήσθημεν)· οὔτε γὰρ θέρους εὔοσμα οὔτε χειμῶνος οὔτε ἦρος, ἀλλὰ μάλιστα τοῦ φθινοπώρου. χειμῶνος μὲν γὰρ ὑγρά, θέρους δ' αὖ[7] ξηρανθέντα (διὸ καὶ μεσημ-

§ 20.4: Xenophon, *Cynegeticus*, iv. 11–viii. 2.

[1] Schneider : γ' ἂν U.
[2] Schneider.
[3] τῷ u aP : το U (τὸ N).
[4] u N (δια U) : κατὰ aP.
[5] u : φύσει U.
[6] u : λαγωιων U : λαγωῶν N aP.

lacking moisture, is weak in scent, like a plant devoid of odour; stock too in fact is not fragrant when the ground is exceedingly dry and thin, and again where the air is extremely hot and torrid,[1] since the air dries it out.

20.3 Whereas tufted thyme, calamint and the like, owing to their dryness, come to have odours that are far too pungent and harsh; whereas under cultivation the odours are milder.

The proper adjustment in the food is evidently responsible for both concoction and fragrance: indeed the odour in fragrant plants requires in addition to their natural fragrance a proper tempering of the air, if the odour is to mix with the air and not suffer any impediment.[2]

20.4 It appears in fact that something of the sort affects the aforementioned[3] tracks of hares: the tracks have a good scent neither in summer nor in winter nor in spring, but mainly in autumn; in winter they are too fluid, in summer again they are too dry (which is also why they are worst at noon),

[1] Rose and stock are scentless in Egypt (*HP* 6 8. 5).
[2] For instance by cold: *cf.* *CP* 6 17. 5.
[3] *CP* 6 19. 5.

---

[7] Wimmer : οὐ U.

βρίας χείριστα), τοῦ δ' ἦρος αἱ τῶν ἀνθῶν ὀσμαὶ παροχλοῦσιν·[1] *τὸ δὲ μετόπωρον σύμμετρον ἔχει πρὸς ἅπαντα τὴν κρᾶσιν.*[2]

περὶ μὲν οὖν ὀσμῶν καὶ χυλῶν τῶν ἐν τοῖς φυτοῖς καὶ καρποῖς ἐκ τούτων θεωρητέον.

ὅσα δ' ἤδη κατὰ τὰς μίξεις καὶ τὰ πάθη πρὸς ἄλληλα καὶ τὰς δυνάμεις, ταῦτα καθ' αὑτὰ λεκτέον.[3]

[1] U (-σι N P) : *παρενοχλοῦσι* a.

[2] Itali : *ὄρασιν* U.

[3] subscription in U *θεοφράστου περὶ φυτῶν αἰτιῶν*. The scribe leaves the remaining 27 lines of fol. 269ᵛ blank : U never contained the lost seventh book.

and in spring the odours of the flowers interfere.[1] Autumn on the other hand has the proper tempering in all respects.

Odours then and flavours in plants and fruits are to be studied in the light of this discussion.

All questions however concerning the later stage of their combination, and of their effects and operations on one another, must be treated separately.[2]

[1] *Cf.* Xenophon, *Cynegeticus*, v. 5: "Spring, which is properly blended in its seasonal qualities, makes the trail clear, except in so far as the earth by putting forth flowers interferes with the hounds by combining with the tracks the smells of flowers"; Plutarch, *Quaest. Nat.* xxiii (917 E–F); [Aristotle], *Mir.* lxxxii (836 b 13–19): "In Sicily at the place called Enna is said to be a cave round which they say grows at all seasons a mass of other flowers, but most of all an immense space is filled with violets, which fill the neighbouring country with fragrance, so that huntsmen are unable to track hares, since the hounds are overpowered by the fragrance."

[2] In the lost seventh book (for flavours) and in the book *On Odours.*

APPENDIX I

BOOK VII

In Diogenes Laertius' list of Theophrastus' works (v. 42–50) the following titles occur:

περὶ φυτικῶν ἱστοριῶν α′ β′ γ′ δ′ ε′ [ϛ′][1] ζ′ η′ θ′ ι′
φυτικῶν αἰτιῶν α′ β′ γ′ δ′ ε′ [ϛ′][1] ζ′ η′
περὶ ὀδμῶν α′
περὶ οἴνου καὶ ἐλαίου
περὶ χυλῶν α′ β′ γ′ δ′ ε′
περὶ καρπῶν.

The first three are the surviving *HP*, *CP*, and *De Odoribus*; of the last three we possess no identified

[1] In the surviving works of Aristotle and Theophrastus that consist of six books or more (the *HP* and *CP*; Aristotle's *Topica*, *HA*, *Metaphysics*, *EN* and *Politica*) the books are numbered by the letters of the alphabet, and not by the Greek numerals; so book VI is Z, not ϛ′. The use of the letters is older (it is found in the *Iliad* and the *Odyssey*), and may well go back to Aristotle and Theophrastus themselves. The use of the numerals for this purpose was later all but universal, and accounts for the interpolated ϛ′ in Diogenes Laertius.

fragments. We may guess that *περὶ οἴνου καὶ ἐλαίου* was another version or another name of *CP* VII, and *περὶ καρπῶν* of *CP* V; and that *περὶ χυλῶν*, in five books, was a collection of information that bore the same relation to *CP* VI–VII as *HP* bears to *CP* I–V.[1]

The following references to *CP* VII occur in the *CP* and the *De Odoribus* (there are none in the *HP*):

1. *CP* 2 16. 1 "But odours and flavours must be studied by themselves at greater length later." (A reference to *CP* VI–VII and the *De Odoribus*.)
2. *CP* 6 3. 3: "Savours occur in three things: plants and animals have certain odours as well as savours depending on the tempering of their qualities; again *savours are found in things mixed by some procedure of art*, or else in things that alter spontaneously, sometimes for the better, sometimes, as in decomposition, for the worse ... We must first discuss the natural savours ..."
3. *CP* 6 7. 6 (On the corruption of wine and its occasional recovery): "But such matters are more properly treated in what follows."
4. *CP* 6 11. 2 (Flavours are all in dry [*i.e.* solid] things. But we separate them from these, as with wine and olive oil; and sometimes we do this by infusion, as in obtaining fruit juices or in mak-

[1] See note 2 on *CP* 6 3. 3.

ing them from wheat or barley.) "In all these cases the starting-points and the powers at work are to be sure natural, but the result is rather the achievement of art and of the intelligence that applies it. The products of intention and art, however, must be studied by themselves."

5. *CP* 6 20. 4: "But all matters that pertain to the later stage when they (*sc.* odours and flavours) are mixed with one another, and when they are acted upon by or act on one another, must be treated separately."
6. *De Odoribus*, iii. 7: "We must endeavour to treat of the odours produced by art and ingenuity as we did of the flavours."
7. *De Odoribus*, iii. 8: "We must know what odours mix well together and what odours work well with what things to produce a single odour, as we did with the savours, since there too people seek for this same result who mix and (as it were) confect them."

In his treatise on the blends and powers of the simple medicines Galen often cites Theophrastus (usually with Aristotle) without mentioning any specific work. It is likely that Galen is here following an author who paraphrased some of Theophrastus' views on savours. The following passages have no close parallels in what we have of Theophrastus and have a good chance of ultimately coming from *CP* VII.

APPENDIX I

Galen, *De Simpl.* iv. 3 (vol. xi, p. 629. 4–11 Kühn): ". . . and they (*sc.* most investigators of drugs, who take them not to be composite but homoeomerous) are still more surprised if we venture to say this of vinegar, that it has lost the native heat of wine, and has the heat coming from decomposition, which is the view of Aristotle [Frag. 222 Rose[3]] and Theophrastus. For the vinous parts of wine become chilled in the passage to vinegar, whereas the watery residue acquires on decomposing a certain adventitious heat, just as do all other things that have decomposed."

Galen, *De Simpl.* iv. 11 (vol. xi, p. 654. 4–18 Kühn): "It also appears that must strike us as sweet not only because of its native, but also because of its adventitious, heat. For as Theophrastus and Aristotle [Frag. 226 Rose[3]] said, a good deal of 'afterglow' (as it were) from the heat of the sun is found in grapes and other fruit, by which the watery and as it were half-concocted part is brought to concoction and prepared by assimilation to the agent producing the change, so that if you press out the juice of any ripe fruit whatever you will find that it immediately effervesces, like must, and that the effervescence differs in degree with the difference in heat between one fruit and another. When the watery and half-concocted part has been changed and prepared, and the heat that produces the effervescence has evaporated, the native qualities of the flavours are seen unadulterated, so that we can then tell apart the natural characters of the wines."

Galen, *De Simpl.* iv. 14 (vol. xi, pp. 664. 4–665. 4 Kühn): "... Theophrastus and Aristotle [Frag. 221 Rose[3]] however ... have taught us among many other such things the following point about wines, that they are affected in much the same way as our bodies ... of wines too the naturally hot ones are made to ripen faster by movement that fans them and sunshine that heats them and flames burning near them; whereas those that are colder and more watery are shown for what they are by all such procedures and forced to undergo sooner what was going to happen to them somewhat later. For everything is preserved in its own nature by heat of its own, and corrupted by excess from outside, whether of alien heat or of cold."

Galen, *De Simpl.* iv. 18 (vol. xi, p. 679. 13–15): "... but Plato [*cf. Timaeus*, 65 E 4–66 A 2] pronounces it (*sc.* the pungent) the hottest of all savours and so do Aristotle [not in Rose] and Theophrastus ..."

APPENDIX II

List I (see *CP* 6 4. 1, note 2)

colour	*savour*	*odour*
white	sweet	sweet
yellow	oily	oily
black (grey)	bitter (salty)	odour of decomposition
red	pungent	pungent
violet	dry-wine	dry-wine
green	astringent	astringent
blue	acid	odour of decomposition?

List II (see *CP* 6 9. 2, note 1)

Savours	*Names of Corresponding Odours in*		
	Aristotle	*Theophrastus*	*Galen*
sweet	sweet	sweet	
oily	oily	no	
dry-wine	dry-wine		no
astringent	astringent	no	no
pungent	pungent		pungent
acid			acid
salty		no	no
bitter		no	no